MAMMALIAN DIVERSIFICATION: FROM CHROMOSOMES TO PHYLOGEOGRAPHY

(A Celebration of the Career of James L. Patton)

For Jim and Carol, with many thanks

Mammalian Diversification: From Chromosomes to Phylogeography

(A Celebration of the Career of James L. Patton)

Eileen A. Lacey[1] and Philip Myers[2] (editors)

[1] Museum of Vertebrate Zoology and Department of Integrative Biology of the University of California at Berkeley

[2] Museum of Zoology and Department of Ecology and Evolutionary Biology of the University of Michigan at Ann Arbor

UNIVERSITY OF CALIFORNIA PRESS
Berkeley • Los Angeles • London

University of California Press, one of the most distinguished university presses in the United States, enriches lives around the world by advancing scholarship in the humanities, social sciences, and natural sciences. Its activities are supported by the UC Press Foundation and by philanthropic contributions from individuals and institutions. For more information, visit www.ucpress.edu.

University of California Press
Berkeley and Los Angeles, California

University of California Press, Ltd.
London, England

Library of Congress Cataloging-in-Publication Data
Mammalian diversification : from chromosomes to phylogeography : a celebration of the career of James L. Patton / Eileen A. Lacey and Philip Myers (editors).
    p. cm. -- (University of California publications in zoology ; v. 133)
  Includes bibliographical references.
  ISBN 0-520-09853-6 (pbk. : alk. paper)
  1. Mammals. I. Patton, James L. II. Lacey, Eileen A. III. Myers, Philip, 1947- IV. Series.

QL703.M357 2005
599--dc22

2005018329

Contents

James L. Patton served as Curator of Mammals in the Museum of Vertebrate Zoology (MVZ) and as Professor of Integrative Biology (formerly Zoology) at the University of California, Berkeley, from January, 1969 until June, 2001. During his 32 years as a curator and a member of the Berkeley faculty, Jim made an indelible mark on vertebrate evolutionary biology through his tireless pursuit of excellence in research and teaching. In addition to significantly advancing studies of mammalian evolutionary genetics, systematics, and phylogeography, Jim was instrumental in shaping the careers of vertebrate biologists throughout the Americas. Given the magnitude of his impact on studies of mammals, it seemed only appropriate to celebrate Jim's retirement from the Berkeley faculty by compiling a volume that reflects the breadth of his contributions to vertebrate biology. At the same time, everyone involved in the project agreed that the volume should capture something of Jim, the person. As those of us who have had the privilege of working with him know, Jim is an enthusiastic, generous, no-nonsense individual who doesn't hesitate to support his students and colleagues in any way that he can. Thus, while Jim's intellect and work ethic have made him a successful scientist, it is his personality that has endeared him to so many of his students and colleagues. Here, we try to capture both elements of Jim's career. The result is a series of rigorous, original research papers combined with more informal recollections of Jim's activities as a scholar, mentor, and museum curator. For those readers who have not had the opportunity to interact with Jim, we hope that the following pages will bring to life both the distinguished career and the distinctive personality of this highly respected evolutionary biologist.

As a starting point for this undertaking, we begin with a brief history of the events leading to the production of this volume. The idea of celebrating Jim's long and productive career originated with several of his former graduate students, notably Márcia Lara, Yuri Leite, and Leonora Costa, all of whom had worked with Jim during his extensive studies of the phylogeography of Amazonian mammals. Knowing that he was likely to object to any type of "fuss" on his behalf, Jim was not informed that a celebration was in the works until plans for the event were well underway. Originally envisioned as a small, informal get-together intended to coincide with Jim's retirement from the Berkeley faculty, "Pattonfest" quickly grew into a three-day affair that included oral and poster presentations of original research, a banquet, and the premiere of a video tribute to Jim compiled by Alison Chubb, then a graduate student in the MVZ. In the end, more than 120 of Jim's colleagues (many of whom were former students and postdoctoral scholars) attended the event, which was held 22-24 June, 2001, in the Valley Life Sciences Building on the UC Berkeley campus.

One outgrowth of Pattonfest was the decision to prepare a volume of original research papers authored by individuals who have worked with Jim as students, colleagues, or both. During the planning stages of this endeavor, we (Eileen Lacey and Philip Myers) were enlisted to see the project to fruition by serving as editors of the volume. Needless to say, this would not have been possible without the help of numerous individuals, beginning with the "Shipwreck Committee" - the band of students (most of whom have now fled to other countries) who conceived of, organized, and executed Pattonfest with great care and skill. In particular, we thank Yuri Leite, Leonora Costa, Maria Soares, and Diogo Meyer for putting together an excellent party. Carol Patton was a willing co-conspirator who helped tremendously by making sure that Jim was not off in the field during the big weekend. The faculty and staff of the MVZ – in particular Craig Moritz and Annie Caulfield - provided considerable assistance with the logistic and financial aspects of the celebration. The event itself would not have been such a success without the participation of the many individuals who traveled to Berkeley to help celebrate Jim's career. With regard to the preparation of this volume, we thank the contributors, each of whom was willing to share the results of their work in this format. Javier Rodríguez-Robles acted as an informal third editor, offering excellent advice on numerous points during preparation of the volume Karen Klitz provided invaluable assistance with the preparation of the tables and figures. Mary Beth Rew, Julie Woodruff, and John Wieczorek contributed to the formatting of the volume. For guidance during preparation of the manuscript, we thank John Lynch, Michelle Echenique, and Charles Crumly of the University of California Press. And, finally, thanks (of course) to Jim for simply being himself.

Part I: North America and the Origins of a "Curator's Curator"

James L. Patton began his career at the University of California, Berkeley, in January of 1969. Jim's position at Berkeley consisted of a 50% appointment in the Department of Zoology (now the Department of Integrative Biology) and a 50% appointment in the Museum of Vertebrate Zoology (MVZ). Although officially Jim's responsibilities as a faculty member were equally divided between these units, it was the Museum that quickly became the focal point of his professional activities. Initially, Jim had little formal experience with museum curation – it was his expertise in mammalian cytogenetics that had attracted the attention of MVZ Director Oliver Pearson, leading to the offer of a faculty position. Following his appointment, Jim immersed himself so thoroughly in all aspects of the MVZ and its collections that, at the time of his retirement, he was generally regarded as one of the world's leading curators of mammal specimens.

Jim has made numerous fundamental contributions to museum-based studies of mammalian evolution. To date, he has deposited nearly 20,000 specimens in the MVZ, making him the most prolific collector of mammal specimens in that institution's nearly 100-year history. These specimens have formed the core of his research program, which has generated more than 140 publications, most of which are peer-reviewed contributions to leading professional journals. While Jim embodies the rugged field ethic, enthusiasm for collecting, and broad knowledge of organismal biology traditionally associated with curators of natural history museums, he is also an innovator who does not hesitate to try new technologies if they will improve the quality of his research or the efficiency of his curatorial activities. To this end, Jim was instrumental in establishing facilities for modern molecular genetic research in the MVZ and he was an early proponent of electronic museum data storage, including the development of the Mammal Networked Information System (MaNIS), an internet-accessible network of specimen data from 17 mammal collections in North America. Jim's absolute commitment to natural history research and his intellectual as well as practical contributions to the MVZ have been instrumental in maintaining the international prestige of this institution. The quality and impact of his work with mammal collections have led to Jim's colleagues to refer to him as a "Curator's Curator."

A key component of Jim's ontogeny as a curator was his comprehensive analysis of the evolutionary genetics of pocket gophers (Rodentia: Geomyidae: *Thomomys*), which contributed more than 6,600 specimens to the MVZ's mammal collection. This now-classic work, which marked the first phase of Jim's career at Berkeley, resulted in a substantial revision of geomyid systematics and produced

1

significant new insights into the processes underlying the diversification of vertebrate populations. It also established Jim as an expert on the small mammal fauna of western North America. In recognition of this work and the geographic focus of Jim's early research, the first section of this volume focuses on studies of North American mammals.

The volume opens with a biographical sketch of James L. Patton by Javier Rodríguez-Robles and Harry Greene. This is an informal overview of Jim's life that summarizes his numerous professional accomplishments while at the same time providing insights into the personality underlying these achievements. For those readers who know Jim, many of the quotes contained in the paper will be immediately recognizable. For those who have not had the opportunity to interact with him, this contribution serves to bring to life this highly accomplished researcher who, when asked to characterize his career, tends to respond that he is "just a rat trapper." As Rodríguez-Robles and Greene reveal, there is considerably more to Jim Patton than his ability to catch rats, although, as his collecting record indicates, capturing rodents is something at which he excels.

The second paper in the volume uses recent ecological changes in the San Joaquin Valley of California to explore the role of natural history collections in the preservation of vertebrate biodiversity. A central theme of this contribution by Patrick Kelly and colleagues is the philosophy for natural history research outlined by Joseph Grinnell, the first Director of the Museum of Vertebrate Zoology (1908-1939). Grinnell's prescient writings regarding the importance of collections to conservation biology have significantly influenced many researchers affiliated with the MVZ, including Jim Patton. As Kelly and his co-authors indicate, the need for collections-based research continues and, indeed, the value of existing collections is growing as new technologies increase the ways in which specimens can be used to understand threatened faunal assemblages. Thus, the paper by Kelly et al. provides critical insights into the historical legacy of Jim's position as Curator of Mammals, as well as the role that his contributions to the MVZ are likely to play in future efforts to protect California's vertebrate fauna.

The third paper in this section, co-authored by Hopi Hoekstra and Michael Nachman, uses state of the art genetic technology to explore an old problem in mammalian natural history – the evolution of pelage color variation among the pocket mice (genus *Chaetodipus*) of the southwestern United States. As first reported in the 1920's and 1930's, conspecific pocket mice living on light sand versus dark volcanic substrates differ phenotypically by displaying, respectively, light and dark pelage. By examining specimens from different substrates, Hoekstra and Nachman identify the genes associated with variation in coat color in these animals. Their work represents an important effort to link phenotypic differences to underlying patterns of genotypic variation and their data yield some surprising results regarding the multiplicity of ways in which selection can act to resolve a single adaptive challenge. The authors' use of molecular technology to address a fundamental question in mammalian biology mirrors the integration of organismal and genetic research that has characterized much of Jim Patton's career.

The final contribution in this section examines the effects of large-scale environmental changes on the distribution of mice in Michigan. Using a naturally occurring contact zone between white-footed mice (*Peromyscus leucopus*) and deer mice (*P. maniculatus*), Philip Myers and his co-authors relate changes in the relative abundance and geographic distribution of these species to annual variation in environmental conditions. The authors' synthesis of field observations and museum records parallels the integration of modern and historical data outlined by Kelly et al. and underscores the critical contributions that collections-based researchers such as Jim Patton are making to efforts to understand faunal change. Myers and his colleagues report a striking decline in deer mice in northern Michigan that provides sobering evidence of the potential effects of environmental perturbations - including human-induced climate changes - on future mammalian diversity.

Genes, Rats, and Sinking Boats:
A Biographical Perspective on James L. Patton

Javier A. Rodríguez-Robles and Harry W. Greene

"Every naturalist of great worth has had his own combination of ways by which he has influenced his science and his associates in it."
Alden H. Miller (1964: 235)

Every scientific discipline has a few practitioners whose influence is pervasive and long-lasting. In the case of cytogenetics, systematics, mammalogy, and phylogeography, James Lloyd Patton has been one of those individuals. Accordingly, a brief biography of "Jim" seems in order as an introduction to this volume, which celebrates Jim's career as Professor of Integrative Biology and Curator of Mammals at the Museum of Vertebrate Zoology, at the University of California, Berkeley. We quickly recognized that this account would be much more true to Jim's impact on biology if it revealed something about the lesser-known aspects of his life, rather than just chronicling the most salient features of his distinguished career. That two authors at markedly different stages of their careers (J.A.R.-R. is an assistant professor; H.W.G., a senior faculty member) and who are herpetologists were invited to profile someone widely recognized as the foremost mammalian systematist of his generation may seem peculiar. But, to us, this merely supports our contention that Jim's influence is far-reaching, and that he has been, and continues to be, a role model to junior and senior researchers from diverse organismal disciplines. When Jim learned that we were writing this synopsis, he exclaimed, in his characteristically unpretentious, but candid way, "Why are you wasting your time? Don't you have anything better to do?" Our answer to the latter question was – and still is – a categorical "No." What better way to salute a great scholar, personal hero, and dear friend? Despite his uneasiness, Jim agreed to be interviewed (by J.A.R.-R.) for this account, and what follows is based largely on that three-hour conversation that took place on the morning of December 7, 2001 at Jim's office in the Museum of Vertebrate Zoology.

THE ADVENTURES OF THE YOUNG JIM PATTON

Jim was born in Saint Louis, Missouri, on June 21, 1941. His father, John Franklin Patton II, was an urologist, and his mother, Marjorie Allyn Marquardt, a homemaker. Jim's father was in the United States Army Reserve, but his hospital unit was called into active duty during World War II, and he served in Africa and

Europe. After the war ended, he went back into civilian life, but during the Korean War took a regular Army commission. He retired as Chief of Surgery at Walter Reed Army Medical Center in Washington, D.C. in 1963 (Layne and Hoffmann, 1994). As a result of his father's career, Jim, the younger of two brothers (John Franklin III is senior by two and a half years), spent many of his early years as an "army brat" at military bases in the United States (Denver, Colorado and Washington, D.C.) and Landstuhl, Germany. Jim credits his father with instilling in him a canon by which he has lived, namely "whatever you do, do it right, and have fun doing it." He also gave Jim complete freedom to pursue his own professional interests.

Growing up in Saint Louis in the 1940's, Jim was destined to be a baseball fan. Initially, he favored the Saint Louis Browns (now the Baltimore Orioles), perhaps because at a church function he won a baseball that had been autographed by Bill Veeck, the Browns's colorful owner. However, when Martina Marion – daughter of the great Cardinals shortstop Marty "Hands" Marion – ended up in Jim's third grade class, he immediately became a Cardinals (and a Martina) fan. Another one of Jim's close friends was "Dicky" Musial, son of Cardinals outfielder Stan "The Man" Musial. Because of his friendship with Martina and Dicky, Jim often went to Cardinals games, where he and his friends collected player's autographs. Even though the Patton family moved from Saint Louis in 1952, to this day Jim remains a loyal Cardinals enthusiast.

"Jimmy," as he was known to his family, was a "good kid" who rarely got into trouble or did anything to upset his parents. His relationship with his brother John was a little less tranquil. Their interactions took a contentious turn in 1947 that lasted ten years. For his sixth birthday, Jim had received some toy lead soldiers, and he and his brother were playing with them by the creek behind their house. At one point John demanded that Jim give him one of the soldiers. When Jim refused, John picked up a rock, threw it at Jim and hit him on the head, causing an injury that bled profusely. This incident marked the beginning of a period of intense sibling rivalry that did not end until 1957, when John stayed in Germany to finish his senior High School year. When he reunited with his family in Washington, D.C. five months later, John and Jim suddenly became best of friends, and have been very close ever since.

Jim was always a good student. In high school in Maryland, he was a member of the National Honor Society, and received an honorable mention as a competitor for a National Merit Scholarship. He was well liked by his classmates, and was voted class president in his sophomore year.

Jim is a very shy person. He believes that the event that most helped him to overcome this tendency was his father's return to the military towards the end of the Korean War. That decision forced the family to live in various parts of the world, to meet new people on a regular basis, and to become immersed in different sets of circumstances.

With Jim, the adage "what you see is what you get" has always applied. Since childhood, he has *always* told it like it is, or at least like *he thinks* it is. Jim was very independent as a kid and seemed to know when he was correct and was not afraid to say so, qualities that he acquired from his father. He was also focused, and once he committed to something, he worked tirelessly to complete it, traits that undoubtedly helped him to attain so many remarkable achievements in his academic career. Indeed, Jim claims that his best quality is his perseverance.

The seeds of many of Jim's most salient personal characteristics were evident early on. He has always been very good with his hands and enjoys working with them. Wherever his family lived, Jim usually set up a workshop in his bedroom with an assortment of tools and paints, and he built model trains, cars, and planes. In high school in Germany, he built a wooden box designed specifically for surgical instruments, complete with drawers and individual compartments for each. During his first years as an Assistant Professor at U.C. Berkeley, he built wooden furniture for his home and for the laboratories at the Museum of Vertebrate Zoology. Jim's manual dexterity is patently visible in the thousands of perfect study specimens that he has prepared.

Jim's propensity towards museum curation was also apparent as a child. In Saint Louis, the Pattons lived in a rural area and Jim and his brother spent virtually all their waking hours outdoors. Jim collected fossils, Indian arrowheads, and other artifacts that he found in woods and fields around his house. With a friend, he put the objects on display and invited neighbors to visit their "museum," where the two would talk about the history of their collections. Accordingly, Jim became very interested in archeology and anthropology, read extensively on these subjects throughout junior high and high school, and decided to become a professional anthropologist.

LIFE AS AN ARIZONA WILDCAT

Jim applied to only two colleges - the University of Arizona, Tucson (U of A) and the University of New Mexico, Albuquerque - which represented the two most prestigious anthropology programs in the nation at that time. He was accepted at both institutions and chose to attend the former. Thus, in fall 1959 Jim became an undergraduate anthropology major at U of A.

During his freshman year, Jim took Introductory Anthropology, taught by Frederick S. Hulse. Jim did very well in the class, and Hulse exhorted him to get involved in the department. As a junior and senior Jim took several graduate seminars from Hulse and, as a senior, undertook a research problem (i.e., an undergraduate Honors Thesis) that Hulse supervised. Hulse encouraged Jim to go to graduate school and at least obtain a Master's degree in Anthropology from U of A. During his senior year, Jim took a required two-semester course in scientific illustration that taught students how to draw publication-quality maps and illustrations and take scientific photographs. Since childhood, he had liked drawing, and he thoroughly enjoyed the class and did well in it. When he went to graduate school the following semester, there were few available scholarships, and for his first year he was offered a teaching assistantship in the Scientific Illustration course. Don E. Wilson, now a senior mammalogist at the National Museum of Natural History in Washington, D.C., was in that class and recalls that "Jim was a patient and willing teacher even then, presaging his long and productive career as an educator."

In 1963 Jim received a Bachelor's of Arts degree in Anthropology (with distinction), and he began his graduate studies at U of A that fall. His ability as a draftsman proved extremely valuable, and he supported himself through his first summer in graduate school by working as a scientific illustrator. The textbook *Zoology*, by Cockrum and McCauley (1965), contains many of Jim's drawings, and McCauley referred to Jim as "the master of the fine line technique." As a graduate student, Jim continued earning money by drafting illustrations, including those of the type series of a subspecies of plethodontid salamander (Lowe et al., 1968). Not surprisingly, to this day, Jim draws virtually every illustration and takes all of the specimen photographs for his numerous publications.

In the fall semester of his second year in graduate school, Jim completed a substantial portion of his Master's thesis in Physical Anthropology. At that time he decided that "physical anthropologists ought to know more about mammals other than primates" and, after receiving special permission from his department, he enrolled that spring in E. Lendell Cockrum's Mammalogy course and William B. Heed's course in Human Genetics in the Zoology Department. The Mammalogy laboratory introduced students to mammals of the world, and Jim was fascinated by their diversity. In the second week of class, Alfred L. "Al" Gardner, a graduate student of Cockrum's and already a professional collector, invited Jim to accompany him to trap rodents on the Santa Rita Experimental Range, located approximately 56 km south of Tucson. Jim happily accepted, and went out on a Friday night to set traps for the first time. The Tucson area has an enormous diversity of small mammals and the first trap they checked the next day had a kangaroo rat (*Dipodomys* sp.) in it. Jim recalls that he was "floored by that animal, which [he] had never seen before, and floored by the diversity that [he] saw, and it just really piqued [his] interest."

That semester, Jim had to conduct a laboratory project in his Human Genetics course. One day, Heed walked into the laboratory and challenged his students to develop a technique that would allow a researcher to screen routinely the mitotic chromosomes of any animal. Jim accepted the challenge. Because he was taking Mammalogy concurrently, he decided to use small rodents as his study organisms, and, by the end of the semester, had succeeded in developing such a procedure.

The article that resulted from his work (Patton, 1967b) has since been designated a Science Citation Classic.[1]

Jim immediately realized that his newly developed protocol for chromosome analysis would give him "a window to look at critters in ways that nobody had ever been able to look at them before." In other words, he now could study "all those little furry creatures running around out there and their possible karyotypes." Physical anthropology no longer elicited this same kind of excitement from Jim. In fact, he was somewhat disenchanted with the state of the discipline at the time, and decided to switch fields. Hulse was very supportive, and advised Jim to transfer to the Zoology Department to pursue his newly-developed interest in genetics. Heed was equally encouraging and, although he did not know much about mammals or mammalian cytogenetics (his research focused on chromosome evolution in natural populations of *Drosophila* [desert fruit flies]), he invited Jim to join his laboratory and offered him a research assistantship. Although the assistantship required Jim to conduct research on chromosomal inversion polymorphisms in *Drosophila* (Patton and Heed, 1965), it provided him the opportunity to work in mammalian cytogenetics, as well as welcome financial support. Jim gladly accepted Heed's offer.

For his second Master's Thesis, this one on cytogenetics of *Perognathus* (pocket mice), Jim took a collecting trip to México with Al Gardner and Don Wilson. The three mammalogists worked their way down the western coast of mainland México, collecting *Perognathus* and driving south to San Blás, Nayarit, where their goal was to trap *Desmodus rotundus* (vampire bats) for the Arizona-Sonora Desert Museum. The Museum was going to pay for the bats, which was how Jim and Al were financing the trip. They knew they had to feed the bats on their way back, even though they were planning to drive nonstop. In anticipation of this, they had brought along a huge laboratory rabbit. Jim assured his field companions that he knew how to do a cardiac puncture on the rabbit, which would allow them to extract enough blood to feed the bats a couple of times, without injuring the rabbit. Having captured the bats, the collectors headed for home, and made their first feeding stop in a small glade of tropical forest that reaches its northern limit near San Blás. Al held the rabbit, and Jim carefully inserted a needle straight into its heart. The lagomorph gave one heavy sigh and promptly expired. While Jim looked puzzled and disappointed with the rabbit for giving up so quickly, Al swung into action and, holding the dead animal against a nearby tree trunk, relieved it of its head with a mighty swing of his machete. Exhorting Don to hold a dish under it, the small amount of blood that flowed from the corpse was carefully collected, defibrinated with glass beads, and fed to the bats using an

---

[1] Some workers (e.g., Robert J. Baker, Charles J. Cole, George C. Gorman) credit Jim's technique with allowing them to develop their own productive careers in comparative karyology.

eyedropper. Heading on their way, the young men began worrying almost immediately about how they were going to manage another feeding without their involuntary blood donor. After the idea of using the lone undergraduate on the trip (Don) was reluctantly abandoned by Jim and Al, a complicated plan was hatched (involving impersonating physicians, an elderly swaybacked horse, a cash incentive, and bilingual cursing) that eventually produced enough blood to feed the bats, all of which survived, thereby providing the wherewithal to pay for the trip. More importantly, Jim got the necessary chromosomal material from his pocket mice, and an undergraduate student got a priceless education in field techniques.

The topic of Jim's thesis (Patton, 1965) was a descriptive, comparative karyological study of several species of *Perognathus*, which he interpreted in an evolutionary context (Patton, 1967a, 1967b). One of the species was *Perognathus goldmani* (now *Chaetodipus goldmani*, Goldman's pocket mouse), which consisted of several distinct chromosomal races that were separated geographically. Jim described some of those races in his thesis and he envisioned that his doctoral dissertation would be a continuation of that work, albeit in much greater detail (e.g., exploring the origin of those races, and whether they interbred in sympatry). However, while conducting fieldwork in the Patagonia Mountains of southern Arizona, he discovered a hybrid zone between two species of pocket gophers, *Thomomys bottae* (Botta's pocket gopher) and *T. umbrinus* (southern pocket gopher) and started to devote more and more time to that project. Consequently, he found himself debating whether to emphasize pocket mice or pocket gophers in his dissertation.

Jim never considered going to any institution other than U of A for his doctorate. He was comfortable at U of A, felt confident about his research, was engaged in numerous projects, and, having spent only one year in the Zoology Department, was still learning biology. Further, he was doing collaborative work with other graduate students and even faculty in his new department, specifically Charles J. Cole, John W. Wright, and Charles H. Lowe on karyotypic diversity in *Sceloporus* (spiny lizards; Lowe et al., 1967) and *Aspidoscelis* (formerly *Cnemidophorus*, the whip-tailed lizards), and Robert J. Baker on chromosome variation in bats (Baker and Patton, 1967; Nelson-Rees et al., 1968).

But Jim's research did not consume all of his waking hours. In the spring of 1961 he met Carol Porter, a freshman, at his U of A fraternity. Carol graduated from the university in 1964, and after spending approximately six months in Europe, returned to her native New York. After completing his Master's degree in May 1965, Jim went to Brown University, Rhode Island, in July of the same year to take a two-week workshop on mammalian cytogenetics.[2] Jim contacted Carol to

---

[2]At that workshop, offered by The Wistar Institute of Philadelphia and the Rhode Island Hospital in Providence, Jim met Tao-Chiuh Hsu, who was the dominant

let her know that he was going to be in Providence, and suggested that they could get together one weekend. Jim maintains that he "just wanted to say hi," and that he had no ulterior motives. However, on June 4, 1966, Jim married Carol, "an intrepid trapper and field companion" (*Dedication* of Patton and Smith, 1990), and has been "saying hi" to her ever since.

After a New York wedding, Jim and Carol drove across country and combined their honeymoon with the American Society of Mammalogists annual meeting in Long Beach, California. This was both Jim's first national meeting and the first time that he presented a professional paper. The talk, on the pocket gopher hybrid zone he had recently discovered in southern Arizona, caught the attention of the late Oliver P. "Paynie" Pearson, Director of the Museum of Vertebrate Zoology (MVZ) at U.C. Berkeley. Later that summer, Pearson telephoned Jim's major professor to ask how long it would take Jim to finish his dissertation, and whether Jim would be interested in coming to Berkeley after completing his degree. When told about Paynie's inquiry, Jim's first reaction was along the lines of "Berkeley? Museum of Vertebrate Zoology? Where's that? Who's that?" He would soon find out.

THE BERKELEY AFFAIR

When Seth B. Benson's retirement as an Associate Curator of Mammals in the Museum of Vertebrate Zoology became imminent, Paynie Pearson and Associate Curator of Mammals William Z. "Bill" Lidicker Jr. started to search for possible recruits for Benson's position. Paynie and Bill quickly focused on Jim as a desirable candidate, and on May 10, 1967, a year and a half before he completed his Ph.D. degree, Jim gave a seminar at "MVZ Lunch," an ongoing, long-standing Wednesday noon-time seminar series in evolutionary biology sponsored by the Museum. That seminar marked the beginning of the Museum's successful recruitment efforts.

In the spring of 1968, Jim returned to Berkeley to conduct an official interview as the sole candidate for Benson's position. Despite the fact that he "was scared to death," he was offered the position of Assistant Curator of Mammals, and Assistant Professor in the Department of Zoology, starting July 1968. However, Al Gardner, then a Ph.D. student at Louisiana State University, Baton Rouge, invited Jim to accompany him and fellow graduate student John P. O'Neill to Perú that summer. Jim did not want to miss the opportunity to work both in the Amazon River basin and with a monolingual Indian tribe, the Cashinahua Jívaros. After Pearson agreed to let him postpone his arrival at Berkeley until January 1969, Jim

---

figure in mammalian cytogenetics at the time, and who subsequently (summer 1966) gave Jim a summer research position in his laboratory at the M. D. Anderson Hospital in Houston, Texas (Patton and Hsu, 1967).

accepted the Museum's offer and left for South America. Upon returning from Perú, he realized that he did not have enough time to finish the extensive fieldwork necessary for his study of the pocket gopher hybrid zone or to complete all of the pocket mouse chromosome work. Instead, he was only able to finish his research on karyotypic race evolution in Goldman's pocket mouse. Jim filed his 67-page dissertation (Patton, 1968) days before moving to California, and in June 1969 received his doctorate in Zoology, with a minor in Geochronology.

Recalling his first years at Berkeley, Jim remarked,

> "When I arrived [on campus] I knew that I was in a game that I had never experienced before. I was convinced that Berkeley would see the error of its ways, and would pack me up and I'd be going some place else in the not-too-distant future. And in fact, I think that being convinced of that took all the pressure off of me. I decided to do my own thing because I knew I could not measure up to those guys [his colleagues at the MVZ and the Zoology Department], so I did not even try. And so I never really felt any pressure about being [a non-tenured professor at Berkeley], but I used to avoid like the plague running into [the late] Frank Pitelka[3] any place or having a conversation with him."

Interestingly, Jim did not apply for a National Science Foundation grant during his entire career as an Assistant Professor, because he thought that his publications were the only relevant criterion for advancement. Hence, when he was being considered for tenure, he was surprised to learn that the fact that he did not have extramural research funds was a negative aspect of his dossier. Despite his concerns, Jim was promoted to Associate Professor of Zoology and Associate Curator of Mammals in 1974, and to Professor of Zoology and Curator of Mammals in 1979.

The aspects of academia that Jim likes most are "the challenges that the students provide [and that] they keep [him] alive and fresh; the quality of the undergraduates at Berkeley, even though they do not always live up to their capabilities; the freedom to set one's own program and pursue one's own interests; and the MVZ's policy of encouraging faculty to spend as much time in the field as possible." As anyone who knows Jim would predict, the aspect of academia that he *intensely* dislikes is "all the bureaucratic bullshit you have to put up with, and that gets in the way of us doing whatever it is that we are doing."

In 1987, the Field Museum of Natural History (FMNH) in Chicago, Illinois conducted an international survey to identify the best mammalian systematist in the world, with the goal of offering that individual a senior curatorial position. Jim

---

[3]A renowned ecologist in the Department, famous for his encyclopedic knowledge of biology and propensity for engaging students and colleagues in intellectually challenging exchanges.

was the unanimous choice for the job, and although he seriously entertained the idea of going to FMNH, he ultimately decided to remain at U.C. Berkeley. Greatly concerned about losing Jim to another institution, the University allocated $80,000 for Jim to spend at his discretion, funds that he used to expand the MVZ Genetics Laboratory to include DNA methodologies.

"PEDAL TO THE METAL"

Jim calls himself an "evolutionary systematist who happens to work on mammals," using the term "systematist" in the same manner as the eminent paleontologist George Gaylord Simpson, meaning a student of the diversity of life, not merely of taxonomy and phylogenetics. Jim's primary research interests are population genetics, geographic divergence, systematics, and the biogeography of small, nonvolant mammals, with an emphasis on geomyid and South American rodents, as well as marsupials.

Jim's first publications dealt almost exclusively with evolutionary cytogenetics, and these and subsequent articles established him as a leader in the field (Baker and Patton, 1967; Patton, 1967a, 1967b, 1969a, 1969b, 1970, 1971, 1972a, 1972b, 1973, 1977, 1981; Patton and Hsu, 1967; Patton and Dingman, 1968, 1970; Patton and Gardner, 1971; Patton and Myers, 1974; Gardner and Patton, 1976; Pearson and Patton, 1976; Bush et al., 1977; Patton and Yang, 1977; Reig et al., 1977; Patton and Feder, 1978; Patton et al., 1979; Patton and Sherwood, 1982, 1983; Hafner J. et al., 1983; Barros and Patton, 1985; Hafner M. et al., 1987; Patton and Rogers, 1993). In fact, only two years after completing his doctorate, he had already earned an international reputation as an authority on the application of karyotypic data to systematic research on mammals.

Jim's detailed studies of the evolutionary genetics of *Thomomys* (pocket gophers) have produced a lengthy series of often cited publications on hybrid zone dynamics, genetic demography of local populations, and phylogeographic structure, and on the relationships of these factors to understanding species boundaries and speciation processes (Patton and Dingman, 1968, 1970; Patton, 1970, 1971, 1972b, 1973, 1981, 1985, 1990, 1993c; Patton et al., 1972, 1979, 1984; Patton and Yang, 1977; Patton and Feder, 1978, 1981; Smith and Patton, 1980, 1984, 1988; Patton and Smith, 1981, 1989, 1990, 1993, 1994; Patton and Sherwood, 1982; Sherwood and Patton, 1982; Hafner J. et al., 1983; Smith et al., 1983; Barros and Patton, 1985; Daly and Patton, 1986, 1990; Hafner M. et al., 1987; Patton and Brylski, 1987; Lessa and Patton, 1989; Patterson and Patton, 1990; Ruedi et al., 1997), as well as species accounts and popular articles (Patton, 1984a, 1993a, 1999a, 1999b, 2001). Although only three of his graduate students (María Alicia Barros, Mary Anne Rogers, Patricia A. Garvey-Darda) specifically worked on pocket gophers for their theses, other students (Juliana H. Feder, Margaret F. Smith, Mark S. Hafner, John C. Hafner, Steven W. Sherwood) and postdoctoral associates

(Joanne C. Daly, Enrique P. Lessa, Manuel Ruedi) have coauthored papers with him on these genetically diverse animals.

Together with his collaborators, Jim has published five monographs: *La Clasificación de los Mamíferos de los Aguaruna, Amazonas, Perú* (Berlín and Patton, 1979), *A Review of the* Boliviensis *Group of* Akodon *(Muridae: Sigmodontinae), with Emphasis on Peru and Bolivia* (Myers et al., 1990), *The Evolutionary Dynamics of the Pocket Gopher* Thomomys bottae, *with Emphasis on California Populations* (Patton and Smith, 1990), *Phylogeography and Systematics of the Slender Mouse Opossum,* Marmosops *(Marsupialia, Didelphidae)* (Mustrangi and Patton, 1997), and *Mammals of the Rio Juruá and the Evolutionary and Ecological Diversification of Amazonia* (Patton et al., 2000), which was featured in the *Forum* section of *Trends in Ecology and Evolution* (Sites, 2001). The latter monograph, possibly the most exhaustive study of Brazilian mammals completed during the last 150 years, reports the results of a ten-month expedition (August 1991 to June 1992) to western Brazil to study the genetics, systematics, ecology, and biogeography of small mammals along the Rio Juruá, the largest white-water tributary of the Amazon River with headwaters to the east of the Andes. Jim and three colleagues, Maria N. F. "Lelé" da Silva, Jay R. Malcolm, and Carlos A. Peres, surveyed the mammal fauna of the river system to test Alfred Russel Wallace's riverine barrier hypothesis (Wallace, 1852), which in its modern version states that the degree of differentiation of conspecific populations should increase along both sides of a river, from its headwaters to the mouth, as the barrier widens and the potential for cross-river gene flow diminishes. This elegant research was the subject of reports in *Science* (Morell, 1996a, 1996b) and *Discover Magazine* (Morell, 1997).

Jim and his colleagues have described species of bats (*Mimon*, Gardner and Patton, 1972), marsupials (*Philander*, Gardner and Patton, 1972), and rodents (*Oxymycterus*, Hinojosa et al., 1987; *Akodon*, Myers and Patton, 1989a, 1989b; Myers et al., 1990; *Punomys*, Pacheco and Patton, 1995; *Scolomys*, Patton and da Silva, 1995; *Isothrix*, Vié et al., 1996; *Neacomys, Rhipidomys, Mesomys*, Patton et al., 2000; *Trinomys*, Lara et al., 2002), and conducted revisions of multiple genera (*Proechimys*, Patton and Gardner, 1972; Patton, 1987; *Isothrix*, Patton and Emmons, 1985; *Scolomys*, Patton and da Silva, 1995; *Philander*, Patton and da Silva, 1997) and groups of species within genera (taxa formerly placed in the subgenus *Urosciurus* of *Sciurus*, Patton, 1984b; the *fumeus* and *boliviensis* groups of *Akodon*, Myers and Patton, 1989b; Myers et al., 1990; Peruvian *Akodon*, Patton and Smith, 1992a). Jim has also published checklists and species accounts of mammals from the New World (Patton, 1993a, 1993b; Myers and Patton, in press; Patton and da Silva, in press; Patton and Gardner, in press; Patton and Stein, in press a, b), and more specifically from southeastern Perú (Pacheco et al., 1993), Brazil (Fonseca et al., 1996; Patton et al., 2000), and northwest México (Patton, 1999a, 2000; Patton and Álvarez-Castañeda, 1999). He has co-edited four volumes: *Mamíferos del Noroeste de México* (Álvarez-Castañeda and Patton, 1999), *Mamíferos del Noroeste de México II*

(Álvarez-Castañeda and Patton, 2000), *Life Underground: The Biology of Subterranean Rodents* (Lacey et al., 2000), and *Handbook of South American Mammals, Volume I* (Gardner et al., in press), and is currently working with his co-editors on Volumes II and III of the latter series.

In addition to his extensive studies of rodents and marsupials, Jim has conducted research on fruit flies (Patton and Heed, 1965), lizards (Lowe et al., 1967), caecilians (Wake et al., 1980), tortoises (Marlow and Patton, 1981), finches (Yang and Patton, 1981), bats (Baker and Patton, 1967; Nelson-Rees et al., 1968; Patton and Gardner, 1971), and monkeys (Peres et al., 1996). He has also published on such diverse topics as ethnomammalogy (Patton et al., 1982; Fleck et al., 1999), the origin of eutherian mammals (Lillegraven et al., 1987), and demographic responses to climate change (Lessa et al., 2003).

Jim has been among the leaders in applying new types of data (i.e., karyotypes [Patton, 1969a], allozymes [Patton et al., 1972], mitochondrial DNA sequences [Smith and Patton, 1991b]) to genetic analyses, and has endeavored to understand the nature of molecular markers and their relevance to organismal evolution (Patton and Sherwood, 1982; Smith et al., 1992). He is also responsible for major conceptual advances in our understanding of chromosome evolution and speciation, and for providing a population genetic perspective on the nature of species (Bush et al., 1977; Reig et al., 1977; Patton and Sherwood, 1983; Patton and Smith, 1994). Jim was also one of the first workers to use molecular genetic data to test explicitly hypotheses of evolutionary and ecological diversification (Patton et al., 1990, 1994; Patton and Smith, 1992b). Although his research is centered on the organism, his approach to biology is synthetic (e.g., Patton and Yang, 1977; Patton et al., 1979, 2000; Daly and Patton, 1986; Patton and Brylski, 1987; Patton and Smith, 1990, 1993; Moritz et al., 2000; Figure 1).

Despite this impressive list of accomplishments, Jim does not believe that he has made any major contributions to science. He acknowledges that he is proud of what he has done, and he sees himself as a meticulous worker who generates a lot of data and who is conservative in their interpretation. "In doing what I do, I try to do the best I can in my limited capabilities, intellectually or technically. I've tried to extend beyond the norm." Jim strongly believes that he is "the product of the intellectual development of colleagues (notably David B. Wake) in the atmosphere of the MVZ, and as such [he] epitomizes the case of nurture over nature."

Whatever Jim has gained from his associates, he has repaid them many times over. As a citizen of many constituencies, ranging from close Berkeley colleagues to biologists the world over, Jim is phenomenally responsible, supportive, and inspiring. He has served on countless committees within and outside of the University, and he always contributes beyond the normal call of duty. Within his own discipline, Jim served as President of the American Society of Mammalogists from 1992 to 1994. In 1983 he received the Society's C. Hart Merriam Award, in recognition for his "outstanding contributions to the discipline of mammalogy." In 1998 he received the Society's The Joseph Grinnell Award for Excellence in Education in Mammalogy, and in 2001 was awarded an Honorary Membership. The latter is the highest award given by the Society, for "significant, lifetime achievements in science and service to the discipline of mammalogy."

James L. Patton's Science:
A Model of Holistic Evolutionary Biology

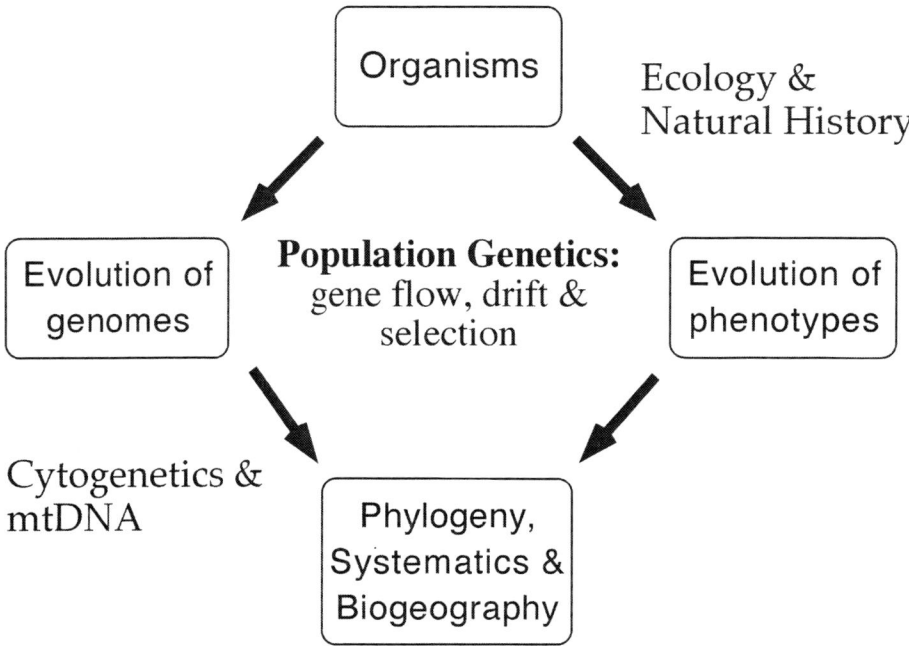

Figure 1. A diagram of James L. Patton's philosophy of science (modified from a talk given by Craig Moritz on June 24, 2001 at the University of California, Berkeley).

Jim never seems to tire of teaching naive but interested herpetologists about tropical rats and desert mice, often on the spur of the moment, and ideally over a tray full of museum specimens. For an academic, he is remarkably free of jealousy toward his closest colleagues, instead regarding their accomplishments with obvious pride. In all aspects of his professional life, Jim leads by example. As David Wake, Director of the Museum of Vertebrate Zoology from 1971 to 1998, said, "The most remarkable thing about Jim is that he does it *all himself*, and that he *masters everything* he does. Jim wakes up in the morning, puts on his pants, and goes to work." Enrique P. Lessa, a former postdoctoral associate of Jim appropriately noted, "I also put on my pants after getting out of bed, but all similarities between Jim and I end right there."

GUIDING GOLDEN BEARS

Jim is an accomplished educator. Despite the fact that one of his favorite mottos is "the only good student is a *stressed* student," both undergraduate and graduate disciples find his teaching style engaging and thought-provoking. At U.C. Berkeley, he has taught undergraduate courses in Introductory Biology, Cytogenetics, the mammal section of Vertebrate Natural History, and Mammalogy, in addition to a graduate seminar on speciation, and occasional graduate colloquia on topics such as Alfred Russel Wallace and phylogeography. An easily overlooked aspect of Jim's teaching at Berkeley is his joint establishment with Dill Lidicker of the Mammal Discussion Group, which ran from 1973 to 1997. The group provided an intellectual and "social" home for mammalogists with diverse interests and from several departments on campus.

Jim mentored five masters and 31 doctoral students, and sponsored eleven postdoctoral researchers during his tenure at U.C. Berkeley (Table 1). He claims that with respect to his graduate students he "was schizophrenic, and did a very poor job of mentoring them in the most important way: intellectually [=conceptually], because [he is] not very intellectual [himself]." Nevertheless, he believes that he "did a good job of helping [his] students develop self confidence, and giving them freedom to pursue the questions they were interested in. [He] developed a strong sense of collegiality with all [his] students, and they "didn't hesitate in coming to [him] when they had personal or other kinds of problems." On several occasions when his students were experiencing harsh financial times, Jim paid their registration fees, bought them field equipment, computers, and even paid for their laboratory research out of pocket. Jim affirms that what he has always had to offer is a "strong push" and a work ethic that demands that he tries to do anything to the best of his abilities - if he does not accomplish his goal it is not for lack of trying. He hopes that if he is going to have any legacy, it is that he instilled in his students the canon that "you are never going to get anywhere in life unless you put out the effort personally; push yourself as far as you can go, and rely on others around you to help push you a little bit further." He proudly asserts that *all* of his students completed their degrees and, for the most part, they have gone on to have productive careers. He has coauthored articles with many of his former graduate students, but his associations with Margaret F. "Peg" Smith and Maria N. F. da Silva have been especially prolific (Patton et al., 1979, 1984, 1989, 1990; Smith and Patton, 1980, 1984, 1988, 1991a, 1991b, 1993, 1999; Patton and Smith, 1981, 1989, 1990, 1992a, 1992b, 1993, 1994; Hafner J. et al., 1983; Smith et al., 1983, 1992, 2001; Zink et al., 1985; Hafner M. et al., 1987; Myers et al., 1990; da Silva and Patton, 1993, 1998; Patton et al., 1994, 1996a, 1996b, 1997, 2000; Patton and da Silva, 1995, 1997, 1998, 2001, in press; Lara et al., 1996; Peres et al., 1996; Ruedi et al., 1997; Gascon et al., 2000; Matocq et al., 2000; Geise et al., 2001).

Jim's teaching abilities are well known in México and South America. Perhaps he is most admired in Brazil, as from 1988 to 2001 he advised six doctoral students (Table 1), and since 1989 has trained dozens of Brazilian students and senior researchers in the field and in the laboratory. Every one of them has been fascinated and has gained from his skills as a field biologist and as an educator.

In 1991, Jim received the Distinguished Teaching Award, granted by the Academic Senate of the University of California, Berkeley, in recognition for his excellence in this arena. As noted above, in 1998 the American Society of Mammalogists presented him with The Joseph Grinnell Award for Excellence in Education in Mammalogy "in recognition for outstanding and intense education of graduate and undergraduate students at the University of California, Berkeley, in mammalogy, and for sharing his deep knowledge of mammals willingly with all students." From the extensive documentation for that case and our own years of observation, we know that Jim's success as a teacher stems, as with other activities, from his insistence on high expectations, his understated but profound affection and concern for others, especially for younger people, and from his leadership by example. Jim's class lectures are fast-paced, jam-packed with information, and flawlessly delivered. His praise for students' work is heartfelt and encouraging, and his criticisms are given in a straightforward manner that makes one want to do better. On field trips Jim is always the first one up the hill, and everyone flocks around whenever he stops to examine something. Students revere Jim, and it may not be an exaggeration to say they would walk off a cliff for him – if he went first.

Claudia Luke aptly described the feelings of hundreds of students who have taken Mammalogy from Jim in a letter she wrote to him on July 4, 2001:

> "I can say truthfully that your mammalogy class was the most challenging, rewarding, and exciting class that I've ever taken. It set a standard of excellence for me and gave me skills that I have used over and over again. In the last ten years, I have been learning more how people learn and trying to design field courses based upon those concepts. And even as I break down learning into its component parts, I find that all these pieces naturally came together in your class and teaching style. Your interest and enthusiasm, placing mammalogy as the subject rather than yourself as the expert, providing the materials and experiences that brought students together socially, motivating individuals to find what interests them, and above all expecting only the best from your students and yourself."

Table 1. Graduate students mentored and postdoctoral researchers sponsored by James L. Patton. For graduate students, names are listed in chronological order (starting in the left column), according to the year in which each person completed their degree (shown in parentheses).

Doctoral Students

Philip Myers (1975)
Sally J. Holbrook (1975)
Susan M. Case (1976)
Margaret F. Smith (1978)
Mark S. Hafner (1979)
Donald O. Straney (1980)
John C. Hafner (1981)
Steven W. Sherwood (1983)
Laurence G. Frank (1984)
William E. Rainey (1984)
Elizabeth D. Pierson (1986)
Duke S. Rogers (1986)
Ángel Spotorno O. (Chile) (1986)
Patrick A. Kelly (Ireland) (1990)
Claudia A. Luke (1992)
Hon-Tsen Yu (Taiwan) (1992)

Francis X. Villablanca (1993)
Márcia C. Lara (Brazil) (1994)
Mary Ellen Holden (1995)
Bernard Peyton III (1995)
Maria N. F. da Silva (Brazil) (1995)
Meika A. Mustrangi (Brazil) (1995)
Elizabeth A. Hadly (1995)
Albert D. Ditchfield (Brazil) (1996)
Luis F. García (Colombia) (1999)
Tina M. Hambuch (2000)
Marjorie D. Matocq (2000)
Randall S. Reiserer (2001)
Leonora Pires Costa (Brazil) (2001)
Yuri L. R. Leite (Brazil) (2001)
Steve Takata (2002)

Masters Students

Juliana H. Feder (1977)
Pamela L. Williams (1982)
María A. Barros (Venezuela) (1983)

Mary Anne Rogers (1983)
Patricia A. Garvey-Darda (1988)

Post-Doctoral Scholars

James T. Mascarello
Margaret F. Smith
Joanne C. Daly (Australia)
Enrique P. Lessa (Uruguay)
Liliana I. Apfelbaum (Argentina)
Sergio F. dos Reis (Brazil)

Eileen A. Lacey
Manuel Ruedi (Switzerland)
John Carlos Garza
Nathan R. Lovejoy (Canada)
Maria José de Jesus Silva (Brazil)

SEAMAN PATTON

Jim is notorious for the series of boating incidents – most of them actually accidents – that he and his unsuspecting traveling companions have experienced over the years. For purposes of completeness, we feel compelled to include a brief description of those events in this biographical sketch.

Episode I, The Omen: On November 1, 1954, Jim's mother and her two sons departed from New York aboard the U.S. Navy ship *A. M. Patch* for Bremerhaven, Germany. Halfway across the northern Atlantic the vessel encountered stormy weather, causing the trip to take longer than usual. Jim was completely unaffected by the rugged journey, but John was in misery most of the time, much to his brother's amusement.

Episode II: In early January 1966, Oscar Soule, Carol Patton, and Jim hired a Mexican fisherman to take them to Isla Dátil (=Turners Island), located southeast of Isla Tiburón, in the Gulf of California. On their way to the island they were hit by a sudden "chubasco" winter storm, their 15-footer boat flooded, and they were forced to seek refuge on Isla Tiburón. In the middle of the night the storm intensified and the waves smashed their dinghy against the rocks. As a result, the party, which originally was going to be out for only one night, was stranded for eight days before being rescued by another fisherman.

Episode III: Returning to Costa Rica in May 1972, after spending approximately four weeks conducting field work in the Galápagos Islands, the research vessel that Jim and his colleagues were using caught fire and burned to the water line, forcing all aboard into a crowded life raft. "Fortunately," the boat went down approximately 150 km off the Costa Rican coast, in the middle of a shipping channel for boats exiting the Panamá Canal. After drifting in the eastern Pacific Ocean for some hours, the shipwrecked passengers were rescued by a Mexican vessel en route to Kobe, Japan. The ship's captain received permission to veer into Mexican waters, and dropped off the researchers in Acapulco a few days later. Of all of Jim's boating accidents, this one ultimately proved the most serious, as essentially everything was lost – specimens, data, field notes, equipment, wallets, passports and other personal documents – and all vestiges of what had been an enormously successful trip sank to the ocean floor. The single exception was the captain's personal documents, which he managed to retrieve before boarding the life raft ahead of everybody else.

Episode IV, The Saga Continues: In August 1978, Jim and his colleagues left (by boat) from Huampami, an Aguaruna Jívaro indian village in northern Perú where they had maintained a base camp for two years. Shortly thereafter, their vessel sank in the Río Cenepa, resulting in another loss of specimens and data.

Episode V: When Philip "Phil" Myers, John E. Cadle, Carol, and Jim were working on the Río Alto Madre de Dios in southern Perú in May 1984, their boat, piloted by two rather intoxicated individuals, ran out of gas while traveling upriver, in rapids, and in near darkness. They almost capsized and had to struggle to get the boat to a bank of rough cobbles. They managed to preserve their belongings, but their situation was precarious for a few hours.

Episode VI, The Inevitable: In late July and early August 1989, Maria da Silva, Jay Malcolm, Carlos Peres, and Jim were in the Brazilian Amazon planning their survey of small mammals along the Rio Juruá. Jim pointed out that they should not begin their surveys in 1990, as that was the year of his next "scheduled" boating accident (notice the periodicity of Jim's boating accidents; Figure 2). As it turned out, it took two years to secure the necessary funding, and thus they began their expedition in August 1991. On February 3, 1992, their research vessel, the *Coró-Coró*, sank in the headwaters of the Rio Juruá following a collision with a submerged log (Patton et al., 2000: 29).

Prompted by these events, when asked to give advice to graduate students planning to accompany Jim to the field, Paynie Pearson immediately stated, "Don't go!" Urged to elaborate, he added "But if you must, find out whether there is any large body of water close to the field site, and if that's the case, make sure that you bring a life jacket, and take heavy insurance against freshwater and marine disasters."

VIGNETTES

Jim is widely known for his loud voice, tenacious – but not dogmatic – opinions, and fondness for punctuating his remarks with colorful expletives. He used to get mad with relative ease, and the warning "Stay away from Jim today!" was often heard in the hallways and graduate student offices of the old MVZ. Although his temperament has calmed in recent years, the possibility of witnessing an episode of rage still exists. Not long ago, his lack of success at figuring out how to get a new water cooler to provide him with hot water for his morning instant coffee unleashed Jim's dormant wrath in an extremely audible manner.

Among his personal heroes, Jim counts the celebrated naturalist Alfred Russel Wallace, the father of zoogeography and codiscoverer of the theory of evolution by natural selection; Theodore Roosevelt, the 26th President of the United States, a man of impressive intellectual achievements and an eminent naturalist whose presidential policies advanced conservation causes; George S. Patton Jr. (no close relation to Jim), probably the most admired and controversial of all United States generals in World War II, a superb military tactician, but also an impetuous, demanding, and ruthless character; and renowned evolutionary biologist David Wake, Jim's colleague in the MVZ since 1969, and his intellectual role model.

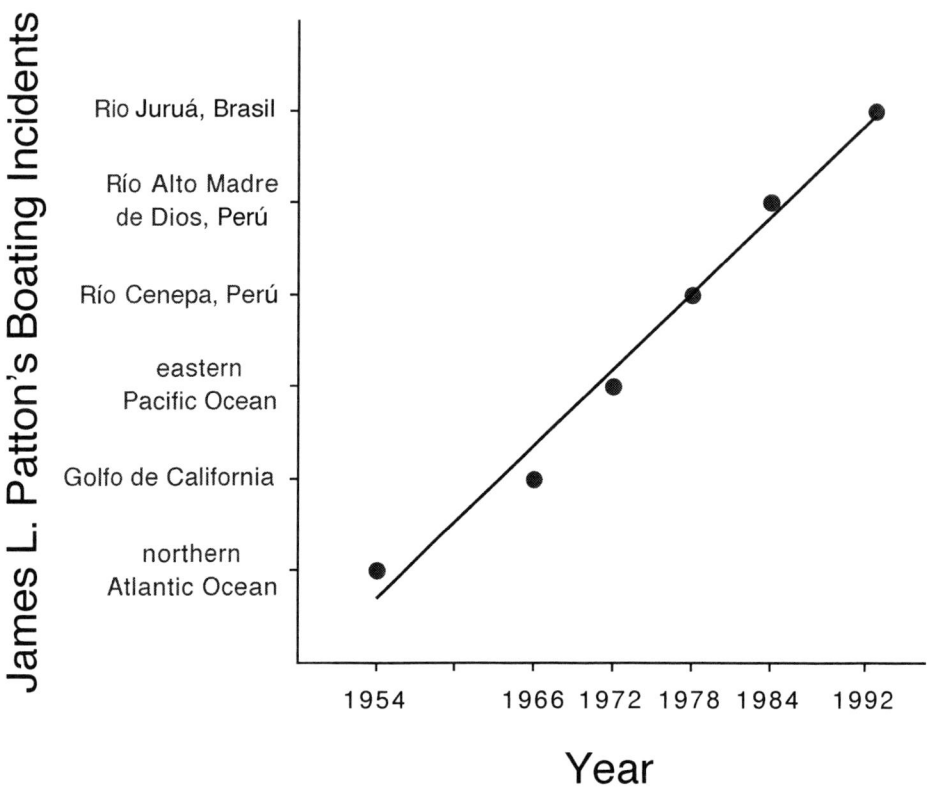

Figure 2. Chronology of James L. Patton's boating incidents (adjusted $r^2 = 0.98$, $F_{(1,4)} = 246.7$, $p < 0.0001$).

Jim is in excellent physical condition, as he maintains a regular exercise program to keep in shape for conducting fieldwork. Indeed, he routinely outperforms younger colleagues on long, arduous hikes. His nephew, John (30 years old as of January 2005), who sees Jim as a role model, still marvels that he has never been able to beat "Uncle Jimmy" in arm-wrestling.

Jim is a superb field biologist. He thoroughly enjoys conducting fieldwork in remote places and under the most demanding conditions. Even the simplest commodity is resented and considered excessive in the field, a sentiment that his field companions do not always share. He likes to go to places that most people do not, and he enjoys the physical challenges that this entails. Since 1963, he has conducted fieldwork in México, the Galápagos Islands, Perú, Venezuela, Argentina, Brazil, Colombia, Taiwan, Vietnam, Iran, and Cameroon. As of March 24, 2004, he had entered 20,810 specimens into his catalogue (including one bird, two snakes, nine salamanders, and 88 frogs). He insists that one must eat what one traps, and, on numerous occasions, after preparing a study specimen, he has eaten the muscle and tissue that would normally be discarded.

For Jim, one of the most important things in life for a field biologist is to identify specimens correctly. By early 1980, one of us (H.W.G.) felt that things had become somewhat stale in the MVZ, and decided to lighten up the mood by renting a gorilla suit. On April 1 (April Fool's Day), Harry put on the costume and visited various biology classrooms, "saying hi" as a gorilla. In the afternoon, he returned to the Museum. At that time, when you entered the front office of the MVZ, the Administrative Assistant's office was immediately to the right. Harry entered the Museum to find Jim leaning across her desk, banging his fist, his voice elevated and angry (some of us know what that can be like). The lady was looking Jim in the eye, wagging a finger in his face, and practically yelling at him. Harry entered the office and stood quietly to the side. After a few seconds, she looked up and exclaimed, "Oh my God, Dr. Patton, there's a bear in here!" Jim quickly looked over, snapped his head back around and immediately replied, "THAT'S NOT A BEAR, IT'S A GODDAMNED GORILLA!" and went right back to banging his fist on the desk and continuing his argument.

Jim is a devoted fan of the 1990 science fiction movie *Tremors*, about the small, isolated town of Perfection, Nevada (14 inhabitants) that finds itself under siege by four monstrous, subterranean wormlike creatures (dubbed "Graboids") that hunt by sensing vibrations in the ground. The plot revolves around the townspeople trying to outwit and escape the creatures, which are tearing the town out from under them. The movie received excellent reviews when it was released, became a "cult hit," and, according to Jim, is the most entertaining movie Hollywood has ever produced. Rumor has it that he *still* watches it two or three times a month, and that he has worn out several copies. Jim also used to watch the television show *Buffy the Vampire Slayer*, in which a teenager girl learns that she belongs to a lineage of ancient vampire killers. After training under the watchful eye of a mentor, she became a spandex-wearing, kung-fu kicking, stake-stabbing slayer of roaming vampires and all sorts of demonic monsters. Jim claims that he watched the show for "total escapism."

CONCLUDING REMARKS

On June 23, 2001, during the banquet to celebrate Jim's retirement, David Wake, acting on behalf of the Chancellor of the University of California, Berkeley, presented Jim with The Berkeley Citation for Distinguished Achievement. This award, equivalent to an honorary degree, is the highest distinction bestowed by the Berkeley campus on a faculty member. Deeply moved by this unusual, but most deserved recognition, Jim nervously stood up, and remained speechless, while the audience, composed of his wife Carol, his brother John, and former and present colleagues, students, and coworkers, gave him a sincere, effusive, and lengthy ovation. John later asserted, "It was truly incredible to see the significant impact Jim has had on the lives of so many, and the affection and admiration that many have for him. I am extremely proud of my little brother!"

Jim's "retirement" became effective on July 1, 2001, but there has been no diminution in his teaching and research, and he remains Curator of Mammals at the MVZ; only his work in academic committees has substantially decreased. He states that "I have no plans for retirement. I'm a little bit scared of retirement. I know what I'm going to do for the next year or so, but I don't know what I'll be doing ten years from now." We know that Jim will always find interesting projects that will keep him occupied.[4]

ACKNOWLEDGMENTS

We thank Christopher J. Bell, Alison L. Chubb, John W. Fitzpatrick, Yuri L. R. Leite, Márcia C. Lara, Enrique P. Lessa, William Z. Lidicker Jr., Claudia Luke, Jay R. Malcolm, Craig Moritz, Philip Myers, John F. Patton III, Jack W. Sites Jr., Margaret F. Smith, and Don E. Wilson for providing anecdotes and valuable information; Barbara R. Stein and Eileen A. Lacey for numerous, constructive comments on the manuscript; Eileen Lacey for inviting us to write this account; and James Patton for willingly sharing many of his experiences and opinions, and for answering innumerable questions.

LITERATURE CITED

Álvarez-Castañeda, S. T., and J. L. Patton (eds.)

    1999   Mamíferos del Noroeste de México. Centro de Investigaciones Biológicas del Noroeste, La Paz, Baja California Sur, México. Pp. 1–583. (In Spanish and English)

    2000   Mamíferos del Noroeste de México II. Centro de Investigaciones Biológicas del Noroeste, La Paz, Baja California Sur, México. Pp. 584–873. (In Spanish and English)

---

[4]In May 2003, Jim and several colleagues from the MVZ began working on a three-year project to reprise a survey of Yosemite National Park wildlife initially conducted by Joseph Grinnell and Tracey I. Storer between 1914 and 1920. The goal of the project is to gather new baseline data to compare to the Grinnell and Storer surveys to determine whether there have been any changes in the abundance or distribution of mammals, birds, amphibians, and reptiles in the park.

Baker, R. J., and J. L. Patton
   1967   Karyotypes and karyotypic variation of North American vespertilionid bats. J. Mammal. 48:270–286.

Barros, M. A., and J. L. Patton
   1985   Genome evolution in pocket gophers (genus *Thomomys*). III. Fluorochrome-revealed heterochromatin heterogeneity. Chromosoma (Berl.) 92:337–343.

Berlín N. B., and J. L. Patton
   1979   La Clasificación de los Mamíferos de los Aguaruna, Amazonas, Perú. Language Behavior Research Laboratory, Berkeley, California. 95 pp. (In Spanish)

Bush, G. L., S. M. Case, A. C. Wilson, and J. L. Patton
   1977   Rapid speciation and chromosomal evolution in mammals. Proc. Natl. Acad. Sci. USA 74:3942–3946.

Cockrum, E. L., and W. J. McCauley
   1965   Zoology. W. B. Saunders and Company, Philadelphia, Pennsylvania. 705 pp.

Daly, J. C., and J. L. Patton
   1986   Growth, reproduction, and sexual dimorphism in *Thomomys bottae* pocket gophers. J. Mammal. 67:256–265.

   1990   Dispersal, gene flow, and allelic diversity between local populations of *Thomomys bottae* pocket gophers in the coastal ranges of California. Evolution 44:1283–1294.

da Silva, M. N. F., and J. L. Patton
   1993   Amazonian phylogeography: mtDNA sequence variation in arboreal echimyid rodents. Mol. Phylogenet. Evol. 2:243–255.

   1998   Molecular phylogeography and the evolution and conservation of Amazonian mammals. Mol. Ecol. 7:475–486.

Fleck, D. W., R. S. Voss, and J. L. Patton
   1999   Biological basis of saki (*Pithecia*) folk species recognized by the Matses Indians of Amazonian Perú. International J. Primatol. 20:1005–1028.

Fonseca, G. A. B., G. Herrmann, Y. L. R. Leite, R. A. Mittermeier, A. B. Rylands, and J. L. Patton
    1996   Lista anotada dos mamíferos do Brasil. Occ. Pap. Conserv. Biol., Number 4. Conservation International, Washington, D.C. 38 pp. (In Portuguese and English)

Gardner, A. L., S. Anderson, and J. L. Patton (eds).
    In press   Handbook of South American Mammals, vol. I. University of Chicago Press, Chicago, Ill.

Gardner, A. L., and J. L. Patton
    1972   New species of *Philander* (Marsupialia: Didelphidae) and *Mimon* (Chiroptera: Phyllostomatidae) from Peru. Occ. Pap. Mus. Zool., Louisiana State Univ. 43:1–12.

    1976   Karyotypic variation in oryzomyine rodents (Cricetinae) with comments on chromosomal evolution in the neotropical cricetine complex. Occ. Pap. Mus. Zool., Louisiana State Univ. 49:1–48.

Gascon, C., J. R. Malcolm, J. L. Patton, M. N. F. da Silva, J. P. Bogart, S. C. Lougheed, C. A. Peres, S. Neckel, and P. T. Boag
    2000   Riverine barriers and the geographic distribution of Amazonian species. Proc. Natl. Acad. Sci. USA 97:13672–13677.

Geise, L., M. F. Smith, and J. L. Patton
    2001   Diversification in the genus *Akodon* (Rodentia: Sigmodontinae) in southeastern South America: mitochondrial DNA sequence analysis. J. Mammal. 82:92–101.

Hafner, J. C., D. J. Hafner, J. L. Patton, and M. F. Smith
    1983   Contact zones and the genetics of differentiation in the pocket gopher *Thomomys bottae*. Syst. Zool. 32:1–20.

Hafner, M. S., J. C. Hafner, J. L. Patton, and M. F. Smith
    1987   Macrogeographic patterns of genetic differentiation in the pocket gopher *Thomomys umbrinus*. Syst. Zool. 36:18–34.

Hinojosa P. F., S. Anderson, and J. L. Patton
    1987   Two new species of *Oxymycterus* (Rodentia) from Peru and Bolivia. Amer. Mus. Novitates 2898:1–17.

Lacey, E. A., J. L. Patton, and G. N. Cameron (eds.)
    2000   Life Underground: The Biology of Subterranean Rodents. University of Chicago Press, Chicago, Ill.

Lara, M. C., J. L. Patton, and M. N. F. da Silva
    1996   The simultaneous diversification of echimyid rodents (Hystricognathi) based on complete cytochrome b sequences. Mol. Phylogenet. Evol. 5:403–413.

Lara, M. C., J. L. Patton, and E. Hingst-Zaher
    2002   *Trinomys mirapitanga*, a new specis of spiny rat (Rodentia: Echimyidae) from the Brazilian Atlantic forest. Mamm. Biol. (Z. Säugetierkunde) 67:233-242.

Layne, J. N., and R. S. Hoffmann
    1994   Presidents. Pp. 22–70 in Seventy-Five Years of Mammalogy (1919-1994) (E. C. Birney and J. R. Choate, eds.). Special Publication Number 11. American Society of Mammalogists, Provo, Utah.

Lessa, E. P., J. A. Cook, and J. L. Patton
    2003   Genetic footprints of demographic expansion in North America, but not Amazonia, during the Late Quaternary. Proc. Natl. Acad. Sci. USA 100:10331–10334.

Lessa, E. P., and J. L. Patton
    1989   Structural constraints, recurrent shapes, and allometry in pocket gophers (genus *Thomomys*). Biol. J. Linn. Soc. 36:349–363.

Lillegraven, J. A., S. D. Thompson, B. K. McNab, and J. L. Patton
    1987   The origin of eutherian mammals. Biol. J. Linn. Soc. 32:281–336.

Lowe, C. H., C. J. Cole, and J. L. Patton
    1967   Karyotype evolution and speciation in lizards (genus *Sceloporus*) during evolution of the North American desert. Syst. Zool. 16:296-300.

Lowe, C. H., C. J. Jones, and J. W. Wright
    1968   A new plethodontid salamander from Sonora, Mexico. Contrib. Sci., Nat. Hist. Mus. Los Angeles County 140:1–11.

Marlow, R. W., and J. L. Patton
    1981   Biochemical relationships of the Galápagos giant tortoises (*Geochelone elephantopus*). J. Zool. (Lond.) 195:413–422.

Matocq, M. D., J. L. Patton, and M. N. F. da Silva
    2000   Population genetic structure of two ecologically distinct Amazonian spiny rats: separating history and current ecology. Evolution 54:1423–1432.

Miller, A. H.
    1964   Joseph Grinnell. Syst. Zool. 13:235–242.

Morell, V.
    1996a   New mammals discovered by biology's new explorers. Science 273:1491.

    1996b   Amazonian diversity: a river doesn't run through it. Science 273:1496–1497.

    1997   On the origin of (Amazonian) species. Discover Magazine 18(April):56–64.

Moritz, C., J. L. Patton, C. J. Schneider, and T. B. Smith
    2000   Diversification of rainforest faunas: an integrated molecular approach. Annu. Rev. Ecol. Syst. 31:533–563.

Mustrangi, M. A., and J. L. Patton
    1997   Phylogeography and systematics of the slender mouse opossum, *Marmosops* (Marsupialia, Didelphidae). Univ. California Publ. Zool. 130:1–86 + x.

Myers, P., and J. L. Patton
    1989a   A new species of *Akodon* from the cloud forests of eastern Cochabamba Department, Bolivia (Rodentia: Sigmodontinae). Occ. Pap. Mus. Zool., Univ. Michigan 720:1–28.

    1989b   *Akodon* of Peru and Bolivia—revision of the *fumeus* group (Rodentia: Sigmodontinae). Occ. Pap. Mus. Zool., Univ. Michigan 721:1–35.

Myers, P., and J. L. Patton
   In press   Genus *Lestoros* Oehser, 1934. In Handbook of South American Mammals, vol. I (A. L. Gardner, S. Anderson, and J. L. Patton, eds.). University of Chicago Press, Chicago, Ill.

Myers, P., J. L. Patton, and M. F. Smith
   1990   A review of the *boliviensis* group of *Akodon* (Muridae: Sigmodontinae), with emphasis on Peru and Bolivia. Misc. Publ., Mus. Zool., Univ. Michigan 177:1–104 + iv.

Nelson-Rees, W. A., A. J. Kniazeff, R. J. Baker, and J. L. Patton
   1968   Intraspecific chromosome variation in the bat, *Macrotus waterhousii* Gray. J. Mammal. 49:706-712.

Pacheco, V., B. D. Patterson, J. L. Patton, L. H. Emmons, S. Solari, and C. F. Ascorra
   1993   List of mammal species known to occur in the Manu Biosphere Reserve, Peru. Public. Mus. Hist. Nat., Univ. Nacional Mayor de San Marcos (Lima) (Ser. A Zool.) 44:1–12.

Pacheco, V., and J. L. Patton
   1995   A new species of the Puna mouse, genus *Punomys* Osgood, 1943 (Muridae, Sigmodontinae), from the southeastern Andes of Perú. Z. Säugetierkunde 60:85–96.

Patterson, B. D., and J. L. Patton
   1990   Fluctuating asymmetry and allozymic heterozygosity among natural populations of pocket gophers (*Thomomys bottae*). Biol. J. Linn. Soc. 40:21–36.

Patton, J. L.
   1965   Cytotaxonomy of the pocket mice, genus *Perognathus* (Rodentia: Heteromyidae). M.S. Thesis, University of Arizona, Tucson. 96 pp.

   1967a   Chromosomes and evolutionary trends in the pocket mouse subgenus *Perognathus* (Rodentia: Heteromyidae). Southwest. Nat. 12:429–438.

   1967b   Chromosome studies of certain pocket mice, genus *Perognathus* (Rodentia: Heteromyidae). J. Mammal. 48:27–37.

   1968   Chromosome evolution in the pocket mouse, *Perognathus goldmani* Osgood. Ph.D. Thesis, University of Arizona, Tucson. 67 pp.

Patton, J. L.

1969a   Chromosome evolution in the pocket mouse, *Perognathus goldmani* Osgood. Evolution 23:645–662.

1969b   Karyotypic variation in the pocket mouse, *Perognathus penicillatus* Woodhouse (Rodentia-Heteromyidae). Caryologia 22:351–358.

1970   Karyotypic variation following an elevational gradient in the pocket gopher, *Thomomys bottae grahamensis* Goldman. Chromosoma (Berl.) 31:41–50.

1971   Possible genetic consequences of meiosis in pocket gopher (*Thomomys bottae*) populations. Experientia 27:593–595.

1972a   A complex system of chromosomal variation in the pocket mouse, *Perognathus baileyi* Merriam. Chromosoma (Berl.) 36:241–255.

1972b   Patterns of geographic variation in karyotype in the pocket gopher, *Thomomys bottae* (Eydoux and Gervais). Evolution 26:574–586.

1973   An analysis of natural hybridization between the pocket gophers, *Thomomys bottae* and *Thomomys umbrinus*, in Arizona. J. Mammal. 54:561–584.

1977   B-Chromosome systems in the pocket mouse, *Perognathus baileyi*: meiosis and C-band studies. Chromosoma (Berl.) 60:1–14.

1981   Chromosomal and genic divergence, population structure, and speciation potential in *Thomomys bottae* pocket gophers. Pp. 255–295 in Ecología y Genética de la Especiación Animal (O. A. Reig, ed.). Equinoccio, Editorial de la Universidad Simón Bolívar, Caracas, Venezuela.

1984a   Pocket gophers. Pp. 628–631 in The Encyclopedia of Mammals (D. Macdonald, ed.). Facts on File, New York.

1984b   Systematic status of the large squirrels (subgenus *Urosciurus*) of the western Amazon basin. Stud. Neotrop. Fauna Environ. 19:53–72.

1985   Population structure and the genetics of speciation in pocket gophers, genus *Thomomys*. Acta Zool. Fennica 170:109–114.

Patton, J. L.

1987   Species groups of spiny rats, genus *Proechimys* (Rodentia: Echimyidae).
        Pp. 305–345 in Studies in Neotropical Mammalogy: Essays in Honor of
        Philip Hershkovitz (B. D. Patterson and R. M. Timm, eds.). Fieldiana:
        Zool. (New Ser.), Number 39.

1990   Geomyid evolution: the historical, selective, and random basis for
        divergence patterns within and among species. Pp. 49–69 in Evolution
        of Subterranean Mammals at the Organismal and Molecular Levels (E.
        Nevo and O. A. Reig, eds.). Alan R. Liss, Inc. New York.

1993a   Family Geomyidae. Pp. 469–476 in Mammal Species of the World: A
        Taxonomic and Geographic Reference, 2nd. ed. (D. E. Wilson and D.
        M. Reeder, eds.). Smithsonian Institution Press, Washington, D.C.

1993b   Family Heteromyidae. Pp. 477–486 in Mammal Species of the World:
        A Taxonomic and Geographic Reference, 2nd. ed. (D. E. Wilson and D.
        M. Reeder, eds.). Smithsonian Institution Press, Washington, D.C.

1993c   Hybridization and hybrid zones in pocket gophers (Rodentia,
        Geomyidae). Pp. 290–308 in Hybrid Zones and the Evolutionary
        Process (R. G. Harrison, ed.). Oxford University Press, New York.

1999a   Family Geomyidae. Pp. 321–350 in Mamíferos del Noroeste de México
        (S. T. Álvarez-Castañeda and J. L. Patton, eds.). Centro de
        Investigaciones Biológicas del Noroeste, La Paz, Baja California Sur,
        México.

1999b   [Species accounts for] Botta's pocket gopher | *Thomomys bottae*, Camas
        pocket gopher | *Thomomys bulbivorus*, Wyoming pocket gopher |
        Thomomys clusius, Idaho pocket gopher | *Thomomys idahoensis*, western
        pocket gopher | *Thomomys mazama*, mountain pocket gopher |
        Thomomys monticola, southern pocket gopher | *Thomomys umbrinus*,
        yellow-faced pocket gopher | *Pappogeomys castanops*. Pp. 466–474, 479–
        480, 492–493 in The Smithsonian Book of North American Mammals
        (D. E. Wilson and S. Ruff, eds.). Smithsonian Institution Press,
        Washington, D.C.

Patton, J. L.
    2000   Family Erethizontidae. Pp. 587–589 in Mamíferos del Noroeste de México II (S. T. Álvarez-Castañeda and J. L. Patton, eds.). Centro de Investigaciones Biológicas del Noroeste, La Paz, Baja California Sur, México.

    2001   Pocket gophers. Pp. 662–665 in The New Encyclopedia of Mammals (D. Macdonald, ed.). Oxford University Press, Oxford.

Patton, J. L., and S. T. Álvarez-Castañeda
    1999   Family Heteromyidae. Pp. 351–443 in Mamíferos del Noroeste de México (S. T. Álvarez-Castañeda and J. L. Patton, eds.). Centro de Investigaciones Biológicas del Noroeste, La Paz, Baja California Sur, México.

Patton, J. L., O. B. Berlin, and E. A. Berlin
    1982   Aboriginal perspectives of a mammal community in Amazonian Perú: knowledge and utilization patterns among the Aguaruna Jívaro. Pp. 111–128 in Mammalian Biology in South America (M. A. Mares and H. M. Genoways, eds.). Special Publication Series, Volume 6, Pymatuning Laboratory of Ecology, University of Pittsburgh, Linesville, Pennsylvania.

Patton, J. L., and P. V. Brylski
    1987   Pocket gophers in alfalfa fields: causes and consequences of habitat-related body size variation. Amer. Nat. 130:493–506.

Patton, J. L., and M. N. F. da Silva
    1995   A review of the spiny mouse genus *Scolomys* (Rodentia: Muridae: Sigmodontinae) with the description of a new species from the western Amazon of Brazil. Proc. Biol. Soc. Washington 108:319–337.

    1997   Definition of species of pouched four-eyed opossums (Didelphidae, *Philander*). J. Mammal. 78:90–102.

    1998   Rivers, refuges, and ridges: the geography of speciation of Amazonian mammals. Pp. 202–213 in Endless Forms: Species and Speciation (D. J. Howard and S. H. Berlocher, eds.). Oxford University Press, New York.

Patton, J. L., and M. N. F. da Silva
    2001   Molecular phylogenetics and the diversification of Amazonian
            mammals. Pp. 139–164 in Diversidade Biológica e Cultural da
            Amazônia (I. C. Guimarães Vieira, J. M. Cardoso da Silva, D. C. Oren,
            and M. A. D'Incao, eds.). Museu Paraense Emílio Goeldi, Belém,
            Brazil.

    In press Genus *Philander* Tiedemann, 1808. In Handbook of South
              American Mammals, vol. I (A. L. Gardner, S. Anderson, and J. L.
              Patton, eds.). University of Chicago Press, Chicago, Ill.

Patton, J. L., M. N. F. da Silva, M. C. Lara, and M. A. Mustrangi
    1997   Diversity, differentiation, and the historical biogeography of
            nonvolant small mammals of the neotropical forests. Pp. 455–465 in
            Tropical Forest Remnants: Ecology, Management, and Conservation of
            Fragmented Communities (W. F. Laurance and R. O. Bierregaard Jr.,
            eds.). University of Chicago Press, Chicago, Ill.

Patton, J. L., M. N. F. da Silva, and J. R. Malcolm
    1994   Gene genealogy and differentiation among arboreal spiny rats
            (Rodentia: Echimyidae) of the Amazon Basin: a test of the riverine
            barrier hypothesis. Evolution 48:1314–1323.

    1996a   Hierarchical genetic structure and gene flow in three sympatric species
            of Amazonian rodents. Mol. Ecol. 5:229–238.

    2000   Mammals of the Rio Juruá and the evolutionary and ecological
            diversification of Amazonia. Bull. Amer. Mus. Nat. Hist. 244:1–306.

Patton, J. L., and R. E. Dingman
    1968   Chromosome studies of pocket gophers, genus *Thomomys*. I. The
            specific status of *Thomomys umbrinus* (Richardson) in Arizona. J.
            Mammal. 49:1–13.

    1970   Chromosme studies of pocket gophers, genus *Thomomys*. II. Variation
            in *T. bottae* in the American Southwest. Cytogenetics 9:139–151.

Patton, J. L., S. F. dos Reis, and M. N. F. da Silva
    1996b   Relationships among didelphid marsupials based on sequence
            variation in the mitochondrial cytochrome b gene. J. Mammal. Evol.
            3:3–29.

Patton, J. L., and L. H. Emmons
   1985   A review of the genus *Isothrix* (Rodentia, Echimyidae). Amer. Mus. Novitates 2817:1–14.

Patton, J. L., and J. H. Feder
   1978   Genetic divergence between populations of the pocket gopher, *Thomomys umbrinus* (Richardson). Z. Säugetierkunde 43:17–30.

   1981   Microspatial genetic heterogeneity in pocket gophers: non-random breeding and drift. Evolution 35:912–920.

Patton, J. L., and A. L. Gardner
   1971   Parallel evolution of multiple sex-chromosome systems in phyllostomatid bats, *Carollia* and *Choeroniscus*. Experientia 27:105–106.

   1972   Notes on the systematics of *Proechimys* (Rodentia: Echimyidae), with emphasis on Peruvian forms. Occ. Pap. Mus. Zool., Louisiana State Univ. 44:1–30.

   In press   Family Mormoopidae Saussure, 1860. In Handbook of South American Mammals, vol. I (A. L. Gardner, S. Anderson, and J. L. Patton, eds.). University of Chicago Press, Chicago, Ill.

Patton, J. L., J. C. Hafner, M. S. Hafner, and M. F. Smith
   1979   Hybrid zones in *Thomomys bottae* pocket gophers: genetic, phenetic, and ecologic concordance patterns. Evolution 33:860–876.

Patton, J. L., and W. B. Heed
   1965   Elevational differences in gene arrangements of *D. pseudoobscura* in the Santa Catalina Mountains, Tucson. *Drosophila* Inf. Serv. 40:69–70.

Patton, J. L., and T.-C. Hsu
   1967   Chromosomes of the golden mouse, *Peromyscus* (*Ochrotomys*) *nuttalli* (Harlan). J. Mammal. 48:637–639.

Patton, J. L., and P. Myers
   1974   Chromosomal identity of black rats (*Rattus rattus*) from the Galápagos Islands, Ecuador. Experientia 30:1140–1141.

Patton, J. L., P. Myers, and M. F. Smith

1989　Electromorphic variation in selected South American akodontine rodents (Muridae: Sigmodontinae), with comments on systematic implications. Z. Säugetierkunde 54:347–359.

1990　Vicariant versus gradient models of diversification: the small mammal fauna of eastern Andean slopes of Peru. Pp. 355–371 in Vertebrates in the Tropics (G. Peters and R. Hutterer, eds.). Museum Alexander Koenig, Bonn.

Patton, J. L., and D. S. Rogers

1993　Cytogenetics. Pp. 236–258 in Biology of the Heteromyidae (H. H. Genoways and J. H. Brown, eds.). Special Publication Number 10, American Society of Mammalogists.

Patton, J. L., R. K. Selander, and M. H. Smith

1972　Genic variation in hybridizing populations of gophers (genus *Thomomys*). Syst. Zool. 21:263–270.

Patton, J. L., and S. W. Sherwood

1982　Genome evolution in pocket gophers (genus *Thomomys*). I. Heterochromatin variation and speciation potential. Chromosoma (Berl.) 85:149–162.

1983　Chromosome evolution and speciation in rodents. Annu. Rev. Ecol. Syst. 14:139–158.

Patton, J. L., and M. F. Smith

1981　Molecular evolution in *Thomomys*: phyletic systematics, paraphyly, and rates of evolution. J. Mammal. 62:493–500.

1989　Population structure and the genetic and morphologic divergence among pocket gopher species (genus *Thomomys*). Pp. 284–304 in Speciation and its Consequences (D. Otte and J. A. Endler, eds.). Sinauer Associates, Inc., Publishers, Sunderland, Massachusetts.

1990　The evolutionary dynamics of the pocket gopher *Thomomys bottae*, with emphasis on California populations. Univ. California Publ. Zool. 123:1–161 + xviii.

Patton, J. L., and M. F. Smith

1992a Evolution and systematics of akodontine rodents (Muridae: Sigmodontinae) of Peru, with emphasis on the genus *Akodon*. Mem. Mus. Hist. Nat., Univ. Nacional Mayor de San Marcos (Lima) 21:83–103.

1992b mtDNA phylogeny of Andean mice: a test of diversification across ecological gradients. Evolution 46:174–183.

1993 Molecular evidence for mating asymmetry and female choice in a pocket gopher (*Thomomys*) hybrid zone. Mol. Ecol. 2:3–8.

1994 Paraphyly, polyphyly, and the nature of species boundaries in pocket gophers (genus *Thomomys*). Syst. Biol. 43:11–26.

Patton, J. L., M. F. Smith, R. D. Price, and R. A. Hellenthal

1984 Genetics of hybridization between the pocket gophers *Thomomys bottae* and *Thomomys townsendii* in northeastern California. Great Basin Nat. 44:431–440.

Patton, J. L., and B. R. Stein

In press (a) Genus *Chironectes* Illiger, 1811. In Handbook of South American Mammals, vol. I (A. L. Gardner, S. Anderson, and J. L. Patton, eds.). University of Chicago Press, Chicago, Ill.

In press (b) Genus *Lutreolina* Thomas, 1910. In Handbook of South American Mammals, vol. I (A. L. Gardner, S. Anderson, and J. L. Patton, eds.). University of Chicago Press, Chicago, Ill.

Patton, J. L., and S. Y. Yang

1977 Genetic variation in *Thomomys bottae* pocket gophers: macrogeographic patterns. Evolution 31:697–720.

Pearson, O. P., and J. L. Patton

1976 Relationships among South American phyllotine rodents based on chromosome analysis. J. Mammal. 57:339–350.

Peres, C. A., J. L. Patton, and M. N. F. da Silva

1996 Riverine barriers and gene flow in Amazonian saddle-back tamarins. Folia Primatol. 67:113–124.

Reig, O. A., A. L. Gardner, N. O. Bianchi, and J. L. Patton
  1977   The chromosomes of the Didelphidae (Marsupialia) and their evolutionary significance. Biol. J. Linn. Soc. 9:191–216.

Ruedi, M., M. F. Smith, and J. L. Patton
  1997   Phylogenetic evidence of mitochondrial DNA introgression among pocket gophers in New Mexico (Family Geomyidae). Mol. Ecol. 6:453–462.

Sherwood, S. W., and J. L. Patton
  1982   Genome evolution in pocket gophers (genus *Thomomys*). II. Variation in cellular DNA content. Chromosoma (Berl.) 85:163–179.

Sites, J. W., Jr.
  2001   Speciation in the world's greatest forest. Trends Ecol. Evol. 16:111–112.

Smith, M. F., D. A. Kelt, and J. L. Patton
  2001   Testing models of diversification in mice in the *Abrothrix olivaceus/xanthorhinus* complex in Chile and Argentina. Mol. Ecol. 10:397–405.

Smith, M. F., and J. L. Patton
  1980   Relationships of pocket gopher (*Thomomys bottae*) populations of the lower Colorado River. J. Mammal. 61:681–696.

  1984   Dynamics of morphological differentiation: temporal impact of gene flow in pocket gopher populations. Evolution 38:1079–1087.

  1988   Subspecies of pocket gophers: causal bases for geographic differentiation in *Thomomys bottae*. Syst. Zool. 37:163–178.

  1991a PCR on dried skin and liver extracts from the same individual gives identical products. Trends Genet. 7:4.

  1991b Variation in mitochondrial cytochrome b sequence in natural populations of South American akodontine rodents (Muridae: Sigmodontinae). Mol. Biol. Evol. 8:85–103.

  1993   The diversification of South American murid rodents: evidence from mitochondrial DNA sequence data for the akodontine tribe. Biol. J. Linn. Soc. 50:149–177.

Smith, M. F., and J. L. Patton
   1999   Phylogenetic relationships and the radiation of sigmodontine rodents in South America: evidence from cytochrome b. J. Mammal. Evol. 6:89–128.

Smith, M. F., J. L. Patton, J. C. Hafner, and D. J. Hafner
   1983   *Thomomys bottae* pocket gophers of the central Rio Grande Valley, New Mexico: local differentiation, gene flow, and historical biogeography. Occ. Pap. Mus. Southwest. Biol., Univ. New Mexico 2:1–16.

Smith, M. F., W. K. Thomas, and J. L. Patton
   1992   Mitochondrial DNA-like sequence in the nuclear genome of an akodontine rodent. Mol. Biol. Evol. 9:204–215.

Vié, J.-C., V. Volobouev, J. L. Patton, and L. Granjon
   1996   A new species of *Isothrix* (Rodentia: Echimyidae) from French Guiana. Mammalia 60:393–406.

Wake, M. H., J. C. Hafner, M. S. Hafner, L. L. Klosterman, and J. L. Patton
   1980   The karyotype of *Typhlonectes compressicauda* (Amphibia: Gymnophiona) with comments on chromosome evolution in caecilians. Experientia 36:171–172.

Wallace, A. R.
   1852   On the monkeys of the Amazon. Proc. Zool. Soc. Lond. 20:107–110.

Yang, S. Y., and J. L. Patton
   1981   Genic variability and differentiation in the Galapagos finches. Auk 98:230–242.

Zink, R. M., M. F. Smith, and J. L. Patton
   1985   Associations between heterozygosity and morphological variance. J. Hered. 76:415–420.

Appendix 1. Publications by James L. Patton during his years as a Curator in the Museum of Vertebrate Zoology. This list, which was compiled by the editors of the volume, includes all publications authored or co-authored by Jim between 1969 and 2005. Entires are presented according to the year of publication in order to capture the conceptual, technological, and taxnomic trends that have characterized Jim's career.

1969

Patton, J. L. Chromosome evolution in the pocket mouse, *Perognathus goldmani* Osgood. Evolution 23:645-662.

Patton, J. L. Karyotypic variation in the pocket mouse, *Perognathus penicillatus* Woodhouse (Rodentia-Heteromyidae). Caryologia 22:351-358.

1970

Patton, J. L., Karyotypic variation following an elevational gradient in the pocket gopher, *Thomomys bottae grahamensis* Goldman. Chromosoma (Berlin) 31:41-50.

Patton, J. L. Chromosome studies of pocket gophers, genus *Thomomys*. II. Variation in *T. bottae* in the American southwest. Cytogenetics 9:139-151.

1971

Patton, J. L. Possible genetic consequences of meiosis in pocket gopher (*Thomomys bottae*) populations. Experientia 27:593-595.

Patton, J. L., and A. L. Gardner. Parallel evolution of multiple-sex chromosome systems in the phyllostomatid bats, *Carollia* and *Choeroniscus*. Experientia 27:105-106.

1972

Patton, J. L., R. K. Selander, and M. H. Smith. Genic variation in hybridizing populations of gophers (genus *Thomomys*). Syst. Zool. 21:263-270.

Patton, J. L. A complex system of chromosomal variation in the pocket mouse, *Perognathus baileyi* Merriam. Chromosoma 36:241-255.

Patton, J. L., and A. L. Gardner. Notes on the systematics of *Proechimys* (Rodentia: Echimyidae), with emphasis on Peruvian forms. Occas. Papers Mus. Zool. (Louisiana) 44:1-30.

Gardner, A. L., and J. L. Patton. New species of *Philander* (Marsupialia: Didelphidae) and *Mimon* (Chiroptera: Phyllostomidae) from Peru. Occas. Papers Mus. Zool. (Louisiana) 43:1-12.

Patton, J. L., and J. K. Jones, Jr. First records of *Perognathus baileyi* from Sinaloa, Mexico. J. Mammal. 53:371-372.

Baker, R. J., A. L. Gardner, and J. L. Patton. Chromosomal polymorphism in the phyllostomatid bat, *Mimon crenalatum* Geoffroy. Experientia 28:969-970.

1973

Patton, J. L. Patterns of geographic variation in karyotype in the pocket gopher, *Thomomys bottae* (Eydoux and Gervais). Evolution 26:574-586.

Patton, J. L. An analysis of natural hybridization between the pocket gophers, *Thomomys bottae* and *Thomomys umbrinus*, in Arizona. J. Mammal. 54:561-584.

Patton, J. L. Committee for standardization of chromosomes of *Peromyscus*, Standardized karyotype of deer mice, *Peromyscus* (Rodentia). Cytogenetics 19:38-43.

1974

Warner, J. W., J. L. Patton, A. L. Gardner, and R. J. Baker. Karyotypic analyses of twenty-one species of molosid bats (Molossidae: Chiroptera). Can. J. Gen. Cytol. 16:165-176.

Patton, J.L., and P. Meyers. Chromosomal identity of black rats (*Rattus rattus*) from the Galapagos Islands, Ecuador. Experientia 30:1140-1141.

1975

Patton, J. L., S. Y. Yang, and P. Myers. Genetic and morphologic divergence among introduced rat populations (*Rattus rattus*) of the Galapagos Archipelago, Ecuador. Syst. Zool. 24(3):296-310.

1976

Patton, J. L., H. MacArthur, and S. Y. Yang. Systematic relationships of the four-toed populations of *Dipodomys heermanni*. J. Mammal. 57:159-163.

Gardner, A. L., and J. L. Patton. Karyotypic variations in oryzomyine rodents (Cricetinae) with comments on chromosomal evolution in the Neotropical cricetine complex. Occas. Papers Mus. Zool. (Louisiana) 49: 48 pp.

Patton, J. L. Biosystematics of the rodent fauna of the Galapagos Archipelago. Am. Phil. Soc. Year Book 1975:352-353.

Pearson, O. P., and J. L. Patton. Relationships among South American phyllotine rodents based on chromosome analysis. J. Mammal. 57:339-350.

1977

Patton, J. L. B-chromosome systems in the pocket mouse, *Perognathus baileyi*: meiosis and C-band studies. Chromosoma (Berlin) 60:1-14.

Reig, O. A., A. L. Gardner, N. O. Bianchi, and J. L. Patton. The chromosomes of the Didelphidae (Marsupialia) and their evolutionary significance. Biol. J. Linn. Soc. 49:191-216.

Bush, G. L., S. M. Case, A. C. Wilson, and J. L. Patton. Rapid speciation and chromosomal evolution in mammals. Proc. Natl. Acad. Sci. USA 74:3942-3946.

Patton, J. L., and S. Y. Yang. Genetic variation in *Thomomys bottae* pocket gophers: macrogeographic patterns. Evolution 31:697-720.

1978

Patton, J. L., and J. H. Feder. Genetic divergence between populations of the pocket gopher, *Thomomys umbrinus* Richardson. Z. Säugetierkunde 43:17-30.

1979

Berlin, O. B., and J. L. Patton. La Clasificacion de los mamiferos de los Aguaruna de Amazonas, Peru. Pp. 1-95. Language Behavior Research Laboratory, Berkeley, CA.

Patton, J. L., J. C. Hafner, M. S. Hafner, and M. F. Smith. Hybrid zones of *Thomomys bottae* pocket gophers: genetic, phenetic, and ecologic concordance patterns. Evolution 33:860- 876.

1980

Wake, M. H., J. C. Hafner, M. S. Hafner, L. L. Klosterman, and J. L. Patton. The karyotype of *Typhlonectes compressicauda* (Amphibia: Gymnophiona) with comments on chromosome evolution in caecilians. Experientia 36:171-172.

Straney, D. O., and J. L. Patton. Phylogenetic and environmental determinants of geographic variation of the pocket mouse *Perognathus goldmani* Osgood. Evolution 34:888-903.

Smith, M. F., and J. L. Patton. Relationships of pocket gopher (*Thomomys bottae*) populations of the lower Colorado River. J. Mamm. 61:681-696.

1981

Yang, S. Y., and J. L. Patton. Genic variability and differentiation in the Galapagos finches. The Auk 98:230-242.

Patton, J. L., S. W. Sherwood, and S. Y. Yang. Biochemical systematics of chaetopidine pocket mice, genus *Perognathus*. J. Mammal. 62:477-492.

Patton, J. L., and M. F. Smith. Molecular evolution in *Thomomys*: phyletic systematics, paraphyly, and the rates of evolution. J. Mammal. 62:493-500.

Patton, J. L. Chromosomal and genic divergence, population structure, and speciation potential in *Thomomys bottae* pocket gophers. Ecol. Gen. Spec. (Caracas, VZ):255-295.

Patton, J. L., and J. H. Feder. Microspatial genetic heterogeneity in pocket gophers: non- random breeding and drift. Evolution 35:912-920.

Marlow, R. W., and J. L. Patton. Biochemical relationships of the Galapagos Giant tortoises (*Geochelone elephantopus*). J. Zool., London 195:413-422.

1982

Patton, J. L. Review: Evolution Today. Science 216:287-288.

Patton, J. L., and S. W. Sherwood. Genome evolution in pocket gophers (genus *Thomomys*). I. Heterochromatin variation and speciation potential. Chromosoma (Berlin) 85:149-162.

Sherwood, S. W., and J. L. Patton. Genome evolution in pocket gophers (genus *Thomomys*) II. Variation in cellular DNA content. Chromosoma (Berlin) 85:163-179.

Patton, J. L., O. B. Berlin, and E. A. Berlin. Aboriginal perspectives of a mammal community in Amazonian Perú: knowledge and utilization patterns among the Aguaruna Jívaro. Pp. 111–128 in Mammalian Biology in South America (M. A. Mares and H. M. Genoways, eds.). Special Publication Series, Volume 6, Pymatuning Laboratory of Ecology, University of Pittsburgh, Linesville, Pennsylvania.

1983

Hafner, J. C., D. J. Hafner, J. L. Patton, and M. F. Smith. Contact zones and the genetics of differentiation in the pocket gopher *Thomomys bottae* (Rodentia: Geomyidae). Syst. Zool. 32:1-29.

Smith, M. F., J. L. Patton, J. C. Hafner, and D. J. Hafner. *Thomomys bottae* pocket gophers of the central Rio Grande Valley, New Mexico: local differentiation, gene flow, and historical biogeography. Occas. Papers Mus. Southwestern Biol. No. 2:16 pp.

Patton, J. L., and M. S. Hafner. Biosystematics of the native rodents of the Galapagos Archipelago, Ecuador. Patterns of Evolution in Galapagos Organisms :539-568.

Patton, J. L., and S. Sherwood. Chromosome evolution and speciation in rodents. Ann. Rev. Ecol. Syst. 14:139-158.

Patton, J. L., and M. A. Rogers. Systematic implications of non-geographic variation in the spiny rat genus *Proechimys* (Echimyidae) Sonderdruck aus. Z. Säugetierkunde 48:363-370.

1984

Patton, J. L. Genetical processes in the Galapagos. Biol. J. Linn. Soc. 21:97-111.

Patton, J. L., M. F. Smith, R. D. Price, and R. A. Hellenthal. Genetics of hybridization between the pocket gophers *Thomomys bottae* and *Thomomys townsendii* in northeastern California. Great Basin Naturalist 44:431-440.

Smith, M. F., and J. L. Patton. Dynamics of morphological differentiation: temporal impact of gene flow in pocket gopher populations. Evolution 38:1079-1087.

Patton, J. L. Systematic status of the large squirrels (Subgenus *Urosciurus*) of the western Amazon basin. Studies on Neotropical Fauna and Environment 19:53-72.

Patton, J. L. Biochemical genetics of the Galapagos giant tortoises. Natl. Geo. Soc. Res. Reports 17:701-709.

Patton, J. L. Pocket gophers. Pp. 628-631 in The Encyclopedia of Mammals (D. W. Macdonald, ed.). Facts on File Publications, New York.

1985

Patton, J. L., and L. H. Emmons. A review of the genus *Isothrix* (Rodentia, Echimyidae). Am. Mus. Nov. 2817: 1-14.

Patton, J. L. Population structure and the genetics of speciation in pocket gophers genus *Thomomys*. Acta Zool. Fenn. 170:109-114.

Barros, M. A., and J. L. Patton. Genome evolution in pocket gophers (genus *Thomomys*) III. Fluorochrome-revealed heterochromatin heterogeneity Chromosoma (Berlin) 92:337-343.

Zink, R. M., M. F. Smith, and J. L. Patton. Associations between heterozygosity and morphological variance. J. Heredity 76: 415-420.

1986

Daly, J. C., and J. L. Patton. Growth, reproduction, and sexual dimorphism in *Thomomys bottae* pocket gophers. J. Mammal. 67:256-265.

Sage, R. D., J. R. Contreras, V. G. Roig, and J. L. Patton. Genetic variation in the South American burrowing rodents of the genus *Ctenomys* (Rodentia: Ctenomyidae). Z. Säugetierkunde 51:158-172.

Patton, J. L. Patrones de distribucion y especiacion de fauna de mamiferos de los bosques nublados Andinos del Peru. Ann. Mus. Hist. Nat. Valparaiso 17:87-94.

1987

Hafner, M. S., J. C. Hafner, J. L. Patton, and M. F. Smith. Macrogeographic patterns of genetic differentiation in the pocket gopher *Thomomys umbrinus*. Syst. Zool. 36:18-34.

Lidicker, W. Z., Jr., and J. L. Patton. Patterns of dispersal and genetic structure in populations of small rodents. Pp. 144-161 in Mammalian Dispersal Patterns (B. D. Chepko-Sade and Z. T. Halpin, eds.). University of Chicago Press, Chicago, Illinois.

Patton, J. L., and P. V. Brylski. Pocket gophers in alfalfa fields: causes and consequences of habitat-related body size variation. Am. Nat. 130:493-506.

Lillegraven, J. A., S. D. Thompson, B. K. McNab, and J. L. Patton. The origin of eutherian mammals. Biol. J. Linn. Soc. 32:281-336.

Hinojosa P., Flavio, S. A., and J. L. Patton. Two new species of *Oxymycterus* (Rodentia) from Peru and Bolivia. Am. Mus. Nov. No. 2898:1-17.

Patton, J. L. Species groups of spiny rats, genus *Proechimys* (Rodentia: Echimyidae). Fieldiana: Zoology 39:305-345.

1988

Smith, M. F., and J. L. Patton. Subspecies of pocket gophers: causal bases for geographic differentiation in *Thomomys bottae*. Syst. Zool. 37:163-178.

Cadle, J.E., and J. L. Patton. Distribution patterns of some amphibians, reptiles, and mammals of the eastern Andean slope of Southern Peru. Neotropical Distribution Patterns Biology Workshop, Sao Paulo, Brazil.

1989

Lessa, E. P., and J. L. Patton. Structural constraints, recurrent shapes, and allometry in pocket gophers (genus *Thomomys*). Biol. J. Linn. Soc. 36:349 363.

Patton, J. L., and M. F. Smith. Population structure and the genetic and morphologic divergence among pocket gopher species (genus *Thomomys*). Pp. 284-304 in Speciation and Its Consequences (D. Otte and J. Endler, eds.). Sinauer Associates, Inc., Massachusetts.

Musser, G. G., and J. L. Patton. Systematic studies of oryzomyine rodents (Muridae): the identity of *Oecomys phelpsi* Tate. Am. Mus. Nov. 2961: 6 pp.

Patton, J. L., and O. A. Reig. Genetic differentiation among echimyid rodents, with emphasis on spiny rats, genus *Proechimys*. Adv. Neotrop. Mammal. 1989:75-96.

Patton, J. L., P. Myers, and M. F. Smith. Electromorphic variation in selected South American Akodontine rodents (Muridae: Sigmodontinae), with comments on systematic implications. Z. Säugetierkunde 54:347-359.

Myers, P., and J. L. Patton. A new species of *Akodon* from the cloud forests of eastern Cochabamba Department, Bolivia (Rodentia: Sigmodontinae). Occas. Papers Mus. Zool., Univ. Michigan 720:1-28.

Myers, P., and J. L. Patton. *Akodon* of Peru and Bolivia -- Revision of the *Fumeus* group (Rodentia: Sigmodontinae). Occas. Papers Mus. Zool., Univ. Michigan 721:1-35.

1990

Patton, J. L. Geomyid evolution: the historical, selective, and random basis for divergence patterns within and among species. Pp. 49-69 in Evolution of Subterranean Mammals at the Organismal and Molecular Levels (E. Nevo and O. A. Reig, eds.). Proc. of the Fifth Interna. Theriological Cong., held in Rome, Italy, Aug. 22-29, 1989. Wiley-Liss, New York.

Myers, P., J. L. Patton, and M. F. Smith. A review of the boliviensis group of *Akodon* (Muridae: Sigmodontinae), with emphasis on Peru and Bolivia. Misc. Pubs. Mus. Zool., Univ. Michigan 177:1-104.

Patterson, B. D., and J. L. Patton. Fluctuating asymmetry and allozymic heterozygosity among natural populations of pocket gophers (*Thomomys bottae*). Biol. J. Linn. Soc. 40:21-36.

Abaturov, B. D., S. Isayev, I. Ya. Pavlinov, and J. L. Patton. The Rodent Population of Man-Made Grasslands In Peruvian Tropical Forests. Bull. Mosk. Obshchestva Isptatelei Prirody Otdel Biolog. 95(1):39-44.

Patton, J. L., and M. F. Smith. The evolutionary dynamics of the pocket gopher *Thomomys bottae*, with emphasis on California populations. Univ. Calif. Publs. Zool. 123:1-161.

Daly, J. C., and J. L. Patton. Dispersal, gene flow, and allelic diversity between local populations of *Thomomys bottae* pocket gophers in the coastal ranges of California. Evolution 44:1283-1294.

Patton, J. L., P. Myers, and M. F. Smith. Vicariant versus gradient models of diversification: the small mammal fauna of eastern Andean slopes of Peru. Pp. 355-371 in Vertebrates in the Tropics (G. Peters and R. Hutterer, eds.). International Symposium on Vertebrate Biogeography and Systematics in the Tropics. Alexander Koenig Zoological Institute and Zoological Museum; Bonn, Germany.

1991

Smith, M. F., and J. L. Patton. Variation in mitochondrial cytochrome b sequence in natural populations of South American akodontine rodents (Muridae: Sigmondontinae). Mol. Biol. Evol. 8:85-103.

Smith, M. F., and J. L. Patton. PCR on dried skin and liver extracts from the same individual gives identical products. Trends in Genetics 7:4.

1992

Patton, J. L., and M. F. Smith. mtDNA phylogeny of Andean mice: a test of diversification across ecological gradients. Evolution 46:174-183.

Smith, M. F., W. K. Thomas, and J. L. Patton. Mitochondrial DNA-like sequence in the nuclear genome of an akodontine rodent. Mol. Biol. Evol. 9:204-215.

Patton, J. L., and M. F. Smith. Evolution and systematics of akodontine rodents (Muridae: Sigmodontinae) of Peru, with emphasis on the genus *Akodon*. Mem. Mus. Hist. Nat., UNMSM (Lima) 21:83-103.

1993

Patton, J. L., and M. F. Smith. Molecular evidence for mating asymmetry and female choice in a pocket gopher (*Thomomys*) hybrid zone. Mol. Ecol. 2:3-8.

da Silva, M. N. F., and J. L. Patton. Amazonian phylogeography: mtDNA sequence variation in arboreal echimyid rodents (Caviomorpha). Mol. Phylog. Evol. 2:243-255.

Pacheco, V., B. D. Patterson, J. L. Patton, L. H. Emmons, S. Solari, and C. F. Ascorra. List of mammal species known to occur in Manu Biosphere Reserve, Peru. Pubs. Mus. Hist. Nat., UNMSM 44:1-12.

Smith, M. F., and J. L. Patton. The diversification of South American muroid rodents: evidence from mitochondrial DNA sequence data for the akodontine tribe. Biol. J. Linn. Soc. 50:149-177.

1993

Patton, J. L., and D. S. Roberts. Biochemical Genetics. Pp. 259-269 in Biology of the Heteromyidae (H. H. Genoways and J. H. Brown, eds.). Special Publication No. 10, The American Society of Mammalogists.

Patton, J.L., and D. S. Rogers. Cytogenetics. In Biology of Heteromyid Rodents (H. H. Genoways and J. H. Brown, eds.). Special Publication No. 10, The American Society of Mammalogists.

Patton, J. L. Hybridization and hybrid zones in pocket gophers (Rodentia, Geomyidae). Pp. 290-308 in Hybrid Zones and the Evolutionary Process (R. G. Harrison, ed.). Oxford Univ. Press.

Patton, J. L. Family Geomyidae. Pp. 469-476 in Mammal Species of the World, 2nd Ed. (D. E. Wilson and D. M. Reeder, eds.). Smithsonian Institution Press.

Patton, J. L. Family Heteromyidae. Pp. 477-486 in Mammal Species of the World, 2nd Ed. (D. E. Wilson and D. M. Reeder, eds.). Smithsonian Institution Press.

1994

Patton, J. L., and M. F. Smith. Paraphyly, polyphyly, and the nature of species boundaries in pocket gophers (genus *Thomomys*). Syst. Biol. 43:11-26.

Patton, J. L., M. N. F. da Silva, and J. R. Malcolm. Gene genealogy and differentiation among arboreal spiny rats (Rodentia: Echimyidae) of the Amazon basin: a test of the riverine barrier hypothesis. Evolution 48:1314-1323.

1995

Patton, James L., and M. N. F. da Silva. A review of the spiny mouse, genus *Scolomys* (Rodentia: Muridae: Sigmodontinae), with the description of a new species from the western Amazon of Brazil. Proc. Biol. Soc. Wash. 108:319-337.

1995

Pacheco, V., and J. L. Patton. A new species of the Puna mouse, genus *Punomys* Osgood, 1943 (Muridae, Sigmodontinae) from the Southeastern Andes of Peru. Z. Säugetierkunde 60:85-96.

1996

Patton, J. L., S. F. dos Reis, and M. N. F. da Silva. Relationships among didelphid marsupials based on sequence variation in the mitochondrial cytochrome b gene. J. Mammal. Evol. 3:3-29.

Lara, M. C., J. L. Patton, and M. N. F. da Silva. The simultaneous diversification of South American echimyid rodents (Hystricognathi) based on complete cytochrome b sequences. Mol. Phylog. Evol. 5:403-413.

Patton, J. L., M. N. F. da Silva, and J. R. Malcolm. Hierarchical genetic structure and gene flow in three sympatric species of Amazonian rodents. Mol. Ecol. 5:229-238.

de Fonseca, G. A. B., G. Herrmann, Y. L. R. Leite, R. A. Mittermeier, A. B. Rylands, and J. L. Patton. Lista Anotada dos Mamíferos do Brasil. Occas. Papers Conserv. Biol. 4:1-38.

Vié, J. C., V. Volobouev, J. L. Patton, and L. Granjon. A new species of *Isothrix* (Rodentia: Echimyidae) from French Guiana. Mammalia 60:393-406.

Peres, C. A., J. L. Patton, and M. N. F. da Silva. Riverine barriers and gene flow in Amazonian saddle-back tamarins. Folia Primatol. 67:113-124.

1997

Mustrangi, M. A., and J. L. Patton. Phylogeography and systematics of the slender mouse opossum Marmosops. Univ. of Calif. Pubs. Zool. 130:1-86.

Patton, J. L., and M. N. F. da Silva. Definition of species of pouched four-eyed opossums (Didelphidae, Philander). J. Mammal. 78:90-102.

1997

Ruedi, M., M. F. Smith and J. L. Patton. Phylogenetic evidence of mitochondrial DNA introgression among pocket gophers in New Mexico (family Geomyidae). Mol. Ecol. 6:453-462.

Patton, J. L., M. N. F. da Silva, M. C. Lara, and M. A. Mustrangi. Diversity, differentiation, and the historical biogeography of nonvolant small mammals of the neotropical forests. Pp. 455-465 in Tropical Forest Remnants (W. F. Laurance and R. O. Bierregaard, Jr., eds.). University of Chicago Press, Chicago, Illinois.

Lavergne, A., O. Verneau, J. L. Patton, and F. M. Catzeflis. Molecular discrimination of two sympatric species of opossum (genus *Didelphis*: Didelphidae) in French Guiana. Mol. Ecol. 6:889-891.

1998

da Silva, M. N. F., and J. L. Patton. Molecular phylogeography and the evolution and conservation of Amazonian mammals. Mol. Ecol. 7:475-486.

Patton, J. L., and M. N. F. da Silva. Rivers, refuges, and ridges: the geography of speciation of Amazonian mammals. Pp. 202-213 in Endless Forms: Species and Speciation (D. J. Howard and S. H. Berlocher, eds.). Oxford University Press, New York.

1999

Álvarez-Castañeda, S. T., and J. L. Patton (eds.). Mamíferos del Noroeste de Mexico (in Spanish and English). Centro de Investigaciones Biológicas del Noroeste, S.C., La Paz, Baja California Sur, Mexico. 583 pp.

Patton, J. L. Rodentia: Family Geomyidae. Pp. 321-350 in Mamíferos del Noroeste de Mexico (S. T. Álvarez-Castañeda and J. L. Patton, eds.). Centro de Investigaciones Biológicas del Noroeste, S.C., La Paz, Baja California Sur, Mexico.

1999

Smith, M. F., and J. L. Patton. Phylogenetic relationships and the radiation of sigmodontine rodents in South America: evidence from cytochrome b. J. Mamm. Evol. 6:89-128.

Patton, J. L., and S. T. Álvarez-Castañeda. Rodentia: Family Heteromyidae. Pp. 351-442 in Mamíferos del Noroeste de Mexico (S. T. Álvarez-Castañeda and J. L. Patton, eds.). Centro de Investigaciones Biológicas del Noroeste, S.C., La Paz, Baja California Sur, Mexico.

Patton, J. L. Pocket gopher accounts. Pp. 466-480 in The Smithsonian Book of North American Mammals (D. F Wilson and S. Ruff, eds.). Smithsonian Institution Press, Washington D.C.

Granjon, L., J. L. Patton, V. Volobouev, and J. C. Vié. Une nouvelle espèce de Ronguer décrite de Guyane Française, *Isothrix sinnamariensis* (in French). Arvicola XI:5-6.

Fleck, D. W., R. S. Voss, and J. L. Patton. Biological basis of Saki (*Pithecia*) folk species recognized by the Matses Indians of Amazonian Perú. Int. J. Primat. 20:1005-1028.

2000

Patton, J. L., M. N. F. da Silva, J. R. Malcolm. Mammals of the Rio Juruá and the evolutionary and ecological diversification of Amazonia. Bull. Amer. Mus. Nat. Hist., No. 244:1-306.

Matocq, M. D., J. L. Patton, and M. N. F. da Silva. Population genetic structure of two ecologically distinct Amazonian spiny rats: separating history and current ecology. Evolution 54:1423-1432.

Moritz, C., J. L. Patton, C. J. Schneider, and T. B. Smith. Diversification of rainforest faunas: an integrated molecular approach. Ann. Rev. Ecol. Syst. 31:533-563.

2000

Gascon, C., J. R. Malcolm, J. L. Patton, M. N. F. da Silva, J. P. Bogart, S. C. Lougheed, C. A. Peres, S. Neckel, and P. T. Boag. Riverine barriers and the geographic distribution of Amazonian species. Proc. Natl. Acad. Sci., USA 97(25):13672-13677.

Álvarez-Castañeda, S. T., and J. L. Patton, (eds.). Mamíferos del Noroeste de México II (in Spanish and English). Centro de Investigaciones Biológicas del Noroeste, S.C., La Paz, Baja California Sur, Mexico. Pp. 587-873.

Patton, J. L. Family Erethizontidae. Pp. 587-589 in Mamíferos del Noroeste de México II (S. T. Álvarez-Castañeda and J. L. Patton, eds.). Centro de Investigaciones Biológicas del Noroeste, S.C., La Paz, Baja California Sur, Mexico.

Lacey, E. A., J. L. Patton and G. N. Cameron (eds.). Life Underground: The Biology of Subterranean Rodents. The University of Chicago Press, Chicago, Illinois. 449 pp.

Steinberg, E. K., and J. L. Patton. Genetic structure and the geography of speciation in subterranean rodents: opportunities and constraints for evolutionary diversification. Pp. 301-331 in Life Underground: The Biology of Subterranean Rodents (E. A. Lacey, J. L. Patton, and G. N. Cameron, eds.). The University of Chicago Press, Chicago, Illinois.

Lara, M. C., and J. L. Patton. Evolutionary diversification of spiny rats (genus *Trinomys*, Rodentia: Echimyidae) in the Atlantic Forest of Brazil. Zool. J. Linn. Soc. 130:661-686.

2001

Geise, L., M. F. Smith, and J. L. Patton. Diversification in the genus *Akodon* (Rodentia: Sigmodontinae) in southeastern South America: mitochondrial DNA Sequence Analysis. J. Mammal. 82:92-101.

Smith, M. F., D. A. Kelt, and J. L. Patton. Testing models of diversification in mice in the *Abrothrix olivaceus/xanthorhinus* complex in Chile and Argentina. Mol. Ecol. 10:397-405.

2001

Patton, J. L., and M. N. F. da Silva. Molecular Phylogenetics and the diversification of Amazonian Mammals. Pp. 139-164 in Diversidade Biologica e Cultural da Amazonia (C. G. Vieira, J. M. Cardoso da Silva, D. C. Oren, and M. A. D'Incao, eds.). Mus. Paraense Emilio Goeldi, Belem, Brazil.

da Silva, M. N. F., A. B. Rylands, and J. L. Patton. Biogeografia e Conservação da Mastofauna na Floresta Amazônica Brasileira. Pp. 110-131 in Biodiversidade e Funções Ecológicas dos Ecosistemas.

2002

Lara, M., J. L. Patton, and E. Hingst-Zaher. *Trinomys mirapitanga*, a new species of spiny rat (Rodentia: Echimyidae) from the Brazilian Atlantic. Mamm. Biol. 67:233-242.

Morshed, S., and J. L. Patton. New records of mammals from Iran with systematic comments on hedgehogs (Erinaceidae) and mouse-like hamsters (*Calomyscus*, Muridae). Zool. Mid. East 26:49-58.

Leite, Y. L. R., and J. L. Patton. Evolution of South American spiny rats (Rodentia, Echimyidae): the star-phylogeny hypothesis revisted. Mol. Phylog. Evol. 25:455-464.

2003

Orlando, L., J. F. Mauffrey, J. Cuisin, J. L. Patton, C. Hanni, and F. Catzeflis. Napoleon Bonaparte and the fate of an Amazonian rat: new data on the taxonomy of *Mesomys hispidus* (Rodentia: Echimyidae). Mol. Phylog. Evol. 27:113-120.

Swei, A., P. V. Brylski, W. D. Spencer, S. C. Dodd, and J. L. Patton. Hierarchical genetic structure in fragmented populations of the little pocket mouse (*Perognathus longimembris*) in Southern California. Conser. Gen. 4:501-514.

2003

Costa, L. P., Y. L. R. Leite, and J. L. Patton. Phylogeography and systematic notes on two species of gracile mouse opossums, genus *Gracilinanus* (Marsupialia: Didelphidae) from Brazil. Proc. Biol. Soc. Wash. 116:275-292.

Lessa, E. P., J. A. Cook, and J. L. Patton. Genetic footprints of demographic expansion in North America, but not Amazonia, during the Late Quaternary. Proc. Natl. Acad. Sci., USA 100:10331-10334.

2004

Álvarez-Casteñeda, S. T., and J. Patton. Geographic genetic architecture of pocket gopher (*Thomomys bottae*) populations in Baja California. Mol. Ecol. 13:2287-2301.

2005

Patton, J. L. Species and speciation: changes in a paradigm through the career of a rat trapper. Pp. 263-276 in Going Afield: The Making of North American Mammalogists and Their Science (C. J. Phillips and C. Jones, eds.). Museum of Texas Tech University, Lubbock, TX.

Documenting Ecological Change in Time and Space: The San Joaquin Valley of California

Patrick A. Kelly, Scott E. Phillips, and Daniel F. Williams

The collections and journal archives of the Museum of Vertebrate Zoology (MVZ) at the University of California, Berkeley, provide a unique historical database that can be used in conjunction with new technologies in genetics and spatial analysis to address serious challenges to the conservation of biological diversity. Through direct reference to the journal entries of Joseph Grinnell and other MVZ biologists of the early 1900s and quantitative analyses of land use changes, we document the tempo and scale of land conversion in the San Joaquin Valley of California during the 20th century. We discuss the impacts of landscape level habitat changes for populations of selected mammalian species, most notably the endemic San Joaquin kangaroo rat (*Dipodomys nitratoides*). Of the three described subspecies of *D. nitratoides*, two (Tipton, *D. n. nitratoides*; Fresno, *D. n. exilis*) are listed as endangered under the California and U.S. endangered species acts, and the third (short-nosed, *D. n. brevinasus*) is a California species of special concern. Despite intensive field surveys begun in 1992, we have been unable to locate a population of Fresno kangaroo rats. This is particularly troubling because analysis of cytochrome-b DNA sequences, which were developed from museum specimen tissue samples, has shown that the Fresno kangaroo rat is unique and strongly differentiated from the other two subspecies. As the population of California continues to grow, the assault on biological diversity will continue. Analyses of the unique historical data provided by the MVZ and other natural history museums using the tools of modern molecular genetics and spatial analysis are essential to addressing these threats and halting or reversing the decline of biological diversity.

DOCUMENTING ECOLOGICAL CHANGE IN TIME AND SPACE: THE SAN JOAQUIN VALLEY OF CALIFORNIA

Since its establishment, the Museum of Vertebrate Zoology (MVZ) at the University of California, Berkeley, has provided remarkable intellectual leadership in the ecology and evolution of terrestrial vertebrates, with emphasis on western North America. When Annie Alexander sponsored and established the MVZ in 1908, she was in pursuit of excellence in research and scholarship, and chose Joseph Grinnell to lead the museum towards that goal, largely because of their deeply shared interest in field biology and mutual commitment to the study of natural history (Stein, 1997, 2001). Under the direction of Grinnell, work at the MVZ was exclusively in the realm of field biology. He set the example for museum scientists with extensive field research throughout California (Grinnell, 1911, 1927, 1928a,b, 1937; Grinnell and Linsdale, 1936; Grinnell and Miller, 1944; Grinnell and Storer, 1924; Grinnell and Wythe, 1927; Grinnell et al., 1930, 1937).

Today, MVZ scientists continue to work in various ecosystems throughout the western United States, Central and South America, and elsewhere. When they are not in the field, MVZ scientists are conducting genetic and other analyses in the laboratory to support their field studies. They continue to address problems that were of interest to Grinnell and other early MVZ scientists. Had Grinnell lived longer, he would have more vigorously pursued his keen interest in pocket gophers (Grinnell, 1927). His goal was achieved 50 years later when two other MVZ scientists, James Patton and Margaret Smith, published their ca 20-year study of the evolutionary dynamics of pocket gophers in California (Patton and Smith, 1990). One of Grinnell's major achievements is that the MVZ became one of the premier institutions for the study of vertebrate zoology. Few other institutions have amassed such a body of scholarship and knowledge about the natural world or have worked as assiduously to promote field research as has the MVZ.

An appreciation for and desire to conserve the natural world has been a central theme in the MVZ's long record of research and service, even if that appreciation was not always explicitly expressed in terms that we might use today. The discipline we call conservation biology is the modern philosophical incarnation of a particular ecological worldview, one that evolved from economic zoology and more recently, from disciplines such as wildlife management, population ecology, behavioral ecology, and other fields (Hall, 1939a,b; Grinnell, 1940; Linsdale, 1942). MVZ scientists have played a large role in conservation biology throughout the museum's history. To understand this role, we provide a retrospective using extensive references to the work and writings of Joseph Grinnell and other principal figures in the early days of the museum.

Almost immediately upon his appointment as the first Director of the MVZ in 1908, Joseph Grinnell took a deep personal interest in Yosemite National Park (Runte, 1990). Grinnell and other MVZ scientists conducted the key surveys that documented the vertebrate fauna of Yosemite (Grinnell and Storer, 1924). From this foundation of intensive field research, Grinnell played a critical role in the evolution of the management policies not only for Yosemite, but also for the National Park Service (Runte, 1990). He wrote thousands of letters and memos to park officials (Runte, 1990, p. 127). This dedication was not out of a sense of duty to assist in public administration. Rather, through his unfading commitment to science, Grinnell gently urged, prodded, and guided park officials towards more scientific management policies, directed primarily towards the conservation of native species (Runte, 1990, p. 129). Through his commitment to public education, he also fostered and encouraged public support for science-based management principles and practices in the parks (Runte, 1990, p. 111-112). Today, Yosemite is one of the crown jewels of the U.S. National Park System and is appreciated annually by millions of visitors from all over the world.

When Joseph Grinnell passed away prematurely in 1939, his legacy was measured by more than the prodigious numbers of study skins, journal pages, and publications he produced during his 21-year tenure as director; his commitment to conservation had become woven into the fabric of the MVZ. Over the past century, MVZ scientists and students have continued to follow Grinnell's philosophy through their profound commitment to preserving as well as understanding the natural world. This commitment is needed more today than ever before.

> It is probable that "Californians Incorporated," a commercial agency whose efforts are expended vigorously toward securing congestion of human population in the San Francisco Bay region, is right now the greatest single enemy of wild animal life in west-central California. The slogan "where life is better" is a curious perversion: it has sinister portents for even man himself.

> Joseph Grinnell (1928a; p. 204 in Grinnell, 1943)

When Joseph Grinnell wrote these prophetic words, California had a human population of less than 6,000,000. By 2000, more than 33,000,000 persons called California home. This more than five-fold increase in population has resulted in very serious consequences for biological diversity throughout California. Further losses in biological diversity, sometimes referred to as natural capital (UNEP-WCMC, 2000), are inevitable. The State of California is expected to have a population of about 46,000,000 people by 2020 (California Department of Finance, 2001). The impacts of this high rate of population growth are being felt throughout California but they have been most profound in four of the State's ten bioregions: South Coast, Central Coast, Bay Delta, and San Joaquin Valley (California Biodiversity Council, 1991). In this analysis, we report on landscape change in the San Joaquin Valley over the past century and describe the consequences of these changes for biological diversity in this region. MVZ scientists have worked in the San Joaquin Valley since 1911. Their collections and journal archives provide a unique historical database that can be explored using new technologies in genetics and spatial analysis to address significant challenges to the conservation of biological diversity in the San Joaquin Valley. We report on an initial exploration of this body of work and make recommendations for future study.

THE SAN JOAQUIN VALLEY: A CENTURY OF CHANGE

March, 1911, saw the arrival of the first MVZ expedition in the San Joaquin Valley (Grinnell, 1911). Joseph Grinnell, Harry Swarth and other MVZ biologists spent much of March, April, and May, 1911, collecting at various localities throughout the San Joaquin Valley and the Carrizo Plain. This was followed by much more work by Grinnell, Swarth, Joseph Dixon, Ward Russell, Seth Benson and many others in 1912 and subsequent years. The extensive journal notes from these MVZ expeditions not only describe the fauna of the region but also paint a detailed and graphic picture of a changing landscape:

> The surrounding country is flat and mostly farmed (wheat and orchards); unless it has been graded the surface shows the queer hummocky condition know locally as hog-wallow land *[vernal pool ecosystem]* of clayey, "hard pan".

> Joseph Grinnell, Lane Bridge, 10 mi. N Fresno, Fresno Co., 6 April 1911.

> Proceeding to Goshen this afternoon, the country is observed from the train. Practically every rod from Berenda *[Madera Co.]* to Selma *[Fresno Co.]* is under close cultivation in grain, alfalfa, raisins, and orchards.

> Joseph Grinnell, Goshen, Fresno Co., 23 April 1911.

> To the north and northwest the county is pretty closely farmed, up to the Tule River; but a belt through Tipton and to the south and east, is largely grazing land yet tho there are pumps being put in and it is only a matter of a few years until every rod of ground in under cultivation.

> Joseph Grinnell, Tipton, Tulare Co., 24 April 1911

> Left Berkeley at 8:00 A.M., catching the 8:53 train at Oakland, which reached Bakersfield at 8 P.M. I had thus a good chance to see the whole length of the San Joaquin Valley, on the east side. It appears to be nearly all under cultivation, or else used as pasturage, and open tracts are evidently changing rapidly, being divided into smaller holdings and more intensively cultivated.

> Harry Swarth, traveling from Oakland to Bakersfield by train, 5 May 1911.

Left Bakersfield on the 8:15 A.M. train (which pulled out at 9:30) reaching McKittrick at 11:30. The stretch of country between the two places is not cultivated as I expected to see it. Outside of Bakersfield were long stretches of brush land, and then miles of bottom land, with ditches and sloughs, pretty well grown up with cotton woods and willows. Around Buttonwillow there was a good deal of alfalfa and other hay fields, but from there to McKittrick it was all brush land, much of it quite sandy.

> Harry Swarth, traveling from Bakersfield to McKittick (western Kern Co.) by
> train, 17 May 1911.

The MVZ Journal archives again and again reveal the commitment to detail exhibited by Grinnell and his field teams. They not only surveyed the landscape and collected examples of the fauna, but they also reported on numerous conversations and interviews with local residents, especially 'old-timers.' Grinnell's journal entries in particular are peppered with references to local observations on the presence, or, more usually, absence of kit foxes (*Vulpes macrotis*), "chipmunks" (antelope ground squirrels, *Ammospermophilus nelsoni*), kangaroo rats (*Dipodomys* spp.), bighorn sheep (*Ovis canadensis*), grizzly bears (*Ursus arctos*) and other mammals that were declining in numbers or losing significant amounts of habitat to land conversion by the early 1900s:

One man interviewed said there were "Kangaroo rats" in his place six years ago when he first plowed but that he had seen none since. Several have told us that 20 years or more ago "rattlesnakes and kangaroo rats" abounded in certain places in the vicinity of Fresno. Evidently the mammal and reptile fauna of the region have been as profoundly modified by human settlement as the birds.

> Joseph Grinnell, Clovis, Fresno Co., 11 April 1911

It would seem that this chipmunk *[antelope ground squirrel]* is retreating in range from the east side of the Tulare Valley, as the country settles up (either cultivated or pastured closely) and as the ground squirrel (C. beecheyi) comes in. We are repeatedly told that the latter has only recently come into this belt, and that it is becoming more numerous all the while.

> Joseph Grinnell, Earlimart, Tulare Co., 1 May 1911

> Mr. J.S. Douglas is the superintendent of the ranch here. He has been in the country since the 70's, and is absolutely trustworthy. He tells me as follows: He personally knows of sheep *[bighorn sheep]* in the Sespe Country, where the last one was killed in 1900. On the San Emigdio ranch, there were many sheep in the steep hills in the 70's and early 80's. Many were shot. In 1888 there were fully 150 sheep in 3 flocks.

> Joseph Grinnell, San Emigdio Ranch (Wind Wolves Preserve today), Kern Co.,
> 22 April 1912

The landscape that Grinnell and Swarth described on the east side of the San Joaquin Valley differed significantly from that of the west side. The great water projects—the Central Valley Project and the State Water Project—had not yet been undertaken and much of the west side was still open rangeland characterized by a mix of grassland, alkali sink scrub, and salt bush scrub (Kahrl, 1978; Reisner, 1987; Thelander, 1994; USFWS, 1998; Hundley, 2001). Between 1915 and 1923, Joseph Dixon spent a considerable amount of time in the San Joaquin Valley, much of it on the "plains" on the west side of the valley. There, he documented the detrimental effects of human activity on native fauna such as pronghorn antelope (*Antilocapra americana*) and the kit fox:

> Serious inroads have been made into the population of kit foxes in the San Joaquin Valley, comprising the subspecies mutica. Large numbers of the animals have been caught there for fur in recent years. For instance, in 1919 Arthur Oliver caught 100 foxes in one week on an area 20 miles long and 2 miles wide, on the plains on the west side of the San Joaquin Valley, in Fresno County. (See fig. 162.) On December 3, 1920, 37 steel traps set in that region caught 5 kit foxes in one night.

> Grinnell, Dixon, and Linsdale (1937 p. 418).

Far more significant inroads than those provided by fur trappers, however, were in store for kit foxes. The MVZ photography archive provides an understanding of agricultural development in this region between 1920 and 1937 (Figure 1a, from Grinnell, Dixon, and Linsdale, 1937, p. 419). For this study, on June 3, 2001 one of us (PAK) took a photograph from about the same location and approximately of the same scene as Dixon's 1920 image (Figure 1b). The location of the 2001 photograph (latitude 36.67738 N, longitude 120.62191 W, Datum WGS 84) is 25.1 km southwest (bearing 216º) of Firebaugh, Fresno Co. It is about 5 km from the base of the foothills and, in contrast to 1920, is now completely cultivated.

Today, most of the west side plains of the San Joaquin Valley are cultivated to the base of foothills.

With the progressive construction of water storage and delivery projects over the past century — notably the massive Central Valley and State Water projects (1935 to 1970) — great tracts of formerly uncultivated or rarely cultivated land were converted to intensive agricultural use (Kahrl, 1978; Reisner, 1987; Thelander, 1994; Hundley, 2001). The natural landscape became increasingly fragmented as grasslands, wetlands, shrublands, woodlands, and forests were converted for cultivation (Figures 2a-d). The fragmented mosaic of natural and cultivated land described by Grinnell and Swarth in 1911 and 1912 gave way over time to the vast cultivated landscape evident today (USFWS, 1998).

Effects on Native Mammals

The scale of land conversion in the San Joaquin Valley over the past 100 to 150 years is staggering (Figure 3, Table 1). We estimate that there has been a loss of more than 27,000 km^2 of natural communities in the San Joaquin Valley. About 65% of grasslands, 64% of San Joaquin Valley shrublands, 88% of water and wetlands, and 95% of riparian forest and oak woodland have been converted, mainly to agricultural use (Table 1). The resulting losses in biological diversity are almost incalculable. If we were to assume conservatively that valley grasslands had an average annual small mammal biomass of 0.300 kg/ha (i.e., low density and low diversity situation of 5-10 kangaroo rats or 15-20 'mice' per hectare), the loss of grasslands represented in Table 1 would translate to an annual loss of 507.42 metric tonnes of small mammals. When we consider the interrelated population dynamics of small mammals and the many predatory species that largely depend on them (e.g, carnivores, raptors, owls, snakes), this loss in biodiversity is compounded significantly.

Mammals that have larger area requirements (e.g., tule elk, *Cervus elaphus;* pronghorn, and kit foxes) were quickly impacted by the settlement and development of the San Joaquin Valley. Perhaps less evident were effects on species with seemingly smaller area requirements. Work by MVZ and other researchers, however, indicates that these species also appear to be very susceptible to habitat fragmentation and degradation. We have already noted Grinnell's 1911 observations that kangaroo rats and antelope ground squirrels appeared to be declining in the face of land settlement. Extensive field research throughout the San Joaquin Valley since 1992 by biologists with the California State University, Stanislaus, Endangered Species Recovery Program (ESRP) confirmed Grinnell's fears that kangaroo rats are susceptible to habitat fragmentation (USFWS, 1998; Uptain et al., 1998). A review of the MVZ collections and journal archives helps reveal the extent of the impact to kangaroo rat populations from land conversion.

Figure 1a. Habitat of the San Joaquin kit fox in western Fresno Co. on 3 December 1920. (Adapted from a photograph by Joseph Dixon, Museum of Vertebrate Zoology Archives; no. 3426).

Figure 1b. Photograph taken on 3 June 2001 from approximately the same location as Joseph Dixon's 1920 photograph. (CSU Stanislaus, Endangered Species Recovery Program).

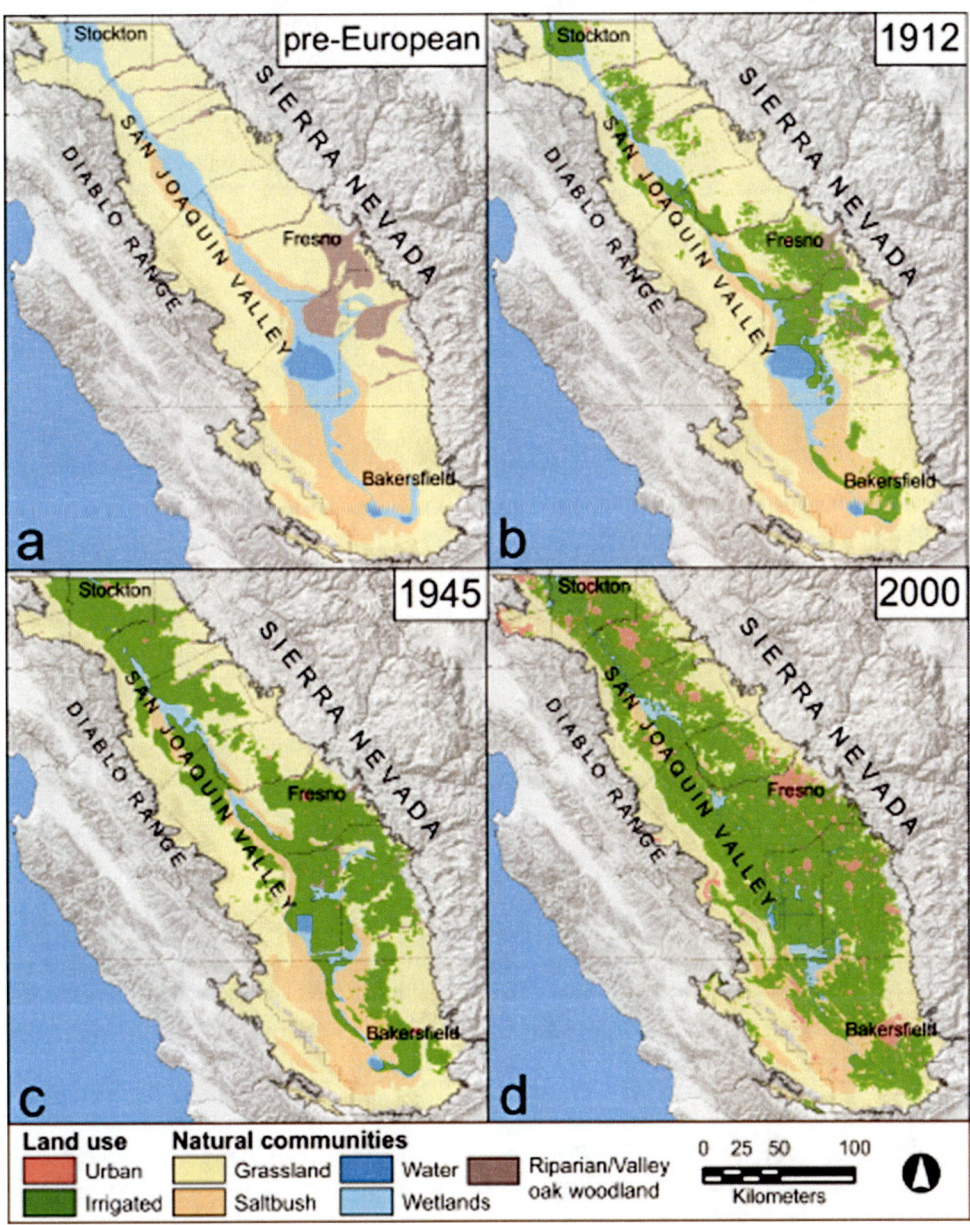

Figure 2. Agricultural land conversion in the area of the Great Valley Region of California (UCSB, 1996) south of latitude 38°N, pre-European settlement to 2000.

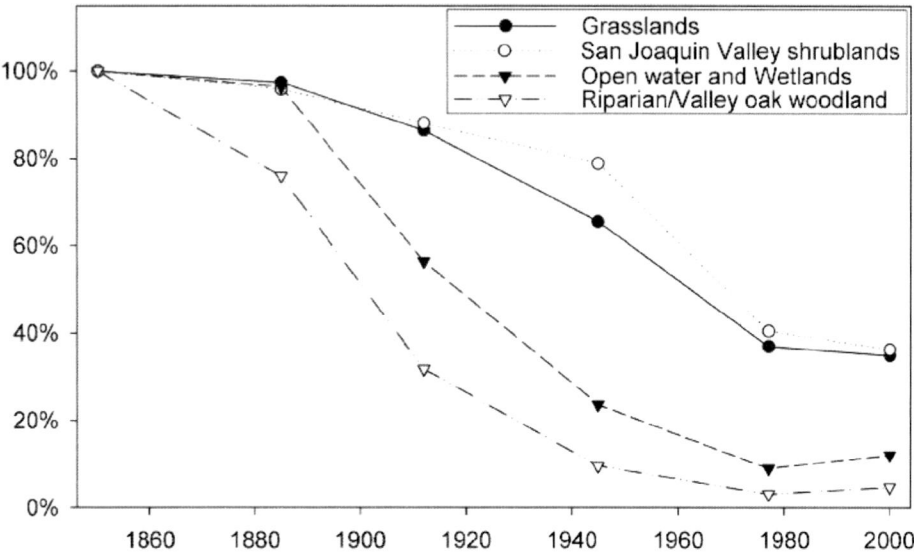

Figure 3. Percent change in natural community cover in the area of the Great Valley Region of California (UCSB, 1996) south of latitude 38°N, pre-European settlement to 2000.

Table 1. Estimated area changes (km²) by major land cover category in the area of the Great Valley Region of California (UCSB, 1996) south of latitude 38°N, pre-European settlement to 2000.

Land Cover Category	Pre-European	2000	Change	%
Developed or degraded	0	27,636	27,636	
Grasslands	25,989	9,075	-16,914	-65.1
San Joaquin Valley shrublands	6,151	2,231	-3,920	-63.7
Open water and Wetlands	5,205	630	-4,575	-87.9
Riparian/Valley oak woodland	2,335	109	-2,225	-95.3

Between 1911 and 1960, MVZ expeditions collected San Joaquin kangaroo rats (*Dipodomys nitratoides*) at 40 localities throughout the San Joaquin Valley (Figure 4). We estimate that *D. nitratoides* is currently extant at probably no more than 18 of these locations, 16 of which are in the band of largely uncultivated rangeland that remains around the margins of the San Joaquin Valley (including the Carrizo Plain National Monument in San Luis Obispo Co.). The remaining two locations are in Tulare County, in the vicinity of the Pixley National Wildlife Refuge and the California Department of Fish and Game's Allensworth Ecological Reserve. These two public land holdings are still occupied by *D. nitratoides*, at least on some land parcels, but most of the valley floor locations trapped by Grinnell and other MVZ researchers have given way to cultivation. Where uncultivated lands remain on the valley floor, they are usually privately owned and closed to trapping surveys.

There are some small populations of *D. nitratoides* at other locations, especially in the southern San Joaquin Valley, that are not represented in the MVZ collections. San Joaquin kangaroo rats however, appear to have largely disappeared from the valley floor, even from most of the larger parcels that remain uncultivated. Populations seem to be in a continuing decline, probably due to the combined effects of habitat conversion, fragmentation, degradation, and other factors (Chesemore and Rhodehamel, 1992; Williams and Kilburn, 1992; Williams and Germano, 1993; Goldingay et al., 1997; USFWS, 1998; Uptain et al., 1998; Kelly 2000). Populations in the more or less continuous band of grassland around the valley periphery are doing better but, even there, San Joaquin kangaroo rats are difficult to find.

THE TOOLS OF CONSERVATION ECOLOGY

Joseph Grinnell was a visionary scientist and conservationist. At the very outset of his MVZ career, he recognized that the true value of the museum would not be gained from the simple accumulation of vertebrate specimens.

> It is quite probable that the facts of distribution, life history, and economic status may finally prove to be of more far-reaching value, than whatever information is obtainable exclusively from the specimens themselves.

> At this point I wish to emphasize what I believe will ultimately prove to be the greatest value of our museum. This value will not, however, be realized until the lapse of many years, possibly a century, assuming that our material is safely preserved. And this is that the student of the future will have access to the original record of faunal conditions in California and the west wherever we now work. He will know the proportional constituency of our faunae by species, the relative numbers of each species and the extent of the ranges of species as they exist to-day.

> Perhaps the most impressive fact brought home to the student of geographical distribution, as he carries on his studies, is the profound change that is constantly going on in the faunal make-up of our country. Right now are probably beginning changes to be wrought in the next few years vastly more conspicuous than those that have occurred in ten times that length of time preceding. The effects of deforestation, of tree-planting on the prairies, of the irrigation and cultivation of the deserts, all mean the rapid shifting of faunal boundaries, the extension of ranges of some animals, restriction in the ranges of others, and, with no doubt whatever, the complete extermination of many others, as in a few cases already on record.

Joseph Grinnell (1910)

The tools that Grinnell employed in his work were his scientific intellect, exceptional observational skills, the thousands of scientific specimens he and his colleagues so painstakingly collected, and, above all else, the detailed journal notes that he accumulated during his surveys of the western United States. A century later, the MVZ has powerful tools that were unavailable to Grinnell. The rapid development of new techniques in genetics, spatial analysis, and landscape ecology that have been pioneered or rapidly adopted by MVZ scientists have provided fresh insights into not only evolutionary questions, but also our most pressing conservation concerns.

Genetics

This application of new technology is exemplified by the work of James Patton (in litt.), who, on the basis of cytochrome-b DNA sequences, has shown that the Fresno kangaroo rat (*D. n. exilis*) is unique and strongly differentiated from the other two subspecies of the San Joaquin kangaroo rat, the Tipton and short-nosed kangaroo rats (*D. n. nitratoides* and *D. n. brevinasus*, respectively). This finding is noteworthy because both the Fresno and Tipton subspecies are listed as endangered by the State and Federal governments. Despite intensive surveys and trapping efforts conducted throughout its former range, not a single Fresno kangaroo rat has been captured since 1992. One individual was captured on the California Department of Fish and Game's Alkali Sink Ecological Reserve (Fresno Co.) on two occasions in Nov. 1992 by one of us (DFW, ESRP data), but Patton's analyses of *D. n. exilis* had to be completed using tissue biopsies taken from study skins in collections at the MVZ and California State University, Fresno. On the basis of this new genetic information, if extant populations of the Fresno kangaroo rat can be located, they are likely to be given the highest priority for conservation by government agencies.

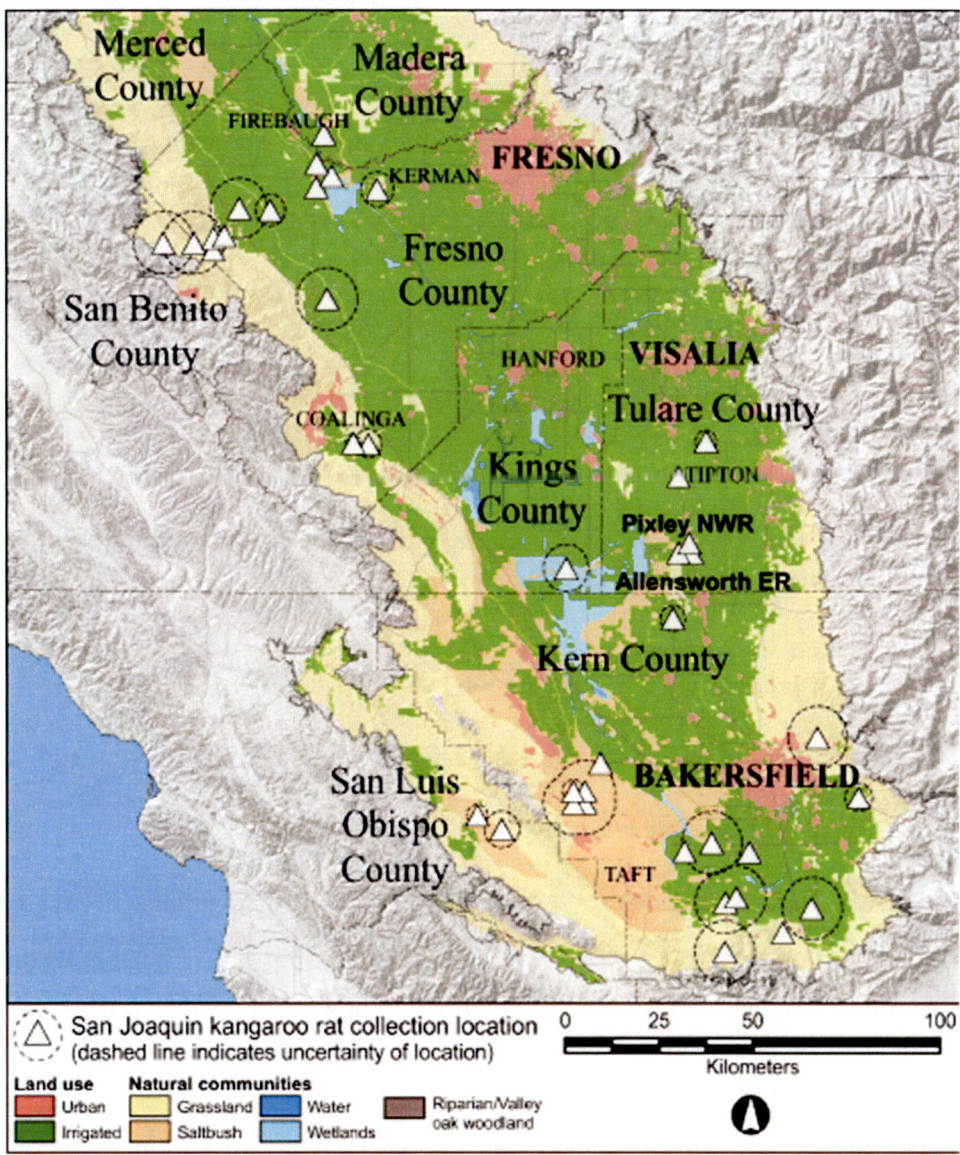

Figure 4. MVZ collection localities for San Joaquin kangaroo rats, 1911 to 1960. (Data provided by the Museum of Vertebrate Zoology, University of California, Berkeley.)

Spatial Analysis

As with genetic analyses, there have been tremendous developments in recent years in computer and database technology. In particular, rapid growth in computer and satellite technologies has provided powerful tools for spatial analyses, notably geographic information systems (GIS) and global positioning systems (GPS). GIS technology allows a user to create databases that store and relate spatial and non-spatial information. Historical map and descriptive data can be combined with current information to assess spatial changes over time. GPS provides researchers with greater capacity to remotely collect and retrieve spatial information. Compact and inexpensive GPS receivers provide the ability to geo-reference collecting and other localities in the field, rather than after the fact. This can be very valuable for relating locations to external spatial data derived from GIS.

In this study, using GIS to associate historical and modern map sources, we have described a striking pattern of habitat loss across the San Joaquin Valley during the 20th century (Figure 2, Table 1). We quantified conversion of four major natural community types from pre-European settlement to 2000 (Table 1); the recent geo-referencing of MVZ museum specimen records (Wieczorek, 2001) allowed us to relate collection locations to external spatial data. The specimen records, along with associated journal entries, genetic and other data, are providing further insights into the dynamics of landscape change in the San Joaquin Valley (see Appendix 1).

THE CHALLENGE

As Joseph Grinnell predicted in 1910, the MVZ collections and journals are needed now more than ever. The persistent growth of the human population of California continues to erode the tremendous, and in many ways unique, biological diversity of the State.

Many of the species that Grinnell inquired about when interviewing local residents of the San Joaquin Valley in 1911 and 1912 are now extinct, rare, or are listed as sensitive, threatened, or endangered by State and Federal governments. Ecosystems throughout California are threatened by urbanization, resource extraction, recreation, and other anthropogenic activities.

The vernal pool ecosystems ("hog-wallow land") that Grinnell commented on in 1911 have been greatly reduced and will likely become further degraded in the coming years. For example, ground breaking for the University of California's 10th campus occurred on Oct. 25, 2002 near the city of Merced, immediately adjacent to the largest remaining expanse of vernal pools in the San Joaquin Valley.

Proposed housing, commercial, highway, and other infrastructure developments present a more routine and pervasive threat to biodiversity. Agency biologists are daily required to provide their professional judgment on the likely impacts of such developments on sensitive species and ecosystems, usually with incomplete knowledge of the distribution, abundance, and natural history of the affected species, and nearly always with no phlyogeographic information for key taxa. It is regrettable that the Fresno kangaroo rat declined to near extinction, if not extinction itself, before a phylogeographic analysis demonstrated its uniqueness. We have no doubt that there are many other taxa with similarly interesting but largely unstudied evolutionary histories that are currently facing uncertain futures in California.

The pressures on the remaining natural land in California are intensifying, suggesting that the decline of ecosystems and their constituent flora and fauna will continue. MVZ scientists and students have at their disposal the tools needed to address these increasingly serious conservation challenges. In particular, they have many of the tools required to identify and to provide a better understanding of major phylogeographic boundaries throughout California.

The combination of the MVZ collections, journal library, genetic laboratories, and GIS technology greatly enhance the analytical power of ecologists, systematists, and conservation biologists. It is our fervent hope that the MVZ will continue its strong leadership role in addressing these conservation challenges in the years ahead.

ACKNOWLEDGEMENTS

The authors would like to extend their sincere gratitude to Jim Patton for his friendship, scholarship, advice, and support over many years of professional and personal interaction. If Joseph Grinnell were alive today, he would be very pleased to know that Jim has so closely followed in his footsteps.

> Our institution is a repository of facts; and no matter what may be said to the contrary by those who undervalue the efforts of the hoarder of facts, it must always be the mass of carefully ascertained facts upon which the valid generalization rests.
>
> Joseph Grinnell (1910)

Jim's accumulation of facts about the ecology and evolution of small mammals throughout the New World has been nothing short of astonishing. Those of us who have had the privilege to work with Jim over the years know that he has also contributed many "valid generalizations."

The biological surveys, genetic studies, and GIS analyses referenced in this chapter were largely supported by the U.S. Bureau of Reclamation, the U.S. Fish and Wildlife Service, and the California Department of Fish and Game. We are grateful for their support. The opinions expressed are those of the authors, not necessarily the opinions of the sponsoring agencies.

LITERATURE CITED

California Biodiversity Council
1991. The Agreement on Biological Diversity. California Resources Agency. September 19, 1991. (http://ceres.ca.gov/biodiv/mou.html)

California Department of Finance
2001 Interim County Population Projections: Estimated July 1, 2000 and Projections for 2005, 2010, 2015, and 2020. State of California, Department of Finance, Demographic Research Unit. June 2001. (5 pages. Taken from the State of California web site March 2002: http://www.dof.ca.gov/HTML/DEMOGRAP/drupubs.htm)

Chesemore, D. L., and W. M. Rhodehamel
1992 Ecology of a vanishing subspecies: the Fresno kangaroo rat (*Dipodomys nitratoides exilis*). Pp. 99-103 in D.F. Williams, S. Byrne, and T.A. Rado, eds. Endangered and sensitive species of the San Joaquin Valley: their biology, management and conservation. (Based on a conference held at California State University, Bakersfield 10-11 December 1987) California Energy Commission.

Endangered Species Recovery Program (ESRP)
2001 Land Use Land Cover in the San Joaquin Valley. 1:100,000. California State University, Stanislaus, Endangered Species Recovery Program, Fresno, California.

Endangered Species Recovery program and U.S. Bureau of Reclamation (ESRP and USBR)
1999 Natural Vegetation of Central California based on Kuchler (1977). 1:1,000,000. California State University, Stanislaus, Endangered Species Recovery Program, Fresno, California.
(http://www.esrp.org/gis/metadata/kuchler_sjv.html)

Goldingay, R., P. Kelly, and D. Williams
1997 The kangaroo rats of California: endemism and conservation of a keystone species. Pacific Conservation Biology 3:47-60.

Grinnell, H. Wood
1940   Joseph Grinnell: 1877-1939. Condor XLII:2-34.

Grinnell, J.
1910   The methods and uses of a research museum. Popular Science Monthly 77:163-169. (Reprinted as Pp. 31-39 in *Joseph Grinnell's Philosophy of Nature*. Univ. of California Press, Berkeley. 1943.)

1911   Field notes from the San Joaquin Valley. Condor XIII:109-111.

1927   Geography and evolution in the pocket gophers of California. Smithsonian Institution Annual Report for 1926.   Pp. 343-355. Washington D.C.

1928a   Presence and absence of animals. Univ. of California Chronicles XXX (October 1928): 429-450. (Reprinted as Pp. 187-208 in Joseph Grinnell's Philosophy of Nature. Univ. of California Press, Berkeley. 1943.)

1928b   A distributional summation of the ornithology of Lower California. University of California Publications in Zoology, Vol. 32, No. 1, 300 pp. Berkeley, California.

1937   Mammals of Death Valley. Proceedings of the California Academy of Sciences 33:115-169.

Grinnell, J., J. Dixon, and J. Linsdale
1930   Vertebrate natural history of a section of northern California through the Lassen Peak region.   University of California Publications in Zoology, Vol. 35, 594 pp. Berkeley, California.

1937   Fur-bearing Mammals of California. 2 Vols. Univ. California Press, Berkeley. 777 pp.

Grinnell, J., and J. Linsdale
1936   Vertebrate Animals of Point Lobos Reserve. Carnegie institution of Washington, Publication No. 481. 159 pp.

Grinnell, J., and A. H. Miller
1944   The Distribution of the Birds of California. The [Cooper Ornithological] Club, Berkeley, California. 608 pp.

Grinnell, J., and T. I. Storer
  1924   Animal Life in the Yosemite: An Account of the Mammals, Birds,
          Reptiles, and Amphibians in a Cross-Section of the Sierra Nevada.
          Univ. California Press, Berkeley. 752 pp.

Grinnell, J., and M. W. Wythe
  1927   Directory to the bird-life of the San Francisco Bay region. The [Cooper
          Ornithological] Club, Berkeley, California. 160 pp.

Hall, E. R.
  1939a   Obituary for Joseph Grinnell. J. Wildl. Manage. 3:366-368.

  1939b   Joseph Grinnell (1877-1939), J. Mammal. 20:408-417.

Hall, W. H.
  1886   Topographic and Irrigation Maps of San Joaquin Valley, Sheets 1-4.
          1:126,720. Obtained from the Water Resources Archives, University of
          California, Berkeley.

Holmes, L. C., E. C. Eckmann, J. W. Nelson, and J. E. Guernsey
  1919   Soil map, reconnaissance survey, Middle San Joaquin Valley sheet,
          California. 126,720. In: Reconnaissance soil survey of the middle San
          Joaquin Valley, California. United States Agriculture Department. Soils
          Bureau; University of California Agricultural Experiment Station. 115
          pp.

Hundley, N.
  2001   The Great Thirst: Californians and Water: A History. University of
          California Press, Berkeley. 800 pp.

Kahrl, W. (ed.)
  1978   California Water Atlas. The State of California. 118 pp.

Kelly, P. A.
  2000   Oh, Rats! BT Journal (Biological and Technical Journal for Wildlife
          Photographers) V:11-13.

Linsdale, J. M.
  1942   In Memorium: Joseph Grinnell. Auk 59:268-285.

Nelson, J. W., W. C. Dean, and E. C. Eckmann
    1921   Soil map, reconnaissance survey, Upper San Joaquin Valley sheet,
            California. 126,720. In: Reconnaissance soil survey of the Upper San
            Joaquin Valley, California. United States Agriculture Department. Soils
            Bureau; University of California Agricultural Experiment Station. 116
            pp.

Nelson, J. W., J. E. Guernsey, L. C. Holmes, and E. C. Eckmann
    1918   Soil map, reconnaissance survey, Lower San Joaquin Valley sheet,
            California. 126,720. In: Reconnaissance soil survey of the Lower San
            Joaquin Valley, California. United States Agriculture Department. Soils
            Bureau; University of California Agricultural Experiment Station. 157
            pp.

Patton, J. L. (in litt.)
            Genetic diversity in and genetic continuity among populations of the
            kangaroo rat *Dipodomys nitratoides* of the San Joaquin Valley of
            California. (manuscript, 30 pages).

Patton, J. L., and M. F. Smith
    1990   The evolutionary dynamics of the pocket gopher *Thomomys bottae*, with
            emphasis on California populations. University of California
            Publications in Zoology 123:1-161 + xviii.

Piemeisel, R. L., and F. R. Lawson
    1937   Types of Vegetation in the San Joaquin Valley of California and Their
            Relation to Beet Leafhopper. U.S. Department of Agriculture Technical
            Bulletin No. 557, Washington, D.C. 28 pp.

Reisner, M.
    1987   Cadillac Desert: The American West and its disappearing water.
            Penguin Books, USA. 582 pp.

Runte, A.
    1990   Yosemite: the embattled wilderness. Univ. of Nebraska Press, Lincoln.
            271 pp.

Stein, B. R.
    1997   Annie M. Alexander: Extraordinary Person. J. Hist. Biol. 30:243-266.

Stein, B. R.
    2001   On Her Own Terms: Annie Montague Alexander and the Rise of
            Science in the American West. University of California Press, Berkeley,
            California. 380 pp.

Thelander, C. (ed.)
    1994   Life on the Edge: A guide to California's endangered natural resources.
            Heyday Books, Santa Cruz, California. 513 pp.

United Nations Environment Programme-World Conservation Monitoring Centre
(UNEP-WCMC)
    2000   Natural capital indicators for OCD countries.   United Nations
            Environment Programme-World Conservation Monitoring Centre
            (undated, viewed 2001-2003:
            http://www.unep-wcmc.org/species/reports/oecdreport.pdf)

University of California Santa Barbara, California Gap Analysis (UCSB)
    1996   Jepson Regions of California. 1:100,000.
            http://www.biogeog.ucsb.edu/projects/gap/gap_data.html

Uptain, C. E., D. F. Williams, P. A. Kelly, L. P. Hamilton, and M. C. Potter
    1998   The status of Tipton kangaroo rats and the potential for their recovery.
            Trans. West. Sect. Wildl. Soc. 35:1-9.

U.S. Bureau of Reclamation (USBR)
    1949   Water Resources Development. 1:1,000,000. In: Central Valley basin; a
            comprehensive report on the development of the water and related
            resources of the Central Valley basin for irrigation, power production
            and other beneficial uses in California, and comments by the State of
            California and Federal agencies. U.S. Govt. Print. Office, Washington,
            DC. 421 pp.

U.S. Department of Agriculture, Office of Experiment Stations (USDA)
    1912   Irrigation map of Central California: to accompany report on the
            irrigation resources of Central California. 1:500,000. U.S. Department
            of Agriculture, Office of Experiment Stations, Irrigation Investigations.
            Obtained from the Water Resources Archives, University of California,
            Berkeley.

U.S. Fish and Wildlife Service (USFWS)
   1998   Recovery plan for upland species of the San Joaquin Valley, California.
           U.S. Fish and Wildlife Service, Portland, Oregon. 319 pp.

U.S. Geological Survey (USGS)
   1990   Land Use Land Cover (LULC). 1:250,000. U.S. Geological Survey Eros
           Data Center, Soiux Falls, South Dakota.
           (http://edcwww.cr.usgs.gov/doc/edchome/ndcdb/ndcdb.html)

Wieczorek, J.
   2001   Museum of Vertebrate Zoology Collections Information System Re-
           engineering Project.
           (http://www.mip.berkeley.edu/mvz/cis/index.html).

Williams, D. F. and D. J. Germano
   1993   Recovery of endangered kangaroo rats in the San Joaquin Valley,
           California. Trans. West. Sect. Wildl. Soc. 28:93-106.

Williams, D. F. and K. S. Kilburn
   1992   The conservation status of the endemic mammals of the San Joaquin
           Faunal Region, California. Pp. 329-348 in D.F. Williams, S. Byrne, and
           T.A. Rado, eds. Endangered and sensitive species of the San Joaquin
           Valley: their biology, management and conservation. (Based on a
           conference held at California State University, Bakersfield 10-11
           December 1987) California Energy Commission, Sacramento,
           California.

APPENDIX 1

Within the boundaries of the Great Central Valley region of California (UCSB, 1996) south of latitude 38°N, we estimated the extent of historical vegetation types using wetland features from 1885 irrigation maps (Hall, 1886), GIS data derived from a map of potential vegetation of California (ESRP and USBR, 1999), and vegetation data derived from additional map sources (Holmes et al., 1919; Nelson et al., 1918; Nelson et al., 1921; Piemeisel et al., 1937). We scanned the source maps and used image-processing software to align the images to a common map coordinate system.

   To normalize the classification differences between sources, we reclassified map features to four general habitat types: grasslands, shrub lands, riparian/Valley oak woodland, and open water/wetlands. We digitized the reclassified map information to quantify the area of each habitat type within our study area. Using scanned irrigation and land cover maps and GIS data, we estimated the extent of developed land in 1885, 1912, 1940, 1977, and 2000 (ESRP 2001; Hall, 1886; USBR 1949; USDA, 1912; USGS, 1990).

Coat Color Variation in Rock Pocket Mice (*Chaetodipus intermedius*): From Genotype to Phenotype

Hopi E. Hoekstra and Michael W. Nachman

In a series of classic studies in mammalian evolutionary biology, Sumner (1921), Benson (1933), and Dice and Blossom (1937) described striking coat color variation in the rock pocket mouse, *Chaetodipus intermedius*, in the deserts of Arizona and New Mexico. These authors showed that *C. intermedius* coat color typically matches the color of the rocks on which the mice live; the dorsal pelage varies from a light, sandy color for populations found on some granites to a dark, nearly black color for populations found on basalt lava flows. Dice and Blossom (1937) suggested that this crypsis is an adaptation to avoid predation. Motivated by the wealth of data on the genetics, biochemistry, and molecular biology of the pigmentation process, we have used a candidate-gene approach to identify the genetic basis of adaptive coat color variation in *C. intermedius*. We review our recent studies on this topic with emphasis on the following key results: the identification of a single gene (the melanocortin-1-receptor, *Mc1r*) in one population that appears to be largely responsible for color differences, the balance between selection and migration among neighboring melanic and light races, and the finding that melanism has evolved independently on different lava flows through changes at different genes.

ADAPTIVE COAT COLOR VARIATION IN RODENTS

Coat color variation in small mammals is a classic example of phenotypic variation in response to selection in different environments; many species closely match the color of the substrate on which they live. This geographic variation in phenotype is well documented both within and between species. Examples of intraspecific color variation in rodents include the canyon mouse (*Peromyscus crinitus*), the deer mouse (*Peromyscus maniculatus*), the oldfield mouse (*Peromyscus polionotus*), Botta's pocket gopher (*Thomomys bottae*), and the rock pocket mouse (*Chaetodipus intermedius*). Although there has been some controversy concerning the importance of crypsis in maintaining this kind of variation (Sumner, 1921; Dice and Blossom, 1937), studies in *Peromyscus* have clearly demonstrated that owls discriminate between mice that do and do not match the color of their substrate (Dice 1947; Kaufman 1974).

At the molecular level, pigmentation is one of the best-studied phenotypic traits in mammals (Silvers, 1979). We have a reasonable understanding of the developmental pathways involved in the formation of melanocytes, the specialized cells that are the site of melanin production, and of the biochemistry of melanin synthesis (Urabe et al., 1993). Importantly for our work, many of the genes underlying these processes have been identified, primarily from studies in the laboratory mouse. Many coat color mutants are known, and over twenty have now been characterized at the molecular level and associated with specific nucleotide changes. With this information, we have been able to target candidate genes that generate phenotypes similar to those seen in wild populations of rodents in an attempt to identify genes involved in adaptive coat color differences.

Here, we review our recent studies on the genetic basis of coat color variation in the rock pocket mouse, *Chaetodipus intermedius* (Rodentia: Heteromyidae). This species of desert-dwelling rodent displays one of the most striking examples of pigmentation variation in mammals, with light and melanic animals inhabiting granitic and volcanic substrates, respectively. This system offers a unique opportunity to investigate the genetic basis of an adaptive phenotype. First, we describe the natural history of rock pocket mice and observed patterns of phenotypic variation. Next, we review the genetics and biochemistry of mammalian pigmentation pathways. We then focus on several key results from our recent work, namely (1) the discovery of a gene that underlies major color differences in one population, (2) our estimates of the strength of selection on this gene, and (3) our finding that similar melanic phenotypes have evolved independently on different lava flows through changes at different genes.

Phenotypic Variation in *Chaetodipus intermedius*

C. intermedius inhabits rocky areas and desert scrub at low elevations in the southwestern deserts of North America, ranging from southern Utah through most of western and southern Arizona, southern New Mexico, western Texas and adjacent areas in northern Mexico. Rock pocket mice are primarily nocturnal, and show diminished activity from November through February. While pregnant mice have been caught at all times of the year, the peak of reproductive activity is June (Reichman and Van de Graaff, 1973). Females can have as many as three litters per year (Brown and Harney, 1993). Young are born naked, usually attaining their first coat at two to three weeks; the first coat is generally thinner and grayer than the adult pelage (Eisenberg, 1993). Young begin to forage outside the burrow at about one month of age and the molt to adult pelage is thought to begin at 8-10 weeks (Eisenberg, 1963). While there are no studies of predation specifically on *C. intermedius*, it is well established that desert heteromyids, in general, are preyed upon by owls as well as snakes and mammalian carnivores (Brown and Harney, 1993).

Throughout most of their range, adult pocket mice have sandy dorsal pelage and nearly white underbellies. Individual hairs on the dorsum are banded, and the ventral hairs are uniformly light colored, as is typical for many rodent species.

However, several populations of rock pocket mice have been described that have strikingly different pigmentation patterns (Figure 1). Dice (1929) described nearly black (melanic) mice on the Carrizozo lava beds of the Tularosa Basin of central New Mexico (Figure 2). Geographically isolated populations of melanic mice have also been reported on other lava beds (Sumner, 1921; Benson, 1933; Dice and Blossom, 1937). In a comprehensive study, Dice and Blossom (1937) surveyed *C. intermedius* coat color variation throughout Arizona and New Mexico. By quantifying light reflectance of dorsal hairs and the surrounding substrate, they demonstrated a strong correlation between the color of the pelage and the color of the substrate on which the mice live. Experiments conducted by Dice (1947) showed that substrate matching in *Peromyscus maniculatus* protected mice against predation from long-eared owls and barn owls. Although comparable experiments have not been conducted with *C. intermedius*, it seems reasonable to expect similar results, since the degree of color variation is greater in *C. intermedius* than in *P. maniculatus*.

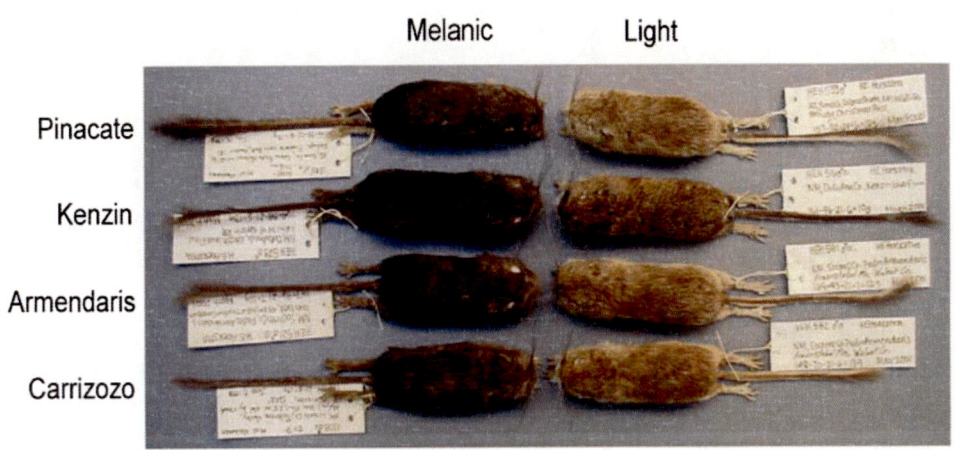

Figure 1. Coat color variation within rock pocket mice, *Chaetodipus intermedius*. The melanic mice in each pair were caught on lava, while the light mice in each pair were caught on nearby light-colored rocks.

While it has been suggested that the color morphs of *C. intermedius* represent different "races" or even subspecies of *Chaetodipus* (Williams et al., 1993), Hoffmeister (1986) noted that phenotypic variation in color is not correlated with morphological variation. In our studies, we found no correlation between coat color variation and mitochondrial DNA variation; instead, we found a strong concordance between mitochondrial phylogeny and geography, independent of coat color (Hoekstra, Krenz and Nachman, 2005). In other words, geographically proximate populations tend to be closely related and share morphological similarities, but can vary dramatically in coat color.

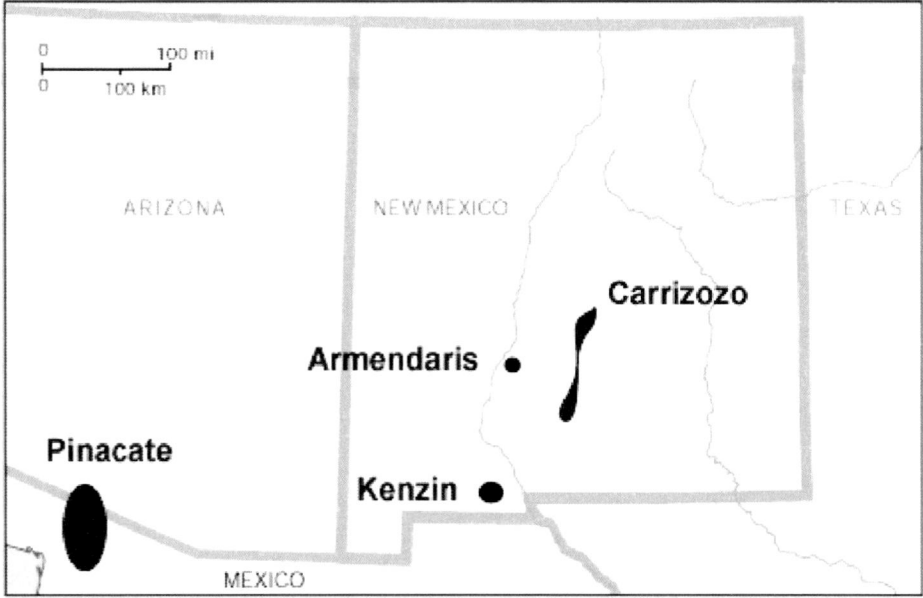

Figure 2. Map of the four lava flows sampled in our studies. Other populations of melanic *C. intermedius* have been identified but are not shown on this map.

The Mammalian Pigmentation Process and Candidate Genes

Pigmentation genes have served as a model system for exploring gene action in a variety of biological processes and for understanding the genetics of mammalian biochemistry, cell biology, and development (reviewed in Silvers, 1979; Jackson, 1994; Barsh, 1996; Jackson, 1997). The deposition of pigment in hair and skin is the culmination of a complex process that involves the coordinated action of many genes and cell types. Melanocytes, the pigment-producing cells, originate in the neural crest and migrate during development throughout the dermis. The melanoblast cell lineage that gives rise to melanocytes is committed early in development, and subsequent expression of many gene products is regulated in a cell-specific manner (Steel et al., 1992; Erickson, 1993; Bronner-Fraser, 1995). Within melanocytes are specialized organelles known as melanosomes (reviewed in Prota, 1992), which are the site of melanogenesis. There are two primary types of melanosomes, and they differ both structurally and biochemically. Eumelanosomes are ellipsoidal and are the site of synthesis of black or brown eumelanin; phaeomelanosomes are spherical and are the site of synthesis of yellow or red phaeomelanin. Once full of melanin, melanosomes are secreted from the melanocyte as pigment granules. Several lines of evidence suggest a close relationship between melanosomes and lysosomes, and it is possible that melanosomes are modified lysosomes (Jackson, 1994, 1997). For example, many mouse mutations, which affect melanosome function, also disrupt lysosome function (e.g. Barbosa et al., 1996; Feng et al., 1997). From an evolutionary perspective, this highlights one of the many ways in which pigmentation mutants may have important pleiotropic consequences. Finally, synthesis of melanin within melanosomes involves the interactions of many loci, and some aspects of melanogenesis are under hormonal regulation.

Pigmentation mutations in the laboratory mouse have been identified in all steps of this complex process (Prota, 1992; Jackson, 1994). For example, there are mutant phenotypes such as *piebald*, *steel*, and *white spotting* that result from improper development or migration of melanocytes, leaving portions of the body without pigment-producing cells. Other mutations, such as *beige* and *pale ear*, interfere with the proper structure and function of melanosomes. Some mutations, such as *albino*, *brown*, or *slaty*, interfere directly with proteins involved in synthesis of melanin. Finally, mutations at the *agouti*, *extension*, and *mahogany* loci disrupt the control and regulation of melanogenesis. Approximately 80 genes have been identified that affect coat color in the laboratory mouse and approximately one-quarter of these have been cloned, sequenced, and characterized at the molecular level (Jackson, 1997). The availability of genes known to affect each of the steps leading to the deposition of melanin in the laboratory mouse presents an opportunity to investigate the genetic changes underlying phenotypic differences in pigmentation in natural populations of rodents.

Several loci involved in pigmentation are particularly likely candidates for the coat color differences seen among populations of *C. intermedius*. A key distinction in melanogenesis is the production of eumelanin (black or brown pigment) versus phaeomelanin (yellow or red pigment). This difference is controlled in large part by the interaction of three proteins: the Agouti signaling protein, alpha-melanocyte stimulating hormone (α-MSH) and melanocortin-1 receptor (MC1R) (Figure 3). MC1R is a transmembrane G-protein-coupled receptor that is highly expressed in melanocytes. Melanocyte-stimulating-hormone activates MC1R and results in elevated levels of cAMP and increased production of eumelanin. Agouti is an antagonist of MC1R; local expression of Agouti results in suppression of synthesis of eumelanin and increased production of phaeomelanin. In the laboratory mouse, multiple alleles have been identified at both *Agouti* and at *Mc1r*, and mutations at both loci produce a range of phenotypes from dark to light color. Dominant *Agouti* mutations result in increased Agouti expression and largely yellow phenotypes. In contrast, recessive, loss-of-function *Agouti* mutations result in nonagouti, all black

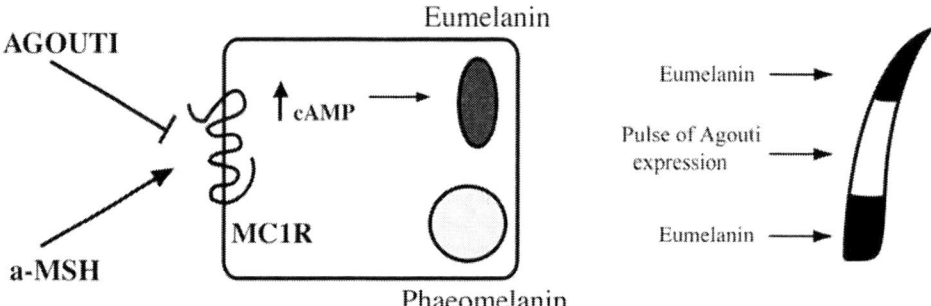

Figure 3. Regulatory control of mammalian melanogenesis. Alpha melanocyte stimulating hormone (α-MSH) signals MC1R, resulting in the up-regulation of cAMP and synthesis of eumelanin. The Agouti protein is an antagonist of MC1R that results in the default pathway of phaeomelanin production. Individual hairs on the dorsal surface of light-colored *C. intermedius* are banded, as shown on the right. In the laboratory mouse, this banded pattern is known to result from a pulse of Agouti expression in individual hair follicles during the middle of the hair cycle. Melanic *C. intermedius* have unbanded dorsal hairs (not shown).

phenotypes. Dominance relationships among *Mc1r* alleles are opposite in effect to those at *Agouti*, meaning that recessive, loss-of-function *Mc1r* mutations typically result in yellow phenotypes (although slightly different phenotypically from the dominant yellow of *Agouti*). Agouti expression varies both spatially and temporally (Bultman et al., 1992; Siracusa, 1994). Wild mice have light bellies as a result of constitutive ventral Agouti expression and associated production of phaeomelanin. In contrast, hairs on the dorsum of wild mice have a banded pattern, with a black tip, a middle yellow band, and a black base (the agouti hair, Figure 3). This banding derives from a pulse of Agouti expression during the mid-phase of the hair cycle, resulting in deposition of phaeomelanin during the middle of hair growth and deposition of eumelanin at the beginning and end of hair growth. The temporal and spatial patterns of Agouti expression are under the control of different promoters (Siracusa, 1994; Vrieling et al., 1994). A mouse mutation known as black-and-tan results from a large insertion in the first intron of the *Agouti* gene (Bultman et al., 1992; Bultman et al., 1994). This insertion eliminates dorsal expression of Agouti but has no effect on ventral expression. The phenotype of these mice includes a light belly and all-black, unbanded dorsal hairs. Similarly, some mutations at *Mc1r* in the laboratory mouse result in dark, unbanded hairs on the dorsum but light hairs on the ventral surface (Robbins et al., 1993).

C. *intermedius* display a variety of color variants (Figure 1). Many of the melanic animals show similarities to *Agouti* or *Mc1r* mutants in the laboratory mouse in having dark, unbanded dorsal hairs and light bellies. These similarities suggested to us that *Agouti* or *Mc1r* might be responsible for the observed color variation in natural populations. Moreover, at several localities, both light and dark C. *intermedius* are found together without phenotypic intermediates, suggesting that pigmentation differences may be controlled by a few genes of major effect, although it is important to bear in mind that strong selection against intermediates or positive assortative mating could also produce this pattern.

We have used association studies and a candidate-gene approach to identify the genes underlying color variation in C. *intermedius*. The general strategy is to develop single-nucleotide-polymorphism (SNP) markers for each candidate gene. These SNPs are then surveyed in geographically adjacent populations of light and dark mice. If a strong association between SNP variants and coat color phenotype is found, we sequence the entire gene in all light and dark individuals and test for population structure by looking at additional, unlinked markers. One advantage of association studies over laboratory crosses is that natural populations will have undergone many more generations of recombination; thus if an association is detected, it is likely that the markers are relatively close to the functional sites.

THE ROLE OF *MC1R* ON COLOR PHENOTYPE
IN THE PINACATE LAVA FLOW

We sampled mice from paired localities in four different regions (Tables 1 and 2, Figure 2). Each region contains a lava flow, and in each region we have sampled mice from the dark lava and from adjacent areas with light-colored rocks. Consistent with earlier work by Dice and Blossom (1937) from many of these same localities, we found a strong association between substrate color and coat color. Here, we concentrate on one well-sampled region, the Pinacate lava bed, which is situated in northern Sonora, Mexico, and the adjacent Cabeza Prieta National Wildlife Refuge in southern Arizona (Figure 2). This lava flow is estimated to be approximately 1.7 million years old (Lynch, 1989), and many of the rocky areas are disjunct due to the accumulation of intervening deposits of sand.

In our first study (Nachman et al., 2003), 29 mice were collected from two sites - one on the dark lava and one on the nearest light rocks (the rocky slopes of the O'Neill Hills, approximately four kilometers east of the lava). Of the 18 mice captured on dark rock, 16 were dark or melanic (89%); while 10 of 11 mice (91%) caught on the light rock were light. We found no association between coat color phenotypes and *Agouti* SNPs. However, several *Mc1r* SNPs revealed a perfect association with coat color phenotypes. The association observed at several *Mc1r* SNPs led us to characterize the entire *Mc1r* gene in *C. intermedius*. PCR primers were designed from conserved regions in the alignment of human and *Mus Mc1r* sequences, and then genome-walking was used to capture the 5' and 3' ends of the gene. *Mc1r* has a single coding exon of 954bp (318 amino acids); this simple gene structure is conserved between birds and mammals. We cloned and sequenced the 58 alleles from our sample of 18 melanic and 11 light mice.

Table 1. Sample sizes of *C. intermedius* at each of the four lava flows in Figure 2.

| | Mouse coat color | |
	Light	Melanic
Pinacates	123	102
Armendaris	12	8
Carrizozo	4	8
Kenzin	6	14

Table 2. Estimated age of oldest lava flow, largest contiguous area of lava, and distance to nearest light rock (Lynch, 1989; Hoffer and Corbitt, 1991).

	Age (yrs)	Area (km²)	Distance (km)
Pinacates	1,700,000	3500	4
Armendaris	760,000	435	5
Carrizozo	1,000	329	20
Kenzin	530,000	142	7

Several lines of evidence suggest that mutations at *Mc1r* are responsible for the observed phenotypic differences in color in mice from the Pinacate Lava Beds (Nachman et al., 2003). First, a perfect association is seen between four amino acid polymorphisms and coat color. These four amino acid polymorphisms are in complete linkage disequilibrium with one another and thus create a single haplotype (which we refer to here as the melanic allele). Mice with one or two copies of the melanic allele are melanic, while mice without this allele are light (Table 3). This perfect association is highly significant (Fisher's Exact test, $p <$ 0.0001). Such an association is expected if *Mc1r* is responsible for the observed coat color differences. However, such an association might also arise if a gene tightly linked to *Mc1r* is responsible for the phenotype, rather than *Mc1r* itself. Genomic sequences and genetic maps from *Mus* and humans suggest that there are few genes neighboring *Mc1r*. Moreover, there are no known genes involved in pigmentation adjacent to *Mc1r* in either species. To further test the hypothesis that *Mc1r*, rather than a linked region, is contributing to the observed phenotypic variation, we are currently using genome-walking to determine the extent of linkage disequilibrium surrounding *Mc1r*. This will delimit the genomic region associated with the phenotype. Preliminary results indicate that linkage disequilibrium decays within 2 kb both upstream and downstream of *Mc1r*, thus ruling out the involvement of a linked locus (Hoekstra et al, in prep). An additional concern with association studies comes from the potential for population structure to create spurious associations between genotypes and unrelated phenotypes (Lander and Schork, 1994; Lynch and Walsh, 1998). The complete ND3 and COIII mitochondrial DNA genes were sequenced in all 29 animals to test for population structure. Patterns of variation at these loci were consistent with a single interbreeding population of light and melanic mice, suggesting that population structure is not responsible for the strong association between *Mc1r* alleles and coat color phenotypes.

A second observation that suggests a direct role for *Mc1r* is that all four amino acid mutations on the melanic allele result in a change of charge and occur at nucleotide sites that may be important for receptor function. These four mutations are derived relative to *Mc1r* sequences in the sister species, *C. penicillatus*, which is light in color. Two mutations are found in extracellular regions and may be important for ligand binding, while two are found in intracellular loops and may be important for G-protein interactions (Figure 4). Previous studies in other species have identified single amino acid changes at *Mc1r* that result in darkening mutations; these mutations have been identified primarily in intracellular regions or at the boundary of transmembrane and extracellular loops (Figure 4).

Third, the dominance pattern seen in *C. intermedius* follows the pattern predicted from studies of the laboratory mouse: dark *Mc1r* alleles are dominant over light alleles. Individual *C. intermedius* that are heterozygous or homozygous for the melanic allele have a dark phenotype. In laboratory mice, gain-of-function mutations at *Mc1r* are dominant, whereas loss-of-function mutations at *Mc1r* are recessive (Robbins et al., 1993).

Finally, overall patterns of nucleotide variability at *Mc1r* show evidence of recent, strong directional selection (Nachman et al., 2003). Comparison of the level of nucleotide variability among the melanic alleles to the level of variability among the light alleles shows a significant reduction in polymorphism among melanic alleles; there is only a single polymorphic nucleotide segregating among the melanic alleles while there are thirteen variable sites among the light alleles. The average level of nucleotide heterozygosity, π, is 0.01% for the melanic alleles and 0.21% for the light alleles, representing a twenty-fold difference. This pattern is consistent with positive, directional selection having acted recently to raise the frequency of the melanic allele. The absence of genetic variation is consistent with hitchhiking effects on linked, neutral sites. Further evidence of genetic hitchhiking comes from the observation that a single silent site is in complete linkage disequilibrium with the four amino acid mutations on the melanic allele.

Together, these observations strongly implicate *Mc1r* in the melanic phenotype in the Pinacate population, and provide a rare example of the molecular changes underlying an adaptive phenotype in a natural population. It remains unclear whether one or more of the four amino acid variants on the melanic allele contribute to the observed phenotypic differences; this is currently being addressed through *in-vitro* functional studies as described below.

Table 3. Association between *Mc1r* genotype and coat color in *C. intermedius* from the Pinacate lava flow. Melanic allele (*D*) and light allele (*d*).

| | Phenotype | |
Mc1r genotype	Light	Melanic
DD	0	11
Dd	0	6
dd	12	0

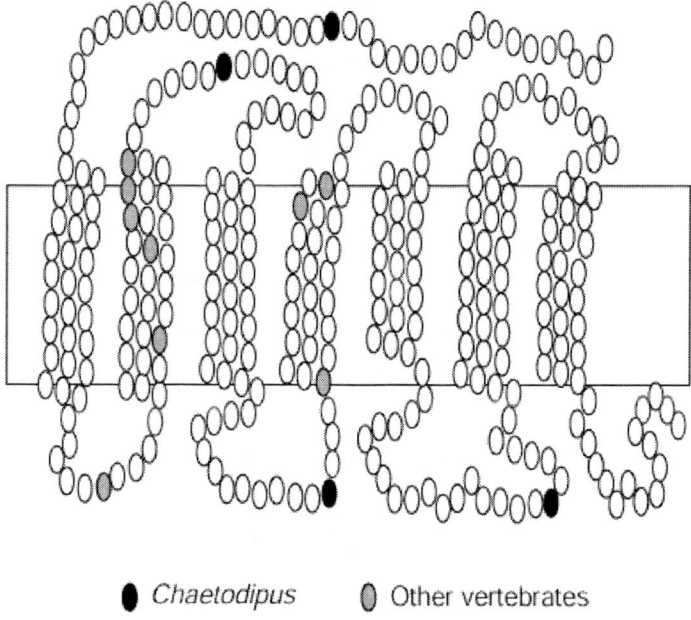

● *Chaetodipus*   ◐ Other vertebrates

Figure 4. Structure of the MC1R protein. The receptor consists of an extracellular N-terminal domain, seven transmembrane regions separated by intracellular and extracellular loops, and a C-terminal intracellular domain. Gray circles identify amino acid positions at which darkening mutations have been identified in other vertebrate species (mouse, fox, dog, cow, sheep, chicken). Black circles identify the four amino acid positions that differ between light and melanic alleles in *C. intermedius*.

The Balance Between Migration and Selection

Identification of a gene underlying phenotypic variation in pocket mice from the Pinacate Lava Flow allowed us to investigate spatial variation in genotype frequencies and to estimate the strength of selection from a simple model of migration-selection balance. We investigated the pattern of genotypic and phenotypic change across geography by conducting a 30 kilometer transect across both light and dark rocks in the Pinacate region (Figure 5) (Hoekstra et al., 2004). This transect included three sites on dark rock and three sites on light rock. The average distance between sites was approximately six kilometers. Except on the

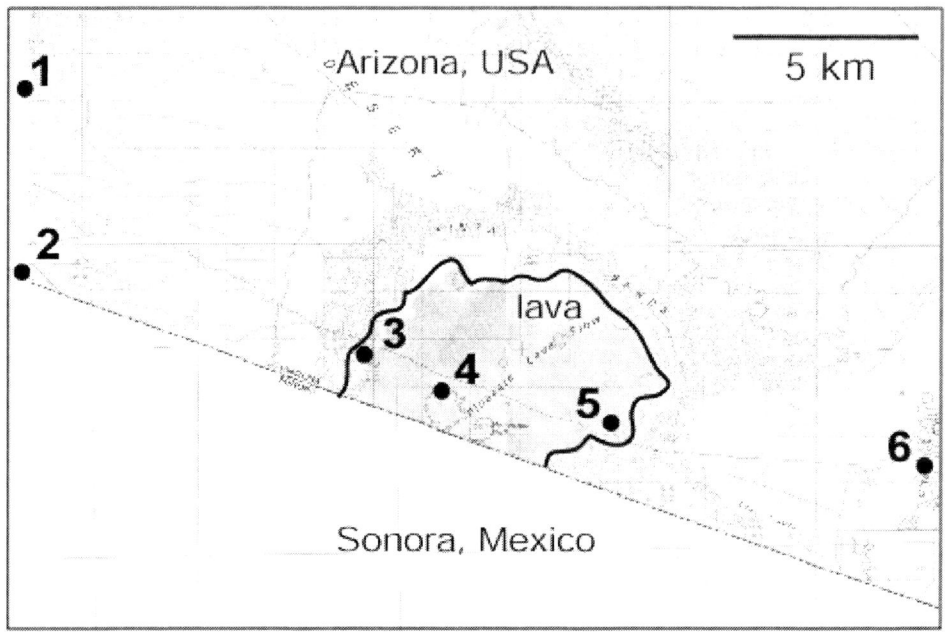

Figure 5. Transect across the Pinacate lava flow. Six sampling sites are indicated: three on volcanic basalt, and three on light-colored granite. Each site is separated by approximately six kilometers.

lava, much of the habitat between sites was dominated by light-colored sand, which is not suitable for *C. intermedius*. In total, 225 mice were sampled. Fifty-seven mice were captured on dark rock; 94.7% of these had a melanic phenotype. One hundred sixty-eight mice were captured on light rock; 71.4% of these had a light phenotype. Across all sites, we found a strong correlation between habitat color and coat color of mice (Figure 6). Interestingly, the correlation between substrate and coat color was stronger on the western side of the lava flow than on the eastern side. The increased frequency of "mis-matched" mice on the eastern edge of the lava may be due to the closer proximity between light and dark habitats, facilitating migration between them.

To determine the distribution of *Mc1r* alleles across this transect, we sequenced the entire *Mc1r* coding region (954bp) for 202 individuals (Hoekstra et al., 2004). Consistent with our earlier studies based on smaller samples, we found that (1) all melanic individuals had at least one melanic allele with the four amino acid variants, (2) there was a strong correlation between *Mc1r* allele frequency and habitat color, and (3) there was no correlation between habitat color and neutral mtDNA markers. This larger sample further strengthened the association between *Mc1r* genotype and coat color phenotype. The strong correlation between *Mc1r* alleles and substrate color, in the face of gene flow observed in patterns of mtDNA variation, suggests that differential selection on different substrates is acting to maintain the color polymorphism.

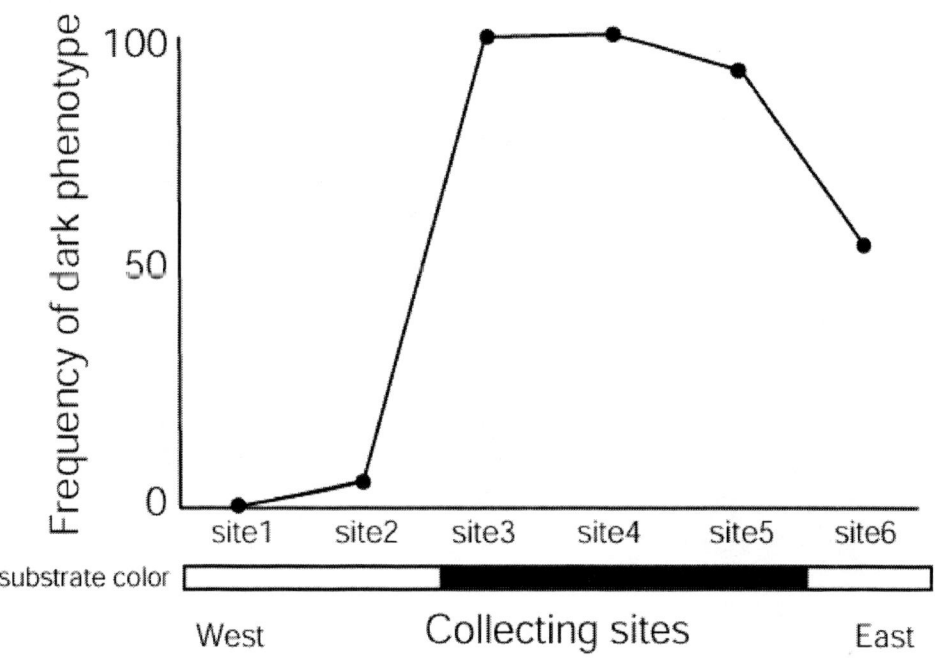

Figure 6. Correlation between coat color of *C. intermedius* and substrate color. Frequency of melanic individuals caught at each of the six sampling sites in Figure 5. Substrate color is shown schematically below.

A simple model of migration-selection balance can be used to estimate the strength of selection in this situation. We assume that the frequency of melanic alleles on light substrate (and the frequency of light alleles on dark substrate) is at steady-state, maintained by the input of new mutations through migration and their elimination by selection. This model further assumes that effective population sizes are equivalent among sites, that mutation is negligible, and that migration occurs between the sites surveyed (as opposed to migration from other sites). Under these assumptions, the equilibrium frequency of a dominant mutation, p, is given by m/s, and the equilibrium frequency of a recessive mutant, q, is given by $\sqrt{m/s}$, where m is the per-generation migration rate and s is the average selection coefficient. By estimating p, q, and m we can then derive a very rough estimate of s (Figure 7). This simplified model ignores emigration and assumes that all *Mc1r* alleles introduced by immigration are deleterious. For a more complete treatment of this topic, see Hoekstra et al. (2004).

Here, we estimate s by considering gene flow between the Pinacate lava and a neighboring light locality, the O'Neill Hills (sites 3 - 5 and 6 in Figure 5). Samples from the three sites on the Lava (sites 3 – 5) were combined because of their close proximity and the lack of population structure between these sites. The total sample contains 77 mice from the light site and 57 mice from the dark site. Thirty-four of 43 mice (79%) from the granite were light, while 55 of 57 mice (96%) from the lava were dark. Gene flow between these sites was estimated by comparing mitochondrial sequence variation for the ND3 and COIII genes using the program MIGRATE (Beerli, 1999; Beerli and Felsenstein, 2001). MIGRATE calculates a maximum likelihood estimate of N_em between populations using a coalescent approach, relaxing the traditional assumption of symmetrical migration. By

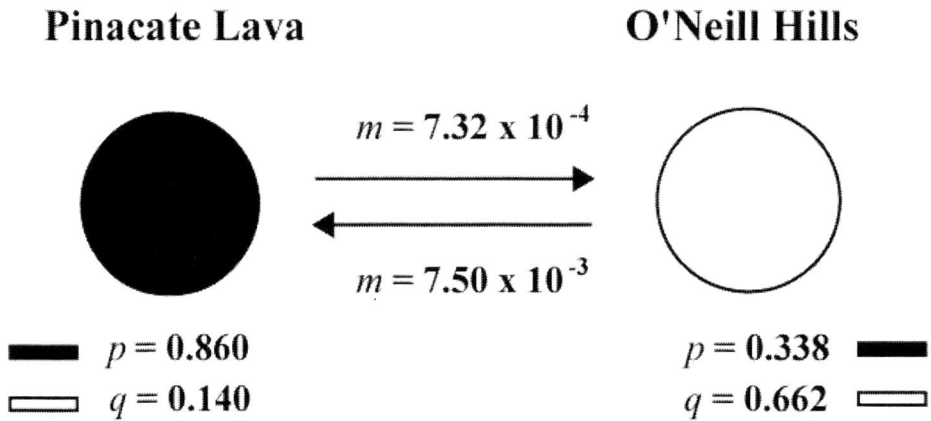

Pinacate Lava   O'Neill Hills

$m = 7.32 \times 10^{-4}$

$m = 7.50 \times 10^{-3}$

$p = 0.860$ $p = 0.338$
$q = 0.140$ $q = 0.662$

Figure 7. Migration-selection balance between populations of *C. intermedius* on dark and light substrates (corresponding to sites 5 and 6 in Figure 5, respectively). Migration rates (m) were estimated from mtDNA. Frequencies of *Mc1r* melanic alleles (p) and *Mc1r* light alleles (q) at each site are shown.

quantifying asymmetric migration rates, we can estimate selection against light mice on dark rocks as well as selection against dark mice on light rocks (Figure 7). The maximum-likelihood estimate of N_em from the Lava to O'Neill Hills was 7.32, while the maximum-likelihood estimate of N_em from O'Neill Hills to the Lava was 75.03. We estimated N_e from the level of mitochondrial nucleotide heterozygosity (π

= 0.01), under the neutral expectation $\pi = N_e\mu$, assuming $\mu = 10^{-6}$ and assuming a sex ratio of one. This suggests that, approximately, $N_e = 10^4$, which is consistent with estimates from other small rodents (e.g. Nachman 1997). This provides an estimate of $m = 7.32 \times 10^{-4}$ from the Lava to O'Neill Hills, and $m = 7.50 \times 10^{-3}$ from O'Neill Hills to the Lava, implying that migration from light rock to lava is higher than migration in the opposite direction. The frequency of the *Mc1r* melanic allele was $p = 0.338$ at O'Neill Hills and the frequency of the light allele was $q = 0.140$ on the Lava. These calculations provide an estimate of $s = 0.002$ for melanic mice on light rocks (O'Neill Hills) and $s = 0.383$ for light mice on dark rocks (Lava). Although these numbers are crude and are based many untested assumptions, they suggest that selection may be stronger against light mice on dark rocks than against melanic mice on light rocks.

Interestingly, the larger estimate of s is associated with the evolution of the novel phenotype; the melanic allele and the dark color are derived relative to the sister species, *C. penicillatus*. Thus, light mice initially encountered and colonized dark rock, and strong selection led to the proliferation of the melanic phenotype. These asymmetrical selection coefficients also parallel observations from peppered moths, in which there was a rapid increase in melanic moths, but a comparatively slow decrease in melanic moths when the habitat became lighter in color as levels of pollution decreased (Kettlewell, 1955). These observations suggest that selection was stronger against light moths on dark background than against dark moths on light background. This pattern may be driven by a bias in avian visual systems in which light objects on dark backgrounds are more conspicuous than the reverse.

Role of *Mc1r* in Other Melanic Populations

A fundamental question in evolution centers on the issue of genetic constraints to adaptive change: given a common evolutionary problem, are there multiple genetic solutions? Having discovered that *Mc1r* appears to play a major role in the color phenotype in the Pinacate population, we were interested in exploring whether similar phenotypes in other populations have evolved independently and, if so, whether they have evolved via changes at the same gene or via changes at different genes (Hoekstra and Nachman, 2003). To explore this issue, we sampled mice from pairs of light and dark localities in four different regions (Table 1, Figure 2). The four different lava flows vary in both size and age (Table 2) and, thus, mice on the lava flows may have very different population histories.

Several observations suggest that similar melanic phenotypes may have evolved independently on different lava flows. First, Hoffmeister (1986) argued there was little gene flow between different melanic races, in part because of the large distances separating many of these populations. Consistent with this, our survey of mitochondrial DNA variation from these four regions revealed a strong correlation between phylogeny and geography, indicating that there has been relatively little gene flow over this geographic scale (Hoekstra et al., 2005). Additionally, subtle differences in phenotype from different lava beds (Dice and Blossom, 1937) suggest that the melanic phenotype may have a different genetic basis. We found that mice from the Pinacates were slightly grayer than populations from Carrizozo and Armendaris, although these differences were quite subtle. There may also be phenotypic differences within lava flows. We observed two types of melanic mice at the Kenzin lava flow, one with completely melanic dorsal hairs and a second type with banded dorsal hairs in which the light band of phaeomelanin was greatly reduced relative to typical light mice. Both limited gene flow among lava flows and phenotypic variation within and among lava flows suggest that there may be multiple genes involved in generating melanic phenotypes, and that melanic color may have evolved independently several times.

To address this question more directly, we sequenced the *Mc1r* coding region in five light and five melanic mice from each of the localities in Figure 2. None of the four *Mc1r* amino acid variants from the Pinacate population were found in melanic mice from any of the other populations. Moreover, no new *Mc1r* mutations showed an association with phenotype in any of the localities (Hoekstra and Nachman, 2003). There is thus no evidence that *Mc1r* is involved in coat color variation in the Pedro Armendaris, Carrizozo, or Kenzin lava flows. These results indicate that a similar melanic phenotype has evolved independently on different lava flows, and has done so via different genetic changes.

FUTURE DIRECTIONS

Our work to date has demonstrated that mutations at *Mc1r* are responsible for adaptive coat color differences in one population of *C. intermedius* and that similar melanic phenotypes have evolved independently in different populations through changes at different genes. This work, which links genotype with phenotype for a trait of clear ecological importance, has raised a number of questions, and our current efforts focus on two of these issues.

First, we are interested in identifying the specific mutation or mutations at *Mc1r* that are responsible for the phenotypic differences observed in the Pinacate population. Because there are four amino acid changes that distinguish all melanic alleles from all light alleles, we do not know whether one or more of these mutations are responsible for the functional differences. It is possible that each mutation contributes a small amount to the observed phenotype, or that two or more mutations interact epistatically to produce the phenotype. Alternatively, some of the mutations may be neutral, having hitchhiked along with the selected mutation(s). To identify the sites of functional importance, we are using an *in-vitro* expression system to measure the function of receptors that have combinations of one or more mutations. Functional assays have been used successfully in *Mus* to show that a single amino acid variant constitutively activates the *Mc1*-receptor in the case of the somber-3J mutation (Robbins et al., 1993). Functional in-vitro studies using *C. intermedius Mc1r* sequences should allow us to identify the specific mutations responsible for adaptive coat color differences in one population of this species.

Second, we are interested in identifying the genes involved in coat color variation in populations where *Mc1r* is clearly not playing a role. Towards this end, we are currently screening SNPs in candidate genes for each of the populations in Figure 2. The presence of several geographically isolated lava flows, and the independent evolution of melanic phenotypes in these areas, provides an excellent opportunity for replicate evolutionary studies. We hope that from these replicates, some generalities about the genetics of adaptation may emerge, including such things as patterns of dominance, epistasis and pleiotropy, the number of mutations contributing to the observed phenotypes, and whether mutations are in coding regions, regulatory regions, or both.

CONCLUSIONS

With the complete DNA sequence of many species now available, it is becoming increasingly possible to make the link between genotype and phenotype for fitness-related traits. In principle, many of the approaches used here for studying pigmentation differences in *C. intermedius* could be extended to other phenotypes of ecological importance in other eutherian mammals. Our work provides evidence for the genetic basis of a key adaptation in a natural population of rodents. It appears that similar melanic phenotypes have arisen independently in *C. intermedius* populations on different lava flows and that these changes have occurred through changes at different genes. In this situation there appears to be more than one genetic solution to a common evolutionary problem.

ACKNOWLEDGEMENTS

We dedicate this work to Dr. James L. Patton. We thank him for the constant inspiration he has provided as a scientist, mentor, and friend. Both authors were undergraduate students of Dr. Patton at UC Berkeley: HEH in 1994 and MWN in 1981. K.E. Drumm and J.A. Kim have provided valuable assistance in both the lab and the field. The work presented here is supported by a National Institutes of Health Postdoctoral Fellowship to HEH and a National Science Foundation grant to MWN.

LITERATURE CITED

Barbosa, M. D. F. S., Q. A. Nguyen, V. T. Tchernev, J. A. Ashley, J. C. Detter, S. M. Blayde, S. J. Brandt, D. Chotai, C. Hodgman, R. C. E. Solari, M. Lovett, and S. F. Kingsmore.
    1996   Identification of the homologous beige and Chediak-Higashi-Syndrome genes. Nature 382:262-265.

Barsh, G. S.
    1996   The genetics of pigmentation: from fancy genes to complex traits. Trends Gen. 12:299-305.

Beerli, P.
    1999   MIGRATE. University of Washington, Seattle, WA.

Beerli, P., and J. Felsenstein.
    2001   Maximum likelihood estimation of a migration matrix and effective population sizes in n subpopulations by using a coalescent approach. PNAS 98:4563-4568.

Benson, S. B.
    1933   Concealing coloration among some desert rodents of the southwestern United States. Univ. Calif. Pub. Zool. 40:1-69.

Bronner-Fraser, M.
    1995   Origins and developmental potential of the neural crest. Exp. Cell Res. 218:405-417.

Brown, J. H., and B. A. Harney.
    1993   Population and community ecology of Heteromyid rodents in temperate habitats. Pp. 618-651 in Biology of the Heteromyidae (H. H. Genoways and J. H. Brown, eds.). Special Publication No. 10, The American Society of Mammalogists.

Bultman, S. J., M. L. Klebig, E. J. Michaud, H. O. Sweet, M. T. Davisson, and R. P. Woychik.
    1994   Molecular analysis of reverse mutations from nonagouti (a) to black-and-tan (a^t) and white-bellied agouti (A^w) reveals alternative forms of agouti transcripts. Genes Dev. 8:481-490.

Bultman, S. J., E. J. Michaud, and R. P. Woychik.
   1992   Molecular characterization of the mouse agouti locus. Cell 71:1195-1204.

Dice, L.
   1929   Description of two new pocket mice and a new woodrat from New Mexico. Occ. Pap. Mus. Zool. Univ. Mich. 203:1-4.

   1947   Effectiveness of selection by owls of deermice (*Peromyscus maniculatus*) which contrast with their background. Contr. Vert. Biol. Lab. Univ. Mich. 34

Dice, L., and P. M. Blossom.
   1937   Studies of mammalian ecology in southwestern North America, with special attention to the colors of desert mammals. Publ. Carn. Inst. Wash. 485:1-25.

Eisenberg, J. F.
   1963   The behavior of heteromyid rodents. Univ. Calif. Publ. Zool. 69:1-100.

   1993   Ontogeny. Pp. 479-490 in Biology of the Heteromyidae (H. H. Genoways and J. H. Brown, eds). Special Publication No. 10, The American Society of Mammalogists.

Erickson, C. A.
   1993   From the crest to the periphery: control of pigment-cell migration and lineage segregation. Pigm. Cell Res. 6:336-347.

Feng, G. H., T. Bailin, J. Oh, and R. A. Spritz.
   1997   Mouse pale ear (ep) is homologous to human Hermasky-Pudlak syndrome and contains a rare AT-AC intron. Hum. Mol. Gen. 6:793-797.

Hoekstra, H. E., K. E. Drumm, and M. W. Nachman.
   2004   Ecological genetics of adaptive color polymorphism in pocket mice: geographic variation in selected and neutral genes. Evol. 58:1329-1341.

Hoekstra, H. E, J. G. Krenz, and M. W. Nachman.
   2005   Local adaptation in the rock pocket mouse (*Chaetodipus intermedius*): natural selection and phylogenetic history of populations. Heredity 94:217-228.

Hoekstra, H. E., and M. W. Nachman.
  2003   Different genes underlie adaptive melanism in different populations of rock pocket mice. Mol. Ecol. 12:1185-1194.

Hoffer, J., and L. Corbitt.
  1991   Evolution of the late cenozoic Jornada volcano, south-central New Mexico. New Mex. Geol. Soc. Guidebook 159-163.

Hoffmeister, D. F.
  1986   Mammals of Arizona. University of Arizona Press, Tuscon, Arizona. 602.

Jackson, I. J.
  1994   Molecular and developmental genetics of mouse coat color. Ann. Rev. Gen. 28:189-217.

  1997   Homologous pigmentation mutations in human, mouse, and other model organisms. Hum. Mol. Gen. 6:1613-1624.

Kettlewell, H. B. D.
  1955   Selection experiments on industrial melanism in the Lepidoptera. Heredity 9:323-342.

Lander, E., and N. Schork.
  1994   Genetic dissection of complex traits. Science 265:2037-2048.

Lynch, D.
  1989   Neogene volcanism in Arizona: the recognizable volcanoes. Pp. 681-700 in Geologic evolution of Arizona (J. Jenny and S. Reynolds, eds.). Arizona Geological Society Digest, Tucson.

Lynch, M., and J. B. Walsh.
  1998   Genetics and Analysis of Quantitative Traits. Sinauer Associates, Inc., Sunderland, MA.

Nachman, M. W., H. E. Hoekstra, and S. D'Agostino.
  2003   The genetic basis of adaptive melanism in pocket mice. Proc. Natil. Acad. Sci USA 100:5268-5273.

Prota, G.
  1992   Melanins and melanogenesis. Academic Press, Inc., San Diego, CA.

Reichman, O. J., and K. M. Van de Graaff.
 1973   Seasonal activity and reproductive patterns of five species of Sonoran desert rodents. Am. Mid. Nat. 90:118-126.

Robbins, L., J. Nadeau, K. Johnson, M. Kelly, L. Roselli-Rehfuss, E. Baack, K. Mountjoy, and R. Cone.
 1993   Pigmentation phenotypes of variant extension locus alleles result from point mutations that alter receptor function. Cell 72:827-834.

Silvers, W. K.
 1979   The Coat Colors of Mice: a Model for Mammalian Gene Action and Interaction. Springer-Verlag, New York.

Siracusa, L. D.
 1994   The agouti gene: turned on to yellow. Trends Gen. 10:423-428.

Steel, K. P., D. R. Davidson, and I. J. Jackson.
 1992   TRP-2/DT, a new early melanoblast marker shows that the steel growth factor (c-kit ligand) is a survival factor. Development 115:1111-1119.

Sumner, F.
 1921   Desert and lava-dwelling mice and the problem of protective coloration in mammals. J. Mamm. 2:75-86.

Urabe, K., P. Aroca, and V. J. Hearing.
 1993   From gene to protein: determination of melanin synthesis. Pigm. Cell Res. 6:186-192.

Vrieling, H., D. M. J. Duhl, S. E. Millar, K. A. Miller, and G. S. Barsh.
 1994   Differences in dorsal and ventral pigmentation result from regional expression of the mouse agouti gene. Proc. Natl. Acad. Sci. USA 91:5667-5671.

Williams, D., H. Genoways, and J. Braun.
 1993   Taxonomy. Pp. 38-196 in Biology of the Heteromyidae (H.H. Genoways and J. Brown, eds.). Special Publication No. 10, The American Society of Mammalogists.

Climate Change and the Distribution of *Peromyscus* in Michigan: Is Global Warming Already Having an Impact?

Philip Myers, Barbara L. Lundrigan, and Robert Vande Kopple

Two species of *Peromyscus*, the woodland deer mouse (*P. maniculatus gracilis*) and the white-footed mouse (*P. leucopus*), are found together in the forests of the northern Lower Peninsula of Michigan. Deer mice have become rare and their rate of decline appears to be accelerating. The relative abundance of *gracilis* in this area, measured as % *gracilis* in collections of *Peromyscus*, has declined from over 40% before 1931 to around 6% in 2003, and in detailed recent studies at one site, from over 50% of *Peromyscus* captures before 1996 to less than 10% in 2003. Numbers of *leucopus* at the University of Michigan Biological Station, northern Lower Peninsula, are related to the length of the winter (measured as the date of ice break-up); it appears that few *leucopus* survive long winters (when ice breaks up in late April or early May). We suggest that a recent tendency for winters to end early (revealed for this area by 100+ years of ice break-up records for Grand Traverse Bay, Lake Michigan) may be responsible for the decline in *gracilis* populations.

ECOLOGY OF *PEROMYSCUS* IN MICHIGAN

Two species of long-tailed, forest-dwelling *Peromyscus* occur together over large areas of the northeastern United States and along the Appalachian Mountains. *Peromyscus leucopus* (the white-footed mouse) is broadly distributed in the eastern and central United States, from southernmost Canada to the Yucatan peninsula and from the Atlantic coast to the western Great Plains. *Peromyscus maniculatus* (the deer mouse) consists of a complex of long-tailed forest and short-tailed grassland "subspecies," which are sometimes found in different habitats of the same region without evidence of interbreeding. Members of this species occur throughout much of the United States and in southern Canada north to Hudson's Bay. Where *leucopus* and long-tailed *maniculatus* occur together, the 2 species may be very similar in appearance and difficult to distinguish, both in the field and in the laboratory (Smith and Speller, 1970; Feldhamer et al., 1983; Long and Long, 1993; Rich et al., 1996; Bruseo et al., 1999).

In Michigan, the forest-dwelling *maniculatus* is *Peromyscus maniculatus gracilis*. It has been found throughout the state's Upper Peninsula and in the northern half of the Lower Peninsula (Figure 1a). Much of the range of *gracilis* lies to the north of Michigan, in Ontario and Quebec. *Peromyscus leucopus*, in contrast, is at the northern limits of its range in this region and in Michigan is restricted to the Lower Peninsula and southernmost Upper Peninsula (Figure 1b). Where *gracilis* and *leucopus* overlap in Michigan, they are both very similar morphologically and are often captured in the same habitats.

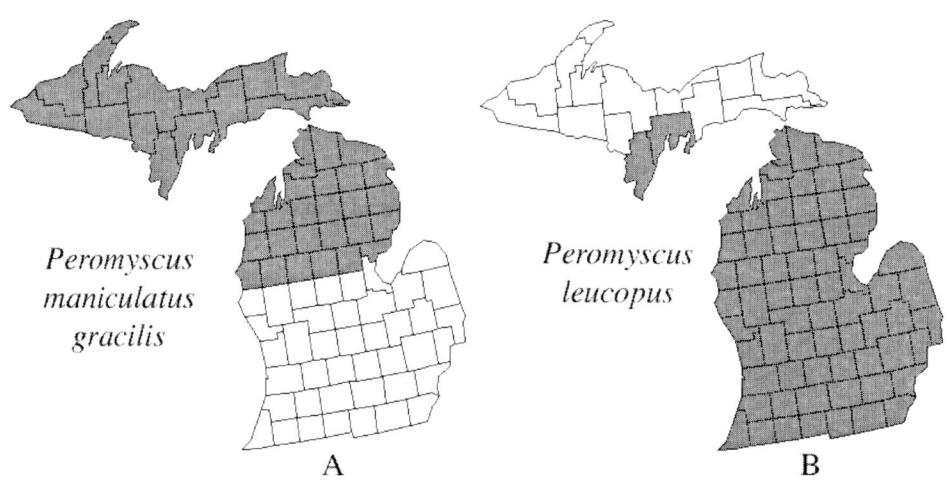

Figure 1. Distribution of *Peromyscus maniculatus gracilis* (a) and *Peromyscus leucopus* (b) in Michigan. Modified from Baker (1983).

Because of their similarity in appearance and habitat use, these two species have been the subject of a number of ecological investigations. Interactions between *leucopus* and *Peromyscus maniculatus nubiterrae* (an Appalachian representative of the forest *maniculatus* complex), for example, have been studied intensively where the two occur together in the mountains of Virginia (Wolff and Hurlbutt, 1982; Wolff et al., 1983; Barry et al., 1984; Wolff, 1985a,b; Wolff et al., 1985; Cranford and Maly, 1986; Wolff, 1986; Wolff and Durr, 1986; Harney and Dueser, 1987; Graves et al., 1988; Dooley and Dueser, 1990). In a summary, Wolff (1985b) reported that *leucopus* and *nubiterrae* "interact ecologically as a single species, though they do not interbreed." Wolff further suggested (1996) that the coexistence of the two species may depend on differences in their winter adaptations, which cause one species (*leucopus*) to flourish when winters are mild and the other (*maniculatus*) to predominate when winters are harsh.

In this paper, we show that the population of *gracilis* in the northern Lower Peninsula of Michigan has declined dramatically since the beginning of the 20th century, when the first extensive surveys of the small mammals of this region were carried out. We combine these observations with data from a recent program of regular censuses of small mammal populations in northern Michigan to suggest that *leucopus* may be sensitive to particular aspects of winter weather. We argue that changes in climate over the last century, but especially in the last 20 years, have favored *leucopus* and may be responsible for their increase in abundance relative to *gracilis*.

ESTIMATING THE RELATIVE ABUNDANCE OF *PEROMYSCUS MANICULATUS GRACILIS* AND *PEROMYSCUS LEUCOPUS*

Trapping surveys provided data concerning the abundance of *Peromyscus*. For records prior to 1985, we relied on specimens and field notes in Museum collections. More recent records were based on our own observations.

Historical records of trapping surveys are difficult to interpret. Even when field notes are available, the reasons why a collector trapped at a particular place are seldom apparent. Habitat descriptions are often absent or are too sketchy to be useful. Further, collectors may not preserve representatives of each species in the ratio in which they were captured. Rare species may be favored, juveniles discarded, and research interests in one species or another may lead to their over-representation in a collection.

Nevertheless, field notes and specimens preserved in collections are usually the only source of information available on the composition of past communities. Extensive collections from Michigan, often with accompanying field notes, are present in the University of Michigan Museum of Zoology and the Michigan State University Museum. They are the work of many collectors. While each collection undoubtedly reflects the interests and biases of the researchers who contributed to it, we shall assume that, in the aggregate, these materials provide a reasonable picture of the composition of small mammal communities in northern Michigan in the past and present.

An additional problem in estimating abundances of *gracilis* and *leucopus* is that these two mice are sometimes very difficult to distinguish, especially in the field. To be certain that we could identify the taxa correctly in northern Michigan, we examined 223 specimens from Michigan whose identity had been confirmed by allozymes (Meagher, 1995). We found that the following field characteristics were useful for distinguishing most Michigan specimens: ear length (longer in *gracilis*); sharpness of the zone of transition between pale and dark areas of the bicolored tail (always sharply distinct in *gracilis*, variable in *leucopus*); degree of hairiness of the tail (longer hairs in *gracilis*); and color of the throat and upper chest (usually white to the base in *leucopus* vs. hairs with a gray base in *gracilis*). No single characteristic worked without error, but when considered together, these characteristics resulted in identifications concordant with electrophoretic evidence in almost all cases. Using these criteria, we re-identified all museum specimens included here. We excluded records based strictly on the field notes of early collectors (i.e., where the mice were not collected) because we could not confirm identifications. We included, however, mice captured and released in our own surveys (described below), because we had the advantage of having examined the electrophoretically-identified specimens before we went into the field.

We documented the relative abundance of *gracilis* and *leucopus* as the ratio (#*gracilis*)/(#*gracilis*+*leucopus*) rather than reporting the actual number of mice captured. We chose not to rely on absolute numbers because trapping effort varied from survey to survey and was often not recorded, and because seasonal fluctuations in population levels of both species were often extreme and would have confounded comparisons of collections made at different times of year. Relative abundance will be misleading if methods used in early surveys (e.g., type of trap, bait, habitats trapped) differed from those used recently and if those methods were biased towards one species. Relative abundance might also be misleading with respect to the absolute abundance of *gracilis* if the total number of *Peromyscus* captured has changed significantly over the last century. There is no evidence that either is the case.

To the museum records, we have added a number of trapping surveys in the area of overlap in the northern Lower Peninsula. The goals of these surveys and the details of the methods used often differed, but they usually involved setting lines of Sherman live-traps, typically a mixture of large and small sizes, and running them for 1 to 3 nights. The number of traps set varied from 30 to over 500. Traps were baited with oats. Trapping was completed by us, by graduate and undergraduate students, and by volunteers. Trappers recorded the identity of each individual captured and its sex, weight, and reproductive condition. We marked each mouse temporarily by clipping a small patch of fur from its rump. Identifications were made by ourselves or by students that we had trained. From 1985 to 2002, we conducted surveys at 118 localities in this region, for a total of over 30,000 trap-nights. Initially, we chose localities that we believed would provide a representation of the diversity of mammals in the area. More recently, we have focused more attention on habitats that we know are used by the less common *gracilis*, as we attempt to find remaining populations of that species.

Records were summarized by county. This was done because very few *Peromyscus* were recorded from many of the individual collection sites, resulting in exaggerated differences in relative abundance unless sites were combined. Counties provided a convenient and unbiased basis for pooling data. Counties with <10 *Peromyscus* captured during a given trapping period were not included in subsequent analyses.

Census Trapping

To examine yearly and seasonal patterns of *Peromyscus* abundance with more precision than was possible with the general survey data, in 1989 we began a biannual census of small mammals at the University of Michigan Biological Station (UMBS), Cheboygan and Emmet Counties, near the northern tip of the Lower Peninsula (Figure 2). Three trap-lines, selected to sample 3 dominant habitat types in the area, were laid out and permanently marked:

(1) Colonial Point: 45° 29.43'N, 84° 41.12'W; mature northern hardwoods with open understory; dominant tree species in the vicinity of the trap-line include sugar and red maple (*Acer saccharum, Acer rubrum*), red oak (*Quercus rubra*), American beech (*Fagus grandifolia*), and white pine (*Pinus strobus*).

(2) Burn Plot: 45° 33.65'N, 84° 42.08'W; mixed hardwoods and pines; understory predominantly huckleberry (*Gaylussacia baccata*)l and ground cover bracken fern (*Pteridium aquilinum*), blueberries (*Vaccinium* sp.), and lichens; dominant tree species in the vicinity of the trap-line include red maple, red oak, aspen (*Populus grandidentata*), white pine, and red pine (*Pinus resinosa*).

(3) Reese's Swamp: 45° 32.87'N, 84° 39.92'W; boreal conifers with ground cover of sphagnum, maple seedlings, and a variety of forbs; dominant tree species in the vicinity of the trap-line include northern white cedar (*Thuja occidentalis*), balsam fir (*Abies balsamea*), red maple, and yellow and white birch (*Betula alleghaniensis, B. papyrifera*).

Each trap-line consists of 20 trapping stations. At each station, 1 large (9 cm x 8 cm x 23 cm) and 2 small (6.5 cm x 5 cm x 16 cm) Sherman traps are set and baited with oats. Trap stations are separated by approximately 25 m (Colonial Point), 20 m (Burn Plot), and 15 m (Reese's Swamp). Each line is trapped for 3 days and nights in early May (before the first litters of the spring are weaned) and again in late September (after weaning of the last litters of the summer).

In addition to these standardized censuses, in 1996 we began regular sampling near the Black River in the Pigeon River State Forest (Otsego Co., 45° 07'N, 84° 25'W; Figure 2), an area that had also been trapped in 1989, 1991, 1992, and 1993 using a protocol similar to the one described above. This area was chosen for additional sampling because of the presence of large numbers of *gracilis*. Approximately 100 traps (small Shermans) are deployed during each trapping session. They are placed in 2 parallel lines of 50 traps each. Because these trap-lines are on public land, they are not permanently marked. The trap-lines always begin at the same points and follow the same compass directions. Traps are placed at intervals of 5-7 m, baited with oats, and run for 1-2 nights during May and September. The trap-lines lie in deciduous forest. Tree species present are primarily sugar maple and American beech, with some red oak and white birch also present. The understory is open and consists primarily of maple and beech seedlings.

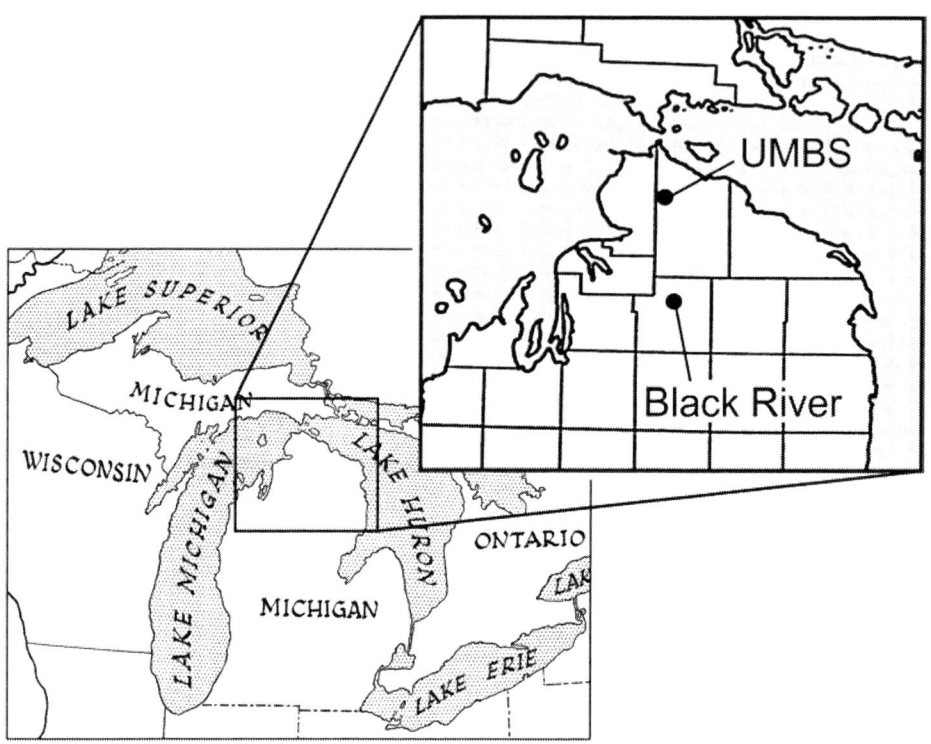

Figure 2. Location of study sites for *Peromyscus leucopus* and *P. maniculatus gracilus* in Michigan.

Mast Data

Acorns have been shown to be an important food resource for *leucopus* in northern New York (Elkinton et al., 1996; Ostfeld et al., 1996). To examine the effects of oak masting on mouse populations in Michigan, we identified years of significant mast production (1988-present) using data from field notes and from the Michigan Department of Natural Resources.

Climate Data

Dates when ice leaves Grand Traverse Bay were obtained from the Traverse City Chamber of Commerce. All other climatic data were recorded at a weather station at UMBS, located within 10 km of the UMBS census sites and approximately 54 km NNW of the Black River site.

DATA ANALYSIS

Analysis of climatic data is problematic because of the large number of climate descriptors available. With a sufficient number of data points, one could record as many weather variables as possible and use multivariate statistical techniques to explore and reduce the variable set. Alternatively, one could carry out a series of univariate tests and adjust significance levels to take into account the number of tests performed. With only 15 points for each spring and autumn (representing the 15 years of UMBS census data), neither approach was feasible. Instead, for these preliminary and exploratory analyses, we chose seven variables that integrate different aspects of the climate experienced by small mammals. We searched for possible effects of winter weather (October-April) on spring populations, and summer weather (May-September) on autumn populations. To represent summer weather, we included total rainfall, total number of cooling degree days (defined as 65°F subtracted from the mean temperature for each day, summed over all days from May 1 through September 30), and the greatest number of consecutive days with no rainfall. For winter weather, we used the total snowfall, the number of heating degree days (defined as the mean temperature for each day subtracted from 65°F, summed over all days from October 1 through April 30), and the minimum winter soil temperature measured at a depth of 50 cm (not available for one year). As an indication of the length of the winter, we used the dates on which ice forms (ice formation) and leaves (ice break-up) Douglas Lake, a large (1520 ha) lake that adjoins UMBS and lies within 10 km of each census site.

Because we did not attempt to adjust significance levels to reflect the number of tests, the patterns reported here should be viewed with caution. Nevertheless, because long-term data such as these are rare and difficult to obtain yet are essential for understanding population processes (Inchausti and Halley, 2001), we present our results to date to point out trends and to encourage and focus future research. Nonparametric statistical tests (Siegel, 1956) were used in the analysis of climatic data.

RESULTS

Are populations of *Peromyscus maniculatus gracilis* declining in the northern Lower Peninsula of Michigan? To examine long-term changes in the relative numbers of *gracilis* and *leucopus*, we divided our records (including general surveys, plus census data from UMBS and Black River) into four time periods, 1901-1930, 1931-1960, 1961-1990, and 1991-2002. For each period, we calculated the %*gracilis* from records for each county within the known range of *gracilis* in the Lower Peninsula. This analysis indicates that the relative abundance of *gracilis* in the northern Lower

Peninsula has declined sharply over the last century (Figure 3; ANOVA, df = 3,27, p = 0.044).

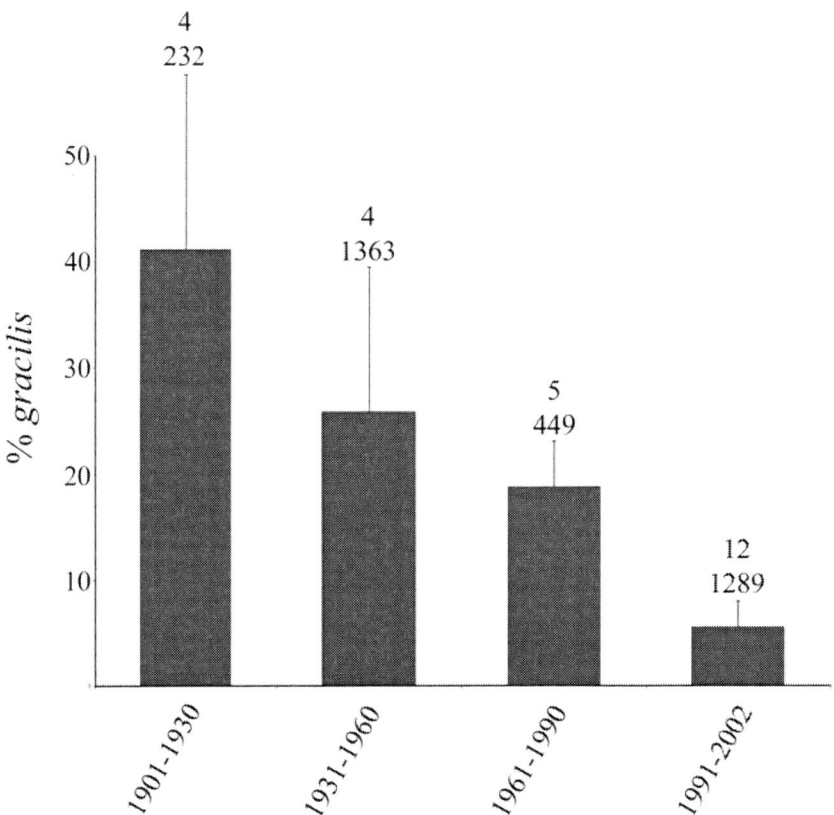

Figure 3. Decline in relative abundance of *P. m. gracilis* in the northern Lower Peninsula of Michigan. The Y-axis shows the relative abundance of *gracilis*, calculated for each time period as the number of *gracilis* captured divided by the total number of *Peromyscus* (*gracilis* + *leucopus*) captured. The error bar represents 1 SE. The upper number above each error bar is the number of counties from which ≥ 10 mice were obtained; the second number is the number of *Peromyscus* (*leucopus* + *gracilis*) in the sample.

Records from UMBS and the Black River provided a more detailed account of recent changes in the relative abundance of *gracilis* and *leucopus*. At UMBS, a few *gracilis* were captured early in the census period, but, by 1993, *gracilis* had disappeared entirely. Both species continued to coexist, however, at the Black River site. Here, we trapped at irregular intervals in 1989, 1991, 1992, and 1993; during that time period, between 50% and 87% of the *Peromyscus* captured were *gracilis*. In 1996, we began a program of biannual trapping at the Black River site (as at UMBS). In the first year of that census, the relative abundance of *gracilis* was similar to previous years (73%), but since 1996, it has fallen sharply to less than 10% (Figure 4).

What are the correlates of fluctuations in *Peromyscus leucopus* populations? At UMBS, 15 years of census data allowed us to examine population fluctuations in light of detailed weather records made very near the census sites. Information on masting patterns was also available. Because *gracilis* was last recorded in 1993, we were unable to study changes in its populations. We could, however, examine records of *leucopus* captures to ask whether the number of animals present in autumn or spring censuses was related to climate or acorn availability.

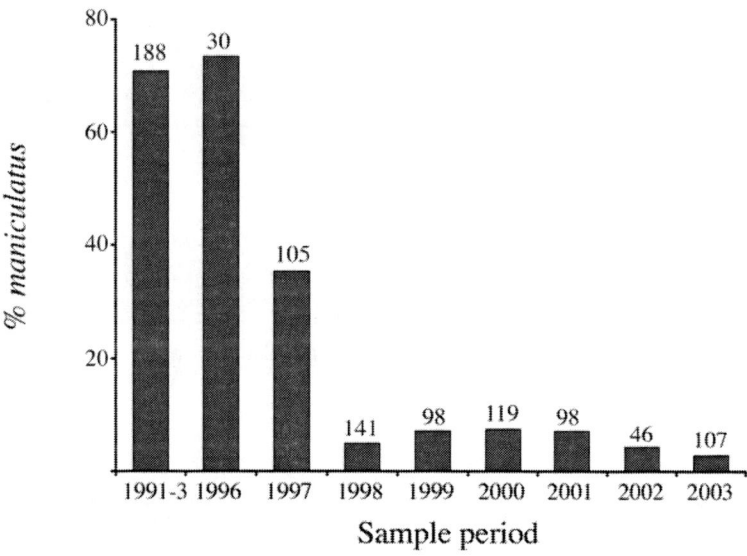

Figure 4. Decline in relative abundance of *P. m. gracilis* in the Black River area in Otsego Co., Michigan. Numbers above the bars indicate total number of *Peromyscus* (*leucopus* + *gracilis*) in the sample; data for 1991-1993 are combined due to small sample sizes for most years. The Y-axis is the same as in Figure. 3.

The abundance of *leucopus* at UMBS varied seasonally and yearly (Figure 5). Not surprisingly, spring populations were always equal to or smaller than populations in the preceding autumn and were usually considerably smaller. Autumn populations ranged from 10 animals in 1996 to 100 in 1993; spring populations fluctuated between 0 mice (1996, 2000) and 22 mice (1990).

Five of the 15 census years were years of substantial mast production (Figure 5). The number of mice captured was unrelated to oak masting. For example, both high (e.g., 1989) and low (e.g., 1996) *leucopus* populations came during the autumns of mast years (the time of year when the acorns drop). Similarly, populations in the spring following a mast were sometimes relatively high (e.g., 1990) and sometimes low (e.g., 1996).

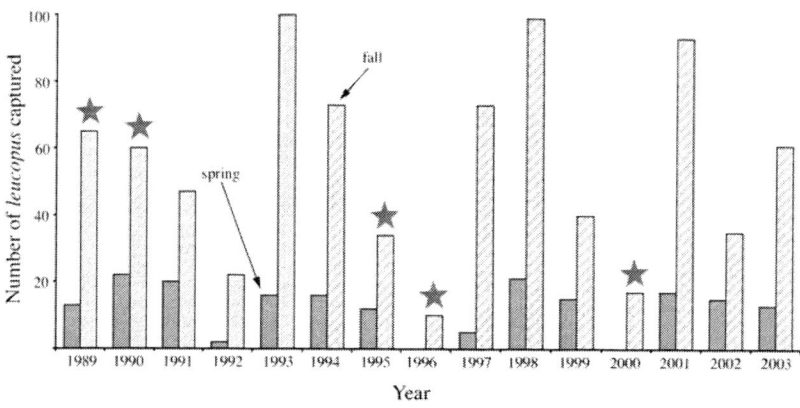

Figure 5. Number of *P. leucopus* captured in semi-annual censuses of small mammals at the University of Michigan Biological Station (UMBS), northern Lower Peninsula. Spring (May) samples are indicated by solid bars; autumn (September) samples by hatched bars. Years of high acorn production are marked by stars.

With respect to climate, northern Michigan experienced a very unusual rainstorm on Jan 1-2, 2000. The hard and prolonged rain collapsed the subnivean space and coated downed branches and tree trunks with a thick layer of ice. *Peromyscus leucopus*, like many other non-hibernating small mammals, relies on these structures for travel and foraging during the winter months. Heavy over-winter mortality in species such as *leucopus*, *Clethrionomys gapperi* (red backed vole*)*, and soricids (*Sorex cinereus*, *Blarina brevicauda*), all of which were characterized by extremely low abundance in spring 2000, was probably a result of this rare storm. We have therefore treated the spring census results of that year as an outlier, labeling them in the following figures and removing them from the statistical analyses of spring populations that follow.

Population size of *leucopus* in the autumn (at the end of the breeding season) was strongly related to the number of mice in the preceding spring (Spearman's rho = 0.65, N = 15, p < 0.01; Figure 6d). It was unrelated to total summer rainfall (Spearman's rho = 0.11, N = 15, p = 0.69), total number of cooling degree days (Spearman's rho = 0.15, N = 15, p = 0.59), or the maximum number of consecutive days with no rainfall (Spearman's rho = -030, N = 15, p = 0.28; Figure 6a-c).

In the spring census, the number of *leucopus* recorded was not related to the number of mice present during the preceding autumn census (Spearman's rho = 0.23, N = 15, p = 0.45; Figure 7d). It was also uncorrelated with total snowfall during the preceding winter (Spearman's rho = -0.32, N = 14, p = 0.27), number of heating degree days (Spearman's rho = -0.38, N = 14, p = 0.19), and minimum soil temperature at a depth of 50 cm (Spearman's rho = -0.05, N =12, p = 0.88; Figure 7a-c). Similarly, mouse population size was unrelated to the date ice formed on Douglas Lake (Spearman's rho = 0.30, N = 14, p = 0.30; Figure 8a). The population size of *leucopus* was, however, correlated with the date of ice break-up (Spearman's rho = -0.63, N = 14, p = 0.02; Figure 8b). When ice lasted late into spring (late April or early May), few *leucopus* survived the winter. When ice left early, in late March or early April, *leucopus* populations were at high levels. The date ice left the lake appeared to be more strongly related to population size than the overall length of the winter (number of days ice covered the lake; Spearman's rho = -0.44, N = 14, p = 0.12; Figure 8c).

DISCUSSION

In the northern Lower Peninsula, *Peromyscus maniculatus gracilis* is now strikingly less common relative to *leucopus* than it was at the beginning of the 20th century. A similar trend has been reported in northern Wisconsin by Long (1996). Over 40% of the *Peromyscus* in collections made in this region of Michigan before 1931 were *gracilis*. Today, that number is less than 6%. Further, because our recent efforts have been aimed at discovering *gracilis* populations and have concentrated on the habitats favored by that species, we strongly suspect that even this number over-represents the frequency of *gracilis*. The population of *gracilis* in northern Lower Michigan has been isolated from other *gracilis* populations for several thousand years and has evolved to become distinct both genetically and morphologically (Meagher, 1995, 1999; Lundrigan, Myers and Meagher, unpublished). The extinction of this unique population is a very real possibility.

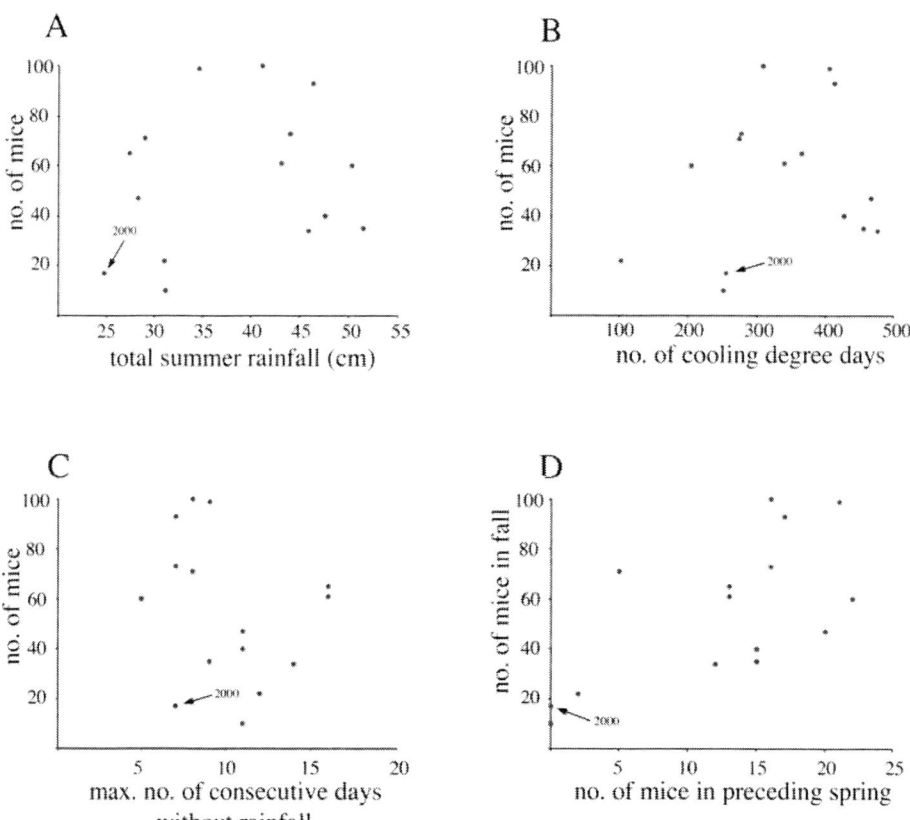

Figure 6. Number of *P. leucopus* captured in each autumn sample at the UMBS (Y axis) vs. (a) total summer (May-September) rainfall, (b) total number of cooling degree days during the summer months, (c) maximum number of consecutive days during the summer with no rainfall, and (d) number of *leucopus* captured during the preceding spring.

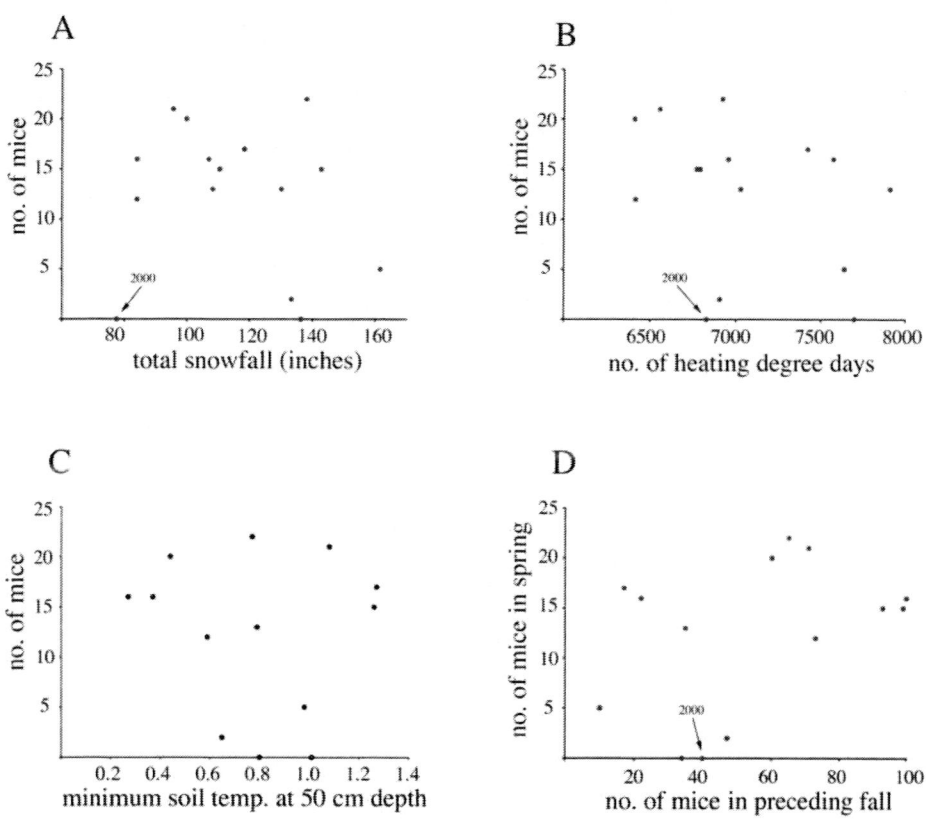

Figure 7. Number of *P. leucopus* captured in each spring sample at UMBS (Y axis) vs. (a) total snowfall during the preceding winter (October-April), (b) total number of heating degree days during the winter months, (c) minimum temperature reached at a soil depth of 50 cm during the winter months, and (d) number of *leucopus* captured during the preceding autumn.

Forest habitats in the northern Lower Peninsula have changed considerably over the last century (Whitney, 1987). Can these changes explain the loss of *gracilis* from this area? *Peromyscus maniculatus gracilis,* like *leucopus,* is found in a variety of habitats and, in the Great Lakes region, *gracilis* is often reported to prefer cooler, more boreal microhabitats (e.g., Dice, 1925; Hooper, 1942; Long, 1996). In the Lower Peninsula, we have found it to be strongly associated with forests, especially northern hardwoods (sugar maple, beech, white and yellow birch) or mixed hardwoods and hemlock (*Tsuga canadensis*), balsam fir, white pine, and red pine.

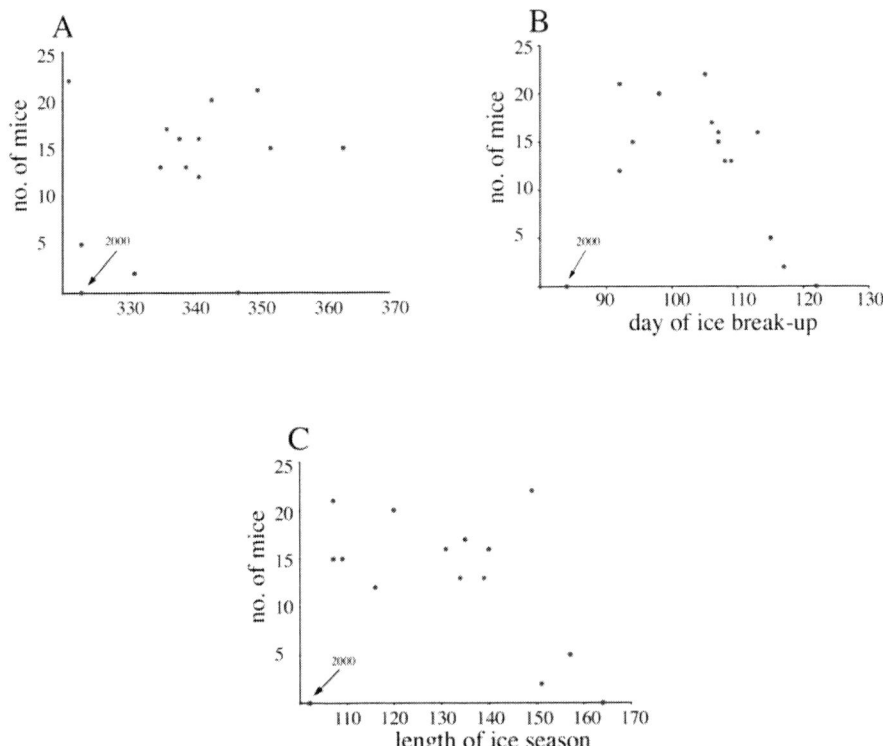

Figure 8. Number of *P. leucopus* captured in each spring sample at UMBS (Y axis) vs. (a) day of the year on which ice first covered Douglas Lake (UMBS) in the preceeding winter, (b) day of the year on which the ice covering Douglas Lake broke up, (c) total number of days ice covered Douglas Lake.

Peromyscus leucopus can also be found in these habitats, sometimes in high numbers. Unlike Lower Peninsula *gracilis*, *leucopus* is also abundant at the edges of fields or even in areas that have been recently logged (*gracilis* is often found in these habitats in parts of the Upper Peninsula where *leucopus* is absent). All of northern Michigan was logged at the end of the 19th and beginning of the 20th centuries and much of the area was subsequently burned in a series of fires that followed logging. Forest habitats largely disappeared. Our early records from this area document collecting soon after logging was completed, yet they contain a high proportion of *gracilis.* Since that time, logging has continued in a more controlled fashion and forests have regenerated to a considerable extent. Based on current knowledge of the habitat preferences of these two species, we would expect conditions for *gracilis* to be improving, not deteriorating.

Additional evidence that the apparent replacement of *gracilis* by *leucopus* is not related to forest regeneration comes from the Black River site. The abundance of *gracilis* at that site has fallen sharply during the last 10 years, yet neither the plant communities nor any other aspect of the habitat has changed appreciably.

With respect to the abundance of *leucopus* at UMBS, the number of years of census data is small. Any statistical analysis of the correlates of mouse population size must be viewed with caution, particularly when, as is the case here, several potential correlates were examined. Nevertheless, even with just 15 census samples, some variables were clearly unrelated to mouse number, while for others, a relationship appeared to emerge.

Oak masting had little if any effect on mouse populations at UMBS, in contrast to its strong influence on *leucopus* populations in New York (Elkinton et al., 1996; Ostfeld et al., 1996). Also, the summer temperature and rainfall variables tested here had no detectable effect. Of the summer variables we examined, the only one affecting population size in the autumn was the number of mice captured during the preceding spring.

The winter pattern was quite different. Autumn population levels had no detectable effect on population levels in the succeeding spring, suggesting that events during the winter months were critical for determining spring (and by correlation, subsequent autumn) population levels. *Peromyscus leucopus* is at the very northern limit of its distribution in this region and it would not be surprising if populations of this species were especially vulnerable to harsh winter weather. Variables such as number of heating degree days and total snowfall, however, bore no relationship to the number of mice remaining in the spring.

Winters in the northern Lower Peninsula are long, as well as cold and snowy. Winter conditions often extend into late April or even early May. Length of winter is difficult to define and even harder to measure. We chose to use dates of ice formation and ice break-up on nearby Douglas Lake to represent the beginning and end of winter conditions, because these dates integrate temperature, snowfall, and other aspects of local weather (Anderson et al., 1996) and because they have been recorded consistently over the years of the study. The number of *leucopus* that successfully over-wintered appeared to be strongly related to the date of ice break-up, slightly less strongly related to the number of days between ice formation and break-up, and unrelated to the date of ice formation. Prolonged winters were always followed by low numbers of *leucopus* in the spring. Shorter winters appear to be more favorable and are generally followed by larger spring populations. Alternatively, they permit other factors to come into play (for example, the unusual rainstorm in January 2000).

Long (1996), reporting on long-term observations of *Peromyscus* populations in Wisconsin, also proposed that winter weather had a greater impact on populations of *leucopus* than on populations of *gracilis*. He suggested that temperature and snow depth might be critical factors (a combination of low temperature and lack of snow being detrimental; see also Long, 1973). Based on reports that *maniculatus nubiterrae* tends to nest in trees during the winter, while *leucopus* tends to move its nests underground (Wolff and Hurlbutt, 1982; Wolff and Durr, 1986), Long (1996) argued that *leucopus* might be more susceptible than *gracilis* to deep frost during severe winters, and that deep frosts were the result of cold temperatures occurring when the ground was unprotected by snow. We found no relationship, however, between the minimum soil temperature reached during the winter and *leucopus* population size the following spring.

Peromyscus leucopus is abundant in southern Michigan, where members of the species begin breeding in March in most years (Baker, 1983). The onset of breeding is probably at least partially determined by photoperiod, and this mechanism is under direct genetic control (Heideman and Bronson, 1991; Heideman et al., 1999). Breeding is a risky and expensive process. One possible explanation for the relationship between winter length and *leucopus* survival reported here is that the timing of breeding in the northern Lower Peninsula, at the northern fringe of the species' range, is influenced by gene flow from southern populations for which a March onset of reproduction is more appropriate. When winter ends early, mice breed successfully. When it ends late, breeding begins before resources are available to sustain the breeding process. If this hypothesis is correct, then (1) winter mortality in *leucopus* populations in the northern Lower Peninsula should be most severe in late winter or early spring, and (2) breeding activity of *leucopus* should begin in March or early April, regardless of weather. We hope to examine both predictions in the future.

Climate and *Peromyscus maniculatus gracilis*

Does weather, especially winter weather, affect *gracilis* in the same way that it affects *leucopus*? Field data on breeding and mortality patterns from the northern Lower Peninsula are lacking. In other areas, however, it has been shown that *gracilis* and the similar *nubiterrae* make better use of torpor than *leucopus* to survive particularly difficult conditions (Tannenbaum and Pivorun, 1988; Pierce and Vogt, 1993; Wolff, 1996). Pierce and Vogt (1993) showed that *gracilis* builds nests that are better insulated than those of *leucopus*, *gracilis* hoards greater quantities of food, and *gracilis* is more likely to go out of breeding condition during the winter months. Unfortunately, these studies involved comparing *gracilis* or *nubiterrae* with *leucopus* from milder climates rather than from syntopic populations. Based on field studies of the two species where they occur together, Wolff (1996) suggested that climatic variation may contribute to the coexistence of these two taxa. *Peromyscus leucopus* should thrive when winters are mild, and *gracilis* populations should rebound when winters are harsh. At UMBS, *leucopus* populations behaved as predicted by Wolff; the number of *leucopus* was high following short winters and low when winters were prolonged. What about *gracilis* populations?

By 1993 *gracilis* had disappeared from the area around the census sites at UMBS. The species still persists, however, at the Black River site. Comparing the number of *gracilis* trapped at the Black River site to climatic conditions was not possible because weather records comparable to those for UMBS were not available, spring samples were available for only seven years, and the number of individuals captured has generally been very low. Nevertheless, we can predict that if *leucopus* consistently experiences higher winter mortality than *gracilis* but recovers during the breeding season, the relative abundance of *gracilis* at the site should be high in the spring and low in the autumn. This was consistently true despite substantial year-to-year variation in the number of individuals of each species captured (resulting from both population fluctuation and from differences in trapping effort; Figure 9).

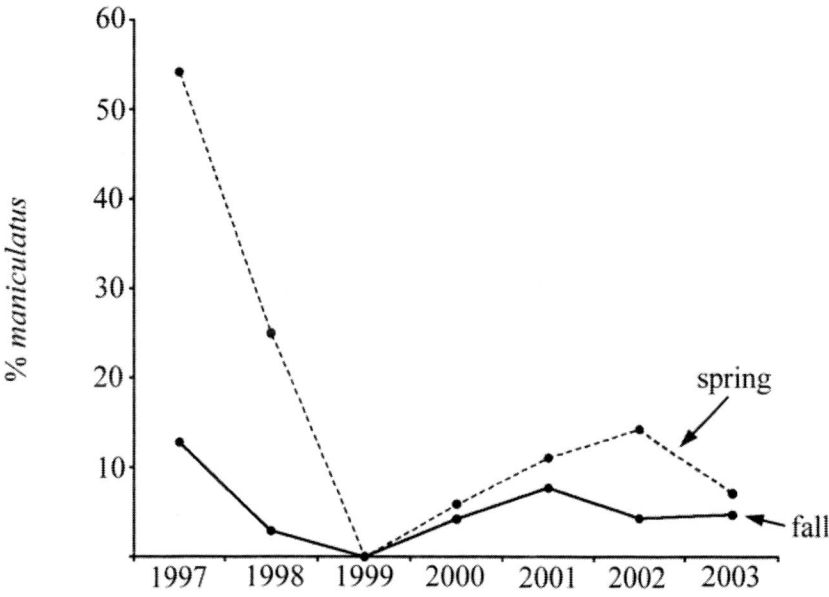

Figure 9. Spring (dashed line) and autumn (solid line) abundance for *P. m. gracilis* at the Black River Study site. Sample sizes (spring/fall): 1997 63/90; 1998 13/200; 1999 72/26; 2000 23/91; 2001 22/135; 2002 26/64; 2003 28/335.

We hypothesize that recent climatic warming may be responsible for the decline in *gracilis* relative to *leucopus* in northern Michigan. In accord with Wolff (1996) and Long (1996), we suggest that *gracilis* survives better than *leucopus* when winters are long and harsh, that *leucopus* has an as-of-yet unknown competitive or reproductive advantage when spring comes early, and that the two species coexist despite their ecological and morphological similarity because of year-to-year variation in the severity of winter. To explain the long-term decline in *gracilis* populations described at the beginning of this paper, however, winters must be shorter now than they were earlier in this century. Further, the Black River data suggest that warming has recently accelerated. Is this the case?

The UMBS ice break-up data do not go far enough back in time to allow us to address this question. A much larger dataset is available, however, for Grand Traverse Bay, on Lake Michigan approximately 100 km SSW of UMBS, where break-up dates have been recorded for over 100 years (Figure 10a). A general decline in break-up date is evident, but there is considerable year-to-year variability. To make long-term trends more obvious, we calculated for each year the average of the break-up date for that year and the preceding 9 years (Figure 10b). These 10-year running averages reveal a striking pattern. During the 20th century, the average time of ice break-up decreased gradually from around the 90th day of the year in the early 1900's to approximately the 75th day of the year in the early 1970's. Since the early 1970's, the average date of break-up has fallen to about day 50 in 2001. The rapid decline of the last 3 decades has been due largely to an increase in the number of years in which ice did not form on the bay (these years were assigned a value of 0, or ice break-up on January 1, for this analysis). Clearly, ice break-up data suggest that winters in this region are ending earlier than in the past.

On average, the earth's climate has warmed 0.3-0.6° C over the last century (Intergovernmental Panel on Climate Change, 1996). This change has not been uniformly distributed. Some areas have experienced much greater warming, while some have actually become cooler (op. cit.). Precipitation patterns have also been altered. Most models predict even greater changes over the next century.

A growing number of studies, like this one in northern Michigan, suggest that recent changes in climate have already had an effect on the organisms inhabiting Earth (McCarty, 2001; Walther et al., 2002). This is true both at the level of individual species (studies showing change in mammal populations include Frey, 1992; Hersteinsson and Macdonald, 1992; Post et al., 1999; Inouye et al., 2000) and at the level of entire communities (e.g., Brown et al., 1997). Taken individually, at best these studies provide evidence of a correlation between climate change and alterations of the biological attributes of species (e.g., distributions, breeding phenology, community structure). Causation is extremely difficult to establish. Nevertheless, as the number of studies that report concordant biological and climate changes grows, the case for global warming as a causal element becomes increasingly compelling.

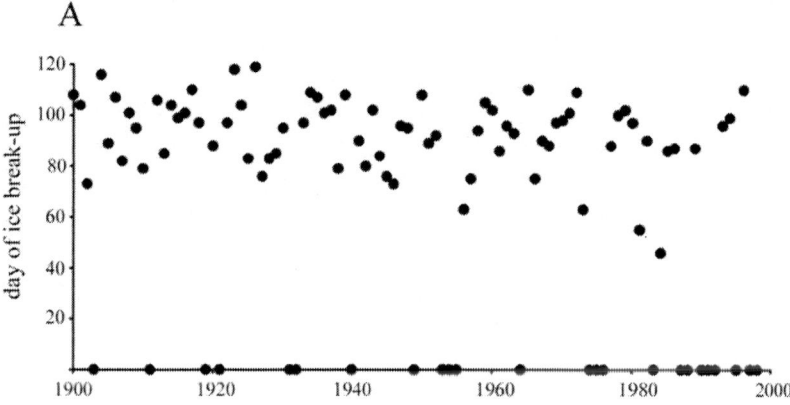

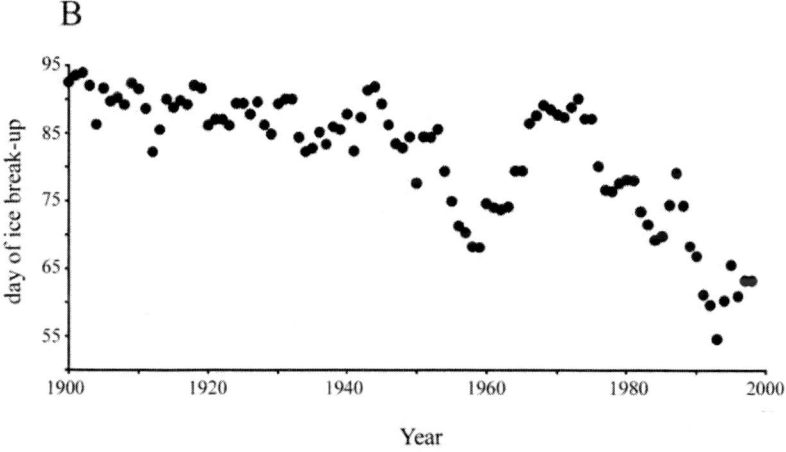

Figure 10. (a) Day of ice break-up on Grand Traverse Bay, Michigan. Years during which continuous ice did not form are recorded as break-up occurring on day 0 (Jan. 1). (b) Day of ice break-up on Grand Traverse Bay, each year calculated as the average of that year and the preceding 9 years.

ACKNOWLEDGMENTS

We are grateful to Jim Patton for his unflagging encouragement and, in particular, for his enthusiasm for an extraordinarily wide variety of research endeavors having to do with mammals. Field surveys were supported in part by a grant from the Michigan Nongame Wildlife Fund, Michigan Department of Natural Resources. The University of Michigan Biological Station has sponsored 14 years of small-mammal censusing; we are very grateful to UMBS and especially to its director during this period, Jim Teeri, for his strong encouragement, support, and advice. Glen Matthews (Michigan Department of Natural Resources) generously gave us information on oak masting, and the Traverse City Chamber of Commerce provided its remarkable records of dates of ice formation and break-up on Grand Traverse Bay. John Megahan gave valuable assistance with figures. Many students and friends have participated in field surveys; they include Rebecca Anderson, Andrew Bunker, Guillermo D'Elia, Toni Gorog, Nedra Klein, Eladio Marquez, Rosa Moscarello, Carol Patton, Jim Patton, Sally Petrella, Dave Storer, Robert Storer, Jerry Svendsen, and the Field Mammalogy class (Biology 453) at UMBS.

LITERATURE CITED

Anderson, W. L., D. M. Robertson, and J. J. Magnuson
    1996   Evidence of recent warming and El Nino-related variations in ice breakup of Wisconsin lakes. Limnology and Oceanography 41:815-821.

Baker, R. H.
    1983   Michigan Mammals. Michigan State University Press, East Lansing, MI., 642 pp.

Barry, R. E., Jr., M. A. Botje, and L. B. Grantham
    1984   Vertical stratification of *Peromyscus leucopus* and *P. maniculatus* in southwestern Virginia. Journal of Mammalogy 65:145-148.

Brown, J. H., T. J. Valone, and C. G. Curtin
    1997   Reorganization of an arid ecosystem in response to recent climate change. Proceedings of the National Academy of Science of America 94:9729-9733.

Bruseo, J. A., S. H. Vessey, and J. S. Graham
    1999   Discrimination between *Peromyscus leucopus noveboracensis* and *Peromyscus maniculatus nubiterrae* in the field. Acta Theriologica 44:151-160.

Cranford, J. A., and M. S. Maly
    1986   Habitat associations among small mammals in an old-field community on Butt Mountain Virginia, USA. Virginia Journal of Science 37:172-176.

Dice, L. R.
    1925   A survey of mammals of Charlevoix County, Michigan and vicinity. Occasional Papers of the University of Michigan Museum of Zoology 159:1-33.

Dooley, J. L., Jr., and R. D. Dueser
    1990   An experimental examination of nest-site segregation by two *Peromyscus* species. Ecology 71:788-796.

Elkinton, J. S., W. M. Healy, J. B. Buonaccorsi, G. H. Boettner, A. M. Hazzard, H. R. Smith, and A. M. Liebhold
    1996   Interactions among gypsy moths, white-footed mice, and acorns. Ecology 77:2332-2342.

Feldhamer, G.A., J. E. Gates, and J. H. Howard
    1983   Field identification of *Peromyscus maniculatus* and *P. leucopus* in Maryland: reliability of morphological comparisons. Acta Theriologica 28:417-423.

Frey, J. K.
    1992   Response of a mammalian faunal element to climatic changes. Journal of Mammalogy 73:43-50.

Graves, S., J. Maldonado, and J. O. Wolff
    1988   Use of ground and arboreal microhabitats by *Peromyscus leucopus* and *Peromyscus maniculatus*. Canadian Journal of Zoology 66:277-278.

Harney, B. A., and R. D. Dueser.
    1987   Vertical stratification of activity of two *Peromyscus* species: an experimental analysis. Ecology 68:1084-1091.

Heideman, P. D., and F. H. Bronson
    1991   Characteristics of a genetic polymorphism for reproductive photoresponsiveness in the white-footed mouse (*Peromyscus leucopus*). Biology of Reproduction 44:1189-1196.

Heideman, P. D., T. A. Bruno, J. W. Singley, and J. V. Smedley
    1999   Genetic variation in photoperiodism in *Peromyscus leucopus*: geographic variation in an alternative life-history strategy. Journal of Mammalogy, 80:1232-1242.

Hersteinsson, P., and D. W. Macdonald
    1992   Interspecific competition and the geographical distribution of red and arctic foxes *Vulpes vulpes* and *Alopex lagopus*. Oikos 64:505-515.

Hooper, E. T.
    1942   An effect on the *Peromyscus* rassenkreis of land utilization in Michigan. Journal of Mammalogy 23:193-196.

Inchausti, P., and J. Halley
    2001   Investigating long-term ecological variability using the Global Population Dynamics Database. Science 293:643-650.

Inouye, D. W., B. Barr, K. B. Armitage, and B. D. Inouye
    2000   Climate change is affecting altitudinal migrants and hibernating species. Proceedings of the National Academy of Science of America 97:1630-1633.

Intergovernmental Panel on Climate Change
    1996   Climate change 1995: the science of climate change. Contribution of working group 1 to the second assessment report of the IPCC. Cambridge University Press, New York, 572 pp.

Long, C. A.
    1973   Reproduction in the white-footed mouse at the northern limits of its geographic range. Southwestern Naturalist 18:11-20.

    1996   Ecological replacement of the deer mouse, *Peromyscus maniculatus*, by the white-footed mouse, *P. leucopus*, in the Great Lakes Region. Canadian Field-Naturalist 110:271-277.

Long, C. A., and J. E. Long
    1993   Discriminant analysis of geographic variation in long-tailed deer mice from northern Wisconsin and Upper Michigan. Transactions of the Wisconsin Academy of Arts, Sciences, and Letters 81:107-116.

McCarty, J. P
    2001   Ecological consequences of recent climate change.   Conservation
            Biology 15:320-331.

Meagher, S.A
    1995   Ecology, genetics, and evolution of deer mice (*Peromyscus maniculatus
            gracilis*) and a parasitic nematode (*Capillaria hepatica*).   Ph.D. thesis,
            University of Michigan, 117 pp.

    1999   Genetic diversity and *Capillaria hepatica* (Nematoda), prevalence in
            Michigan deer mouse populations.   Evolution 53:1318-1324.

Ostfeld, R. S., C. G. Jones, J. O. Wolff
    1996   Of mice and mast.   Bioscience 46:323-330.

Pierce, S. S., and F. D. Vogt
    1993   Winter acclimatization in *Peromyscus maniculatus gracilis, P. leucopus
            noveboracensis,* and *P. l. leucopus.*   Journal of Mammalogy 74:665-677.

Post, E., R. O. Peterson, N. C. Stenseth, and B. E. McLaren
    1999   Ecosystem consequences of wolf behavioral response to climate.
            Nature 401:905-907.

Rich, S. M., C. W. Kilpatrick, J. L. Shippee, and K. L. Crowell
    1996   Morphological differentiation and identification of *Peromyscus leucopus*
            and *Peromyscus maniculatus* in northeastern North America.   Journal of
            Mammalogy 77:985-991.

Siegel, S.
    1956   Nonparametric statistics for the behavioral sciences. McGraw-Hill, Inc.,
            New York, 312 pp.

Smith, D. A., and S. W. Speller
    1970   The distribution and behavior of *Peromyscus maniculatus gracilis* and
            Peromyscus leucopus noveboracensis (Rodentia: Cricetidae) in a
            southeastern Ontario woodlot.   Canadian Journal of Zoology 48:1187-
            1199.

Tannenbaum, M. G., and E. B. Pivorun
    1988.   Seasonal study of daily torpor in southeastern *Peromyscus maniculatus* and *Peromyscus leucopus* from mountains and foothills. Physiological Zoology 61:10-16.

Walther, G.-R., E. Post, P. Convey, A. Menzel, C. Parmesan, T. J. C. Beebe, J.-M. Fromentin, O. Hough-Guldberg, and F. Bairlein.
    2002   Ecological responses to recent climate change. Nature 416:389-395.

Whitney, G. G.
    1987   An ecological history of the Great Lakes forest of Michigan. Journal of Ecology 75:667-684.

Wolff, J. O.
    1985a   Comparative population ecology of *Peromyscus leucopus* and *Peromyscus maniculatus*. Canadian Journal of Zoology 63:1548-1555.

    1985b.   The effects of density, food, and interspecific interference on home range size in *Peromyscus leucopus* and *Peromyscus maniculatus*. Canadian Journal of Zoology 63:2657-2662.

    1986   Life history strategies of white-footed mice (*Peromyscus leucopus*). Virginia Journal of Science 37:208-220.

    1996   Coexistence of white-footed mice and deer mice may be mediated by fluctuating environmental conditions. Oecologia 108:529-533.

Wolff, J. O., and B. Hurlbutt.
    1982   Day refuges of *Peromyscus leucopus* and *Peromyscus maniculatus*. Journal of Mammalogy 63:666-668.

Wolff, J. O., M. H. Freeberg, and R. D. Deuser
    1983   Interspecific territoriality in two sympatric species of *Peromyscus* (Rodentia: Cricetidae). Behavioral Ecology and Sociobiology 12:237-242.

Wolff, J. O., R. D. Dueser, and K. S. Berry
    1985   Food habits of sympatric *Peromyscus leucopus* and *Peromyscus maniculatus*. Journal of Mammalogy 66:795-798.

Wolff, J. O., and D. S. Durr

    1986   Winter nesting behavior of *Peromyscus leucopus* and *Peromyscus maniculatus*. Journal of Mammalogy 67:409-412.

Part II: South America and Journeys into the Unknown

During his appointment as a Curator in the Museum of Vertebrate Zoology (MVZ), James L. Patton's vigorous program of field research and numerous influential publications quickly established him as an authority on the mammals of western North America. This geographic emphasis was logical, given both Jim's interest in the evolutionary genetics of pocket gophers (genus *Thomomys*, distributed throughout the western United States into southern Canada and northern Mexico) and the location of the MVZ on the University of California's Berkeley campus. At the same time, Jim nurtured a strong interest in the mammals of South America, as evidenced by his multiple field trips to this region - in particular Peru - during the 1970's and 1980's. This work resulted in several important publications regarding the taxonomy, systematics, and biogeography of akodontine rodents from eastern Peru.

Jim's research took on a decidedly South American focus around 1990, when he embarked on a nearly decade-long exploration of the biogeography of Amazonian mammals. In many ways, this marked the start of a second research career for Jim, one that led him from the relatively well-traveled deserts of North America to the largely unknown jungles of Brazil. As part of this undertaking, Jim spent his 1991-1992 sabbatical leave collecting mammals along the Rio Juruá in the southwestern Amazon. A primary goal of this adventure was to test Alfred Russel Wallace's Riverine Barrier Hypothesis, which predicts that the degree of divergence among organisms on opposite shores of a river should increase as the width of the waterway increases from its head to its mouth. The specimens obtained during this trip included more than a dozen new species of mammals and provided the foundation for Jim's seminal monograph *Mammals of the Rio Juruá and the Evolutionary and Ecological Diversification of Amazonia* (co-authored with Maria da Silva and Jay Malcolm). This work, published in 2000 by the American Museum of Natural History, is widely regarded as the definitive treatment of the small mammal fauna of northern Brazil. It also epitomizes many of the traits that have defined Jim's research career – the integration of rugged fieldwork, careful sampling design, and modern molecular systematics to test fundamental evolutionary hypotheses.

The second part of this volume pays tribute to Jim's contributions to the study of South American mammals. The first two papers in the section use studies of a relatively well-known taxon of South American rodents, the genus *Ctenomys* (the tuco-tucos), to explore patterns of morphological and genetic differentiation within and among closely related species. Given the striking ecological parallels between

Ctenomys and the North American *Thomomys*, these chapters represent a logical conceptual extension of Jim's studies of pocket gophers as well as a reflection of his shift in geographic focus from North to South America. In the first of these contributions, Thales de Freitas examines variation in skull morphology among multiple species of *Ctenomys* and attempts to relate cranial differences to patterns of karyotypic variation in these animals. From this work, Freitas concludes that morphological and karyotypic differentiation in *Ctenomys* are not concordant, which raises intriguing questions concerning the processes generating and maintaining these patterns of variation.

In the second paper regarding ctenomyid diversity, Enrique Lessa and co-authors examine patterns of genetic variation within the Río Negro tuco-tuco (*Ctenomys rionegrensis*), which is distinguished by an unusual pattern of coat color polymorphism. By characterizing genetic differentiation across multiple spatial scales, Lessa and colleagues provide evidence that this species has undergone an historical range expansion followed by differentiation of relatively isolated local populations. By providing one of the few detailed analyses of genetic structure in *Ctenomys*, this study serves as an important contribution to ongoing efforts to understand the evolutionary diversification of this highly speciose lineage.

The next three papers in this section of the volume address questions that have arisen, at least in part, due to Jim's efforts to understand the systematics of Neotropical rodents. The third chapter, by Pablo Gonçalves and colleagues, uses morphological and cytological data to examine the systematic affinities of *Bibimys labiosus*, a bizarre species of sigmodontine rodent that was first described in 1887 but not recaptured until the 1990's. Based on their analyses of historical and newly collected specimens, Gonçalves et al. conclude that *B. labiosus* is a valid species but that placement of *Bibimys* within the Tribe Scapteromyini is problematic given the degree of divergence between this and other scapteromyine genera. The fourth paper in this section, authored by Guillermo D'Elía and colleagues, expands upon this theme to consider the phylogenetic position of *Bibimys* relative to other sigmodontine rodents. Using analyses of mitochondrial DNA sequences, D'Elía et al. provide further evidence against the inclusion of *Bibimys* within the Scapteromyini, but call into question the validity of the three species currently assigned to *Bibimys*. Together, these chapters underscore the critical need for continued taxonomic, systematic, and phylogenetic analysis of Neotropical sigmodontine rodents.

The fifth paper in this section explores the generic-level systematics of one of the most diverse families of Neotropical rodents, the Echimyidae. Focusing on the arboreal members of this family, Louise Emmons uses morphological characters to propose two new genera of echimyids, including the aptly named *Pattonomys*. As with the preceding chapters regarding the genus *Bibimys*, Emmons' work reveals how much remains to be learned regarding the systematics of the mammals of South America. In this regard, Jim Patton's travels along the Rio Juruá and the analyses of mammalian diversity that they have inspired have truly been "journeys into the unknown."

The final two papers in this volume examine the structure and diversification of Neotropical rodent communities. Márcia Lara and co-authors explore phylogeographic patterns among rodents of the Brazilian Atlantic Forest, an increasingly isolated region whose historical connections to Amazonia are still unclear. Comparative analyses of mitochondrial DNA sequences from rodents of the Atlantic Forest lead Lara and colleagues to conclude that diversification of these taxa is not the result of shared environmental pressures but, instead, likely reflects the distinctive phylogeographic and environmental histories of each lineage considered. Their findings have important implications for understanding processes of diversification in not just the Atlantic Forest, but in all tropical ecosystems.

The volume concludes with a study by Jay Malcolm and colleagues that represents one of the few analyses of the community ecology of Amazonian mammals ever undertaken. For many of the species included in this work, these data reflect virtually all that is known regarding their interactions with their environments. This is a particularly appropriate note on which to close the volume; as Malcolm et al. demonstrate, even when the phylogeographic relationships among species are relatively well understood, much remains to be learned regarding other aspects of their biology. Given the role that Jim Patton has played in mentoring the work presented in each of the chapters in this section of the volume, it is clear that he has been instrumental in encouraging new generations of biologists to explore the mammal faunas of Amazonia and other portions of South America.

Analysis of Skull Morphology in 15 Species of the Genus *Ctenomys*, Including Seven Karyologically Distinct Forms of *Ctenomys minutus* (Rodentia: Ctenomyidae)

Thales Renato O. de Freitas

The genus *Ctenomys* presents extensive karyotypic variation, suggesting that chromosomal speciation is the main evolutionary force acting in this genus. Two types of karyotype variation are found: the first type of variation is interspecific and causes diploid numbers ranging from 2n = 10 to 70, while the second is intraspecific, and is pronounced in some species. This work discusses the role of chromosomal speciation in *Ctenomys* using data from two analyses that explore correlations between karyotypic and morphological variation. Skull morphology of 15 species of *Ctenomys* (*C. australis*, *C. haigi*, *C. mendocinus*, *C. argentinus*, *C. sociabilis*, *C. brunneus*, *C. maulinus*, *C. minutus*, *C. torquatus*, *C. lami*, *C. flamarioni*, *C. fulvus*, *C. opimus*, *C. leucodon*, and *C. peruanus*) is analyzed to assess a possible correlation between skull characteristics and diploid numbers. The second analysis investigates the relationships between karyotypic form and skull morphology in 18 populations of *C. minutus* located along a North-South transect in southern Brazil. Sexual size dimorphism in *C. minutus* is also examined. The report shows that there is no evolutionary pattern related to skull morphology in the 15 species studied, although almost each species has its own karyotype. In contrast, *Ctenomys minutus* presents detectable evolutionary relationships between cytotypes and skull morphology.

KARYOTYPIC VARIATION IN *CTENOMYS*

The subterranean rodents of the genus *Ctenomys* exhibit considerable karyotypic diversity. Although the number of species recognized differs between authors (e.g., Reig et al. [1990] indicated 56 species while Woods [1993] considered only 48 forms), it is clear that fundamental number and chromosomal structure differ substantially across species, with almost each species displaying a unique karyotype. This extensive karyotypic variation suggests that chromosomal speciation is the main evolutionary force acting in this genus. Two types of karyotype variation are found. The first is chromosome variation among species, which results in diploid numbers ranging from 2n = 10 to 70. The second type of variation is intraspecific, which is substantial in some species. For example, in Uruguay, *C. pearsoni* displays karyotypes of 2n = 56, 64, and 70 (Kiblisky et al., 1977;

131

Novello and Lessa, 1986), while *C. rionegrensis* presents 2n = 52 and 56 (Ortells et al., 1990). *Ctenomys torquatus* shows 2n = 44 in Uruguay (Kiblisky et al., 1977), but in southern Brazil 2n = 44 and 46 (Freitas and Lessa, 1984). Also in southern Brazil, *C. minutus* has karyotypes with 2n = 42, 45, 46a, 46b, 47, 48, 49 and 50 (Freitas, 1997). In Argentina, karyotypes of *C. perrensis* range from 2n = 50 to 57 (Garcia et al., 2000), while four species from the Mendocinus group, *C. mendocinus, C. porteousi, C. azarae* and *C. chasiquensis,* are characterized by 2n = 46, 47 and 48, respectively (Massarini et al., 1991, 1998). Although this extensive chromosomal variability is well documented, few studies have examined the association between this form of genetic variation and morphology in *Ctenomys*. Freitas and Lessa (1984) showed that specimens of *C. torquatus* with different karyotypes displayed few different skull characteristics. Marinho and Freitas (2000) found minor differences in skull form in a chromosomal hybrid zone in *C. minutus* (2n = 46a, 47 and 48) reported by Freitas (1997). With the exception of these two studies, however, the relationship between karyotypic and morphological variation has not been characterized for this genus.

In the case of *C. minutus,* Freitas (1997) found the highest intraspecific chromosomic variation known for the genus *Ctenomys*, recording seven karyotypes (2n = 42, 45, 46a, 46b, 47, 48, 49 and 50) distributed along a 330 km transect in the Coastal Plain of Rio Grande do Sul and Santa Catarina in the South of Brazil. The diploid numbers 49 and 50 occur in Jaguaruna beach. Karyotype 2n = 46a spreads from the southern banks of Araranguá River to Emboaba Lake. A hybrid population with 2n = 46a, 47 and 48 was found from the eastern banks of Barros Lake to Fortaleza Lake. In Passinhos and Palmares do Sul, only karyotype 2n = 48 was observed. Karyotype 2n = 42 was found in Mostardas while 2n = 46b was observed in Tavares. Freitas (1997) analyzed G-banding patterns in these karyotypes and documented multiple rearrangements. Specifically, the metacentric pair 2 and the acrocentric pairs 16, 17, 19, 20, 22, 23, and 24 from 2n = 50 are found in other karyotypes as chromosomes or chromosome arms. The variant from 2n = 50 is 2n = 49, which has the 20/17 fusion in heteromorphic form. The diploid number 2n = 46a presents two fusions, 20/17 and 23/19, which are also present in the karyotype 2n = 48 in addition to a fission in chromosome 2(2*p* and 2*q*) that is responsible for the heterozygote 2n = 47. The diploid number 2n = 42 has the fusions 20/17 and 23/19, the metacentric pair 2, and two different rearrangements among pairs 16, 24, and 22, forming a new chromosome: a tandem fusion between chromosomes 24 and 16, followed by a centric fusion with chromosome 22.

This work discusses the role of chromosomal speciation in *Ctenomys* using data from two analyses that explore correlations between karyotypic and morphological variation. First, the skull morphology of 15 species of *Ctenomys* is analyzed to determine if there is a correlation between skull characteristics and diploid numbers. Second, relationships between karyotypic form and skull morphology are examined for 18 populations of *C. minutus* located along a North-South transect in southern Brazil. Sexual size dimorphism in *C. minutus* is also examined. Regarding the genus *Ctenomys,* this report shows that there is no evolutionary relationship between chromosomal variation and skull morphology in the 15 species studied, although almost each species has its own karyotype. *Ctenomys minutus,* on the other hand, is characterized by detectable relationships between cytotypes and skull morphology.

MATERIALS AND METHODS

To examine interspecific variation in skull morphology, 337 females from 15 species were studied: *C. australis,* 2n = 48 (*N* = 23); *C. haigi,* 2n = 50 (*N* = 24); *C. mendocinus,* 2n = 48 (*N* = 4); *C. argentinus,* 2n = 44 (*N* = 2); *C. sociabilis,* 2n = 56 (*N* = 8); *C. brunneus,* 2n = 26 (*N* = 3); *C. maulinus,* 2n = 26 (*N* = 10) from Argentina; *C. minutus,* 2n = 42, 45, 46a, 46b, 47, 48, 49 and 50 (*N* = 84); *C. torquatus,* 2n = 44 and 46 (*N* = 50); *C. lami,* 2n = 54, 55, 56, 57 and 58 (*N* = 61); *C. flamarioni,* 2n = 48 (*N* = 35) from Brazil; *C. fulvus,* ?n – 26 (*N* = 3) from Chile; *C. opimus,* 2n = 26 (*N* = 21); *C. leucodon,* 2n = 36 (*N* = 3) and *C. peruanus* from Peru (Figure 1). The animals from Brazil are deposited in the Mammalian Collection of the Department of Genetics, Federal University of Rio Grande do Sul. Tuco-tucos from Argentina, Peru, and Chile are deposited in the Museum of Vertebrate Zoology, University of California, Berkeley.

To avoid the potential confounding effects of variation in skull morphology due to sexual dimorphism, only female specimens were used. The following skull measurements were used: total length of skull, condylobasal length, nasal length, nasal breadth, zygomatic breadth, bimeatal breadth, interorbital breadth, mastoid breadth, rostral breadth, frontal length, and diastema length based on Langguth and Abella (1970).

The analyses of intraspecific variation in skull morphology among different karyotypes of *C. minutus* were performed using a sample of 101 specimens of this species (50 males and 52 females, Table 1) collected in the northeastern coastal plains of the states of Rio Grande do Sul and Santa Catarina, Brazil (Figure 2). Skulls and skins of these animals are deposited in the collection of the Department of Genetics, Federal University of Rio Grande do Sul. To investigate morphological variation among karyotypes 2n = 50, 49, 48, 47, 46a, 45 and 42, the following 20 measurements described by Langguth and Abella (1970) were used: total length of skull, condylobasal length, nasal length, nasal width, zygomatic breadth, bimeatal breadth, interorbital breadth, mastoid breadth, rostral breadth, pre-orbital foramen length, frontal length, condylopremolar length, diastema length, braincase breadth, length of maxillary toothrow, palatal length, IV pre-molar length, auditory bulla width, auditory bulla length, mandibular width, and palatal length.

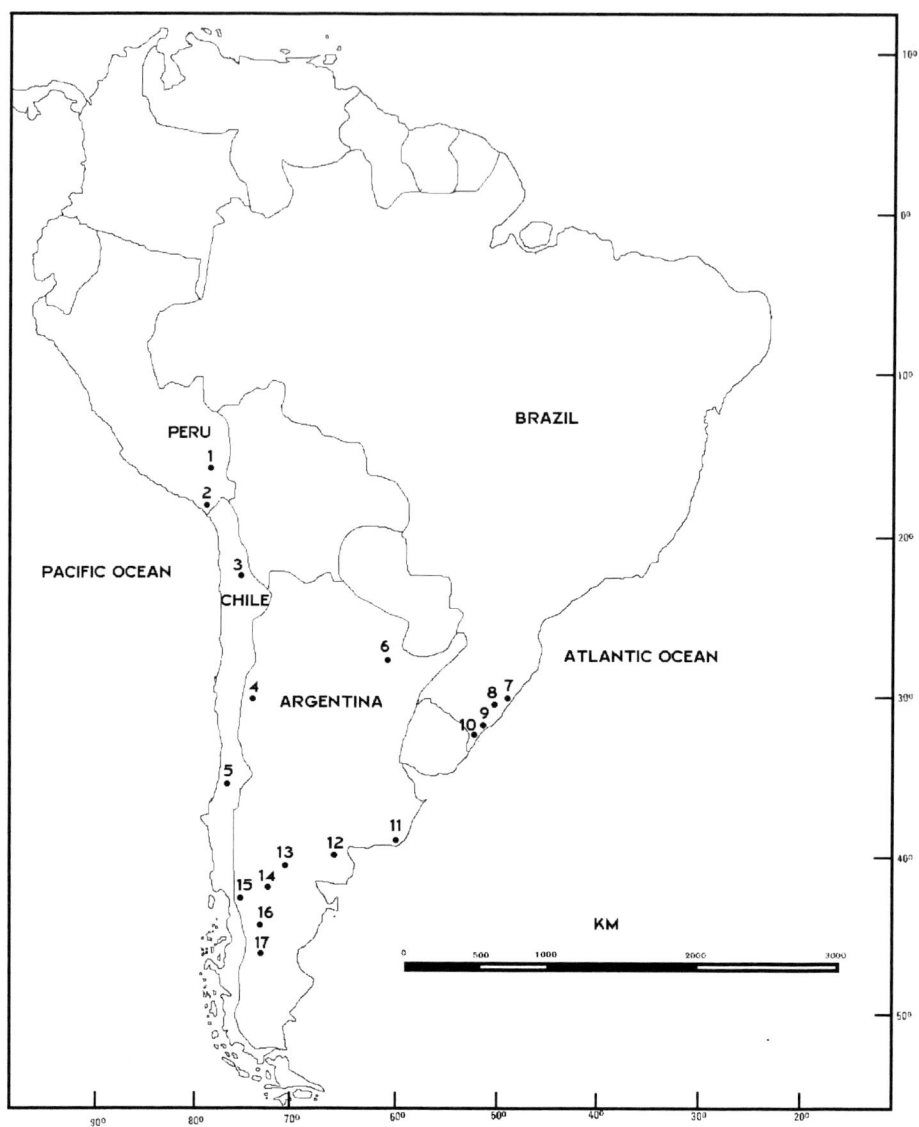

Figure 1. Collecting localities in South America: (1) Puno, (2) Tacna, (3) Calama, (4) Tumuyan, (5) Talca, (6) Brenas Chaco, (7) Santo Antônio da Patrulha, (8) Lami, (9) Pelotas, (10) Taim, (11) Necochea, (12) Rio Colorado, (13) Neuquen, (14) Rio Negro, (15) Bariloche, (16) Nahuel Huapi, and (17) Perito Moreno.

Table 1: *Ctenomys minutus* specimens examined. For each specimen, locality data, sample size (N) and diploid number (2n) are given.

Order number in the N-S transect	Locality	Collection site number	N	2n
1	Jaguaruna (Beach)	36	4	49-50
2	Jaguaruna (Field)	33	7	49-50
3	Morro dos Conventos	37	5	46a
4	Passo das Torres	28	3	46a
5	Torres	27	2	46a
6	Capão Novo	26	3	46a
7	Praia do Barco	24	5	46a
8	Caiera Lake	22	2	46a
9	Traíras Lake	23	5	46a
10	Weber Ranch	71	1	46a
11	Osório-Capivarí	65	6	48
12	Emboaba Lake	19	5	46a
13	Barros Lake	35	6	46a-47-48
14	Osório	31	11	46a
15	S margin of Barros Lake - 1	67	7	47-48
16	S margin of Barros Lake - 2	68	1	48
17	Estância Velha	70	1	46a
18	E Manuel Nunes Lake	78	1	46a
19	Fortaleza Lake 1	29	10	47-48
20	Passinhos	34	4	48
21	Fortaleza Lake 2	80	1	47-48
22	Palmares do Sul	30	8	48
23	Mostardas	38	6	42
24	Tavares	46	3	46b

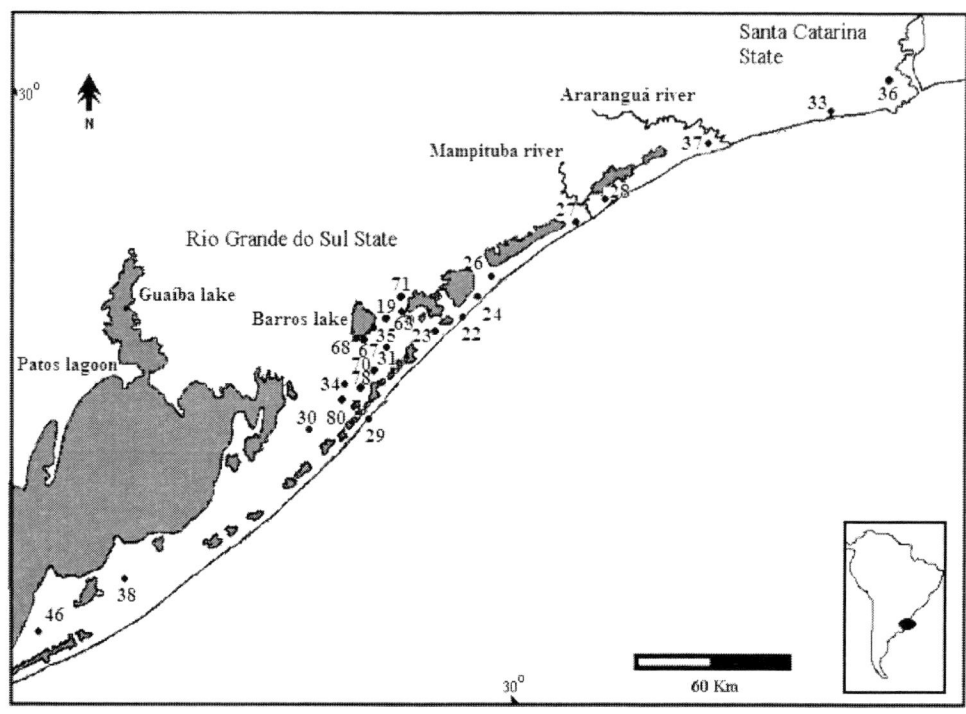

Figure 2. Collecting localities in the Coastal Plain of Rio Grande do Sul and Santa Catarina States: (46) Tavares, (38) Mostardas, (30) Palmares do Sul, (80) Fortaleza Lake-2, (34) Passinhos, (29) Fortaleza Lake-1, (78) Manuel Nunes Lake, (70) Estância Velha, (68) S Barros Lake, (67) S Barros Lake-1, (31) Osório, (35) Barros Lake-2, (19) Emboaba Lake, (65) NW Barros Lake, (71) Weber Ranch, (23) Traíras Lake, (22) Caiera Lake, (24) Praia do Barco, (26) Capão Novo, (27) Torres, (28) Passo das Torres, (37) Morro dos Conventos, (33) Jaguaruna 2, and (36) Jaguaruna 1.

For statistical analyses, measurements of skull characters were first log-transformed. Discriminant Analysis and Principal Components Analysis (PCA) were used to compare the morphology of the 15 *Ctenomys* species. PCA was also used to separate *C. minutus* morphologically by sex and karyotype. Box plots were used to compare the distribution of PCA scores across the 15 species as well as among the different populations of *C. minutus*, including the analysis of sexual dimorphism in skull morphology in this species. Box plots were also used to relate differences in PCA scores to differences in diploid numbers, geographic distribution, and type of chromosomal rearrangement. Number Cruncher Statistical System for Windows, version 2001, was used for all statistical analyses.

RESULTS

Morphological Variation in 15 Species of *Ctenomys*

The Discriminant Analysis based on skull morphology classified only 77% of the specimens correctly (Table 2). Of the 15 species studied, specimens from eight (*C. argentinus*, 2n = 44, *C. australis*, 2n = 48; *C. brunneus*, 2n = 26; *C. fulvus*, 2n = 26; *C. leucodon*, 2n = 36; *C. mendocinus*, 2n = 48; *C. peruanus*; and *C. sociabilis*, 2n = 56) were classified correctly. The classification of specimens from *C. flamarioni*, 2n = 48; *C. haigi*, 2n = 50; *C. lami*, 2n = 54, 55, 56, 57 and 58; *C. maulinus*, 2n = 26; *C. minutus*, 2n = 42, 45, 46a, 47, 48, 49 and 50; *C. opimus*, 2n = 26; and *C. torquatus*, 2n = 44 and 46 was not entirely correct. For example, 19 out of 85 specimens of *C. minutus* were classified as *C. lami*, and three *lami* were classified as *minutus* (Table 2). These results indicate that variability in skull morphology is low among these species and there is no relationship between geographic distribution and diploid number.

Table 2. Classification of 15 species of *Ctenomys* using Discriminant Analysis. The diagonal line shows the exemplars that were correctly classified to species. The sample size for each species is given in the rightmost column.

	arg	aus	bru	fla	ful	hai	lam	leu	mal	men	min	opi	per	soc	tor	Total
C. argentinus	2															2
C. astralis		23														23
C. brunneus			3													3
C. flamarioni			1	29		1	2				1				1	35
C. fulvus					3											3
C. haigi						21				3	1					25
C. lami			1	3			52				3				2	61
C. leucodon								3								3
C. maulinus			3						6	1						10
C. mendocinus										4						4
C. minutus			3	1		1	19		6	1	49				5	85
C. opimus			1	5				1				14	2			23
C. peruanus													6			6
C. sociabilis														8		8
C. torquatus	2	1								1	1				45	50

The first component of Principal Components Analysis (PC1) explained 67.59% of the variation related to the characters skull length, zygomatic breadth, condylobasal length, mastoid breadth, bimeatal breadth, rostral breadth, nasal length and palatal length. The second component (PC2) was related only to frontal length. Figure 3 shows the scores of PC1 and PC2 in females of the 15 species examined. As is evident, the distribution of taxa is the same as that obtained with the Discriminant Analysis (Figure 4) of these animals. The distribution of PC1 scores shows two classes of species, which are also found when analyzing skull size. However, the PC2 scores obtained from these analyses confirm the result obtained with the Discriminant Analysis, namely that PC2 is not informative. The box-plot distributions of PC1 and PC2 scores are provided in Figures 5 and 6. Figure 5 shows that the first axis (PC1) classified the species in two groups: large skull species (*C. australis*, 2n = 48; *C. flamarioni*, 2n = 48; *C. fulvus*, 2n = 26; *C. leucodon*, 2n = 36; *C. opimus*, 2n = 26; and *C. peruanus*) and small skull species (*C. argentinus*, 2n = 44; *C. brunneus*, 2n = 26; *C. haigi*, 2n = 50; *C. lami*, 2n=54, 55, 56, 57 and 58; *C. maulinus*, 2n = 26; *C. mendocinus*, 2n = 48; *C. minutus*, 2n = 42, 45, 46a, 47, 48, 49 and 50; *C. sociabilis*, 2n = 56 and *C. torquatus*, 2n = 44 and 46). The second axis (PC2) in Figure 6 also showed two groups. Specifically, *C. argentinus*, 2n = 44; *C. australis*, 2n = 48; *C. brunneus*, 2n = 26; *C. flamarioni*, 2n = 48 *C. lami*, 2n = 54, 55, 56, 57 and 58; *C. maulinus*, 2n = 26; *C. minutus*, 2n = 42, 45, 46a, 47, 48, 49 and 50; *C. sociabilis*, 2n = 56; and *C. torquatus*, 2n = 44 and 46 form a group with the same skull form, which is distinct from the form of *C. fulvus*, 2n = 26; *C. haigi*, 2n = 50; *C. leucodon*, 2n = 36; *C. mendocinus*, 2n = 48; *C. opimus*, 2n = 26 and *C. peruanus*. This result suggests that, as was found for PCA and Discriminant Analysis, there is no relationship between diploid number, geographic distribution and morphological species groups.

Morphological Analysis of Sexual Dimorphism in *Ctenomys minutus*

In the analyses of male and female specimens of *C. minutus*, the first component of Principal Components Analysis (PC1) explained 68.73% of the variation related to the characters skull length, zygomatic breadth, palate length, condylobasal length, bimeatal breadth, rostral breadth, diastema length, pre-orbital foramen length, condylobasal length, nasal length, nasal breadth, length of maxillary toothrow and mandibular eidth. The second component (PC2), which reflects auditory bulla breadth and length, explains 5.89% of variation. Figure 7 shows the PC1 and PC2 scores for males and females, revealing the size differences between sexes; as is evident, males are bigger than females. PC1 scores for males are statistically different from those for females ($t = 6.34$, $n = 138$, $p < 0.0001$). In contrast, no differences were found in PC2 scores for males and females ($t = 0.48$, n = 138, $p = 0.62$).

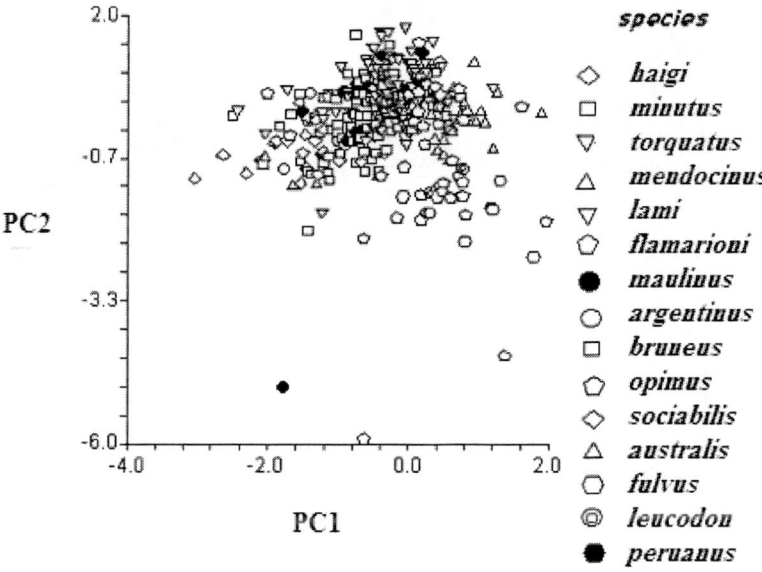

Figure 3.   Distribution of 337 PCA scores obtained from analyses of morphological variation in 15 species of *Ctenomys*.

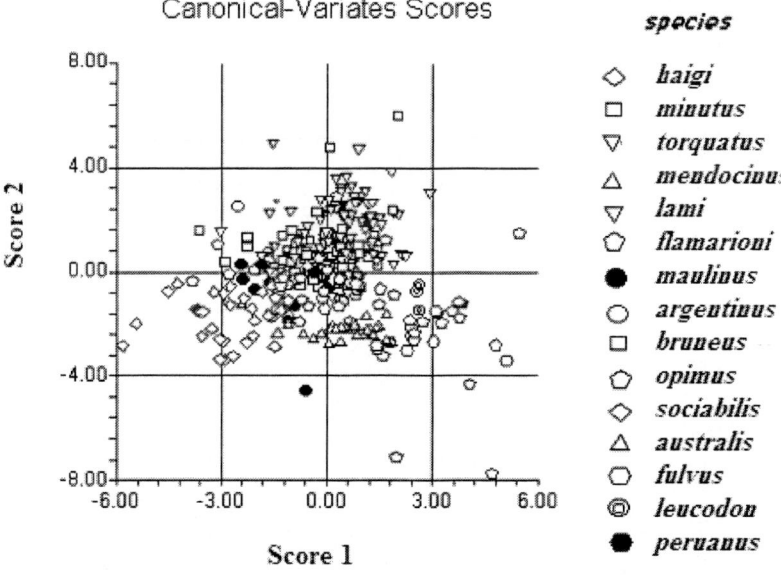

Figure 4. Distribution of 337 canonical scores obtained from analyses of morphological variation in 15 species of *Ctenomys*.

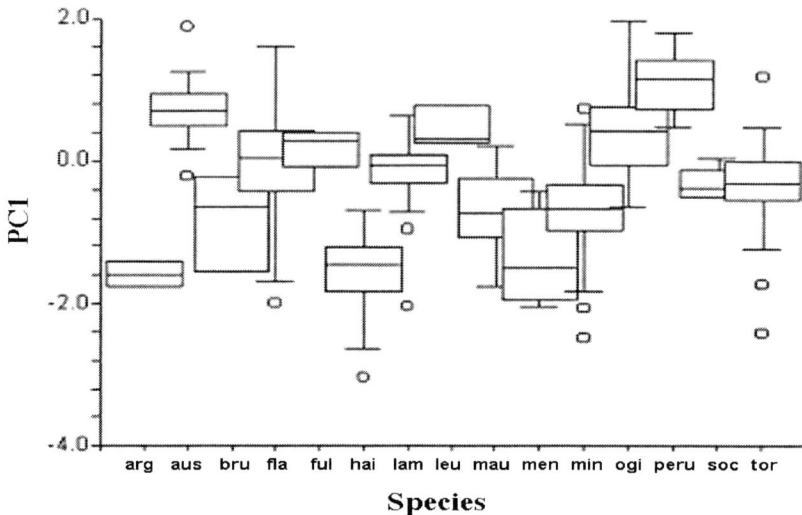

Figure 5. Box plot of PC1 scores for 15 *Ctenomys* species. For each species, the median score is shown, as are the upper and lower quartiles. Vertical lines depict minimum and maximum values falling within 1.5X quartiles of the median. Open circles denote values falling outside of 1.5X quartile values.

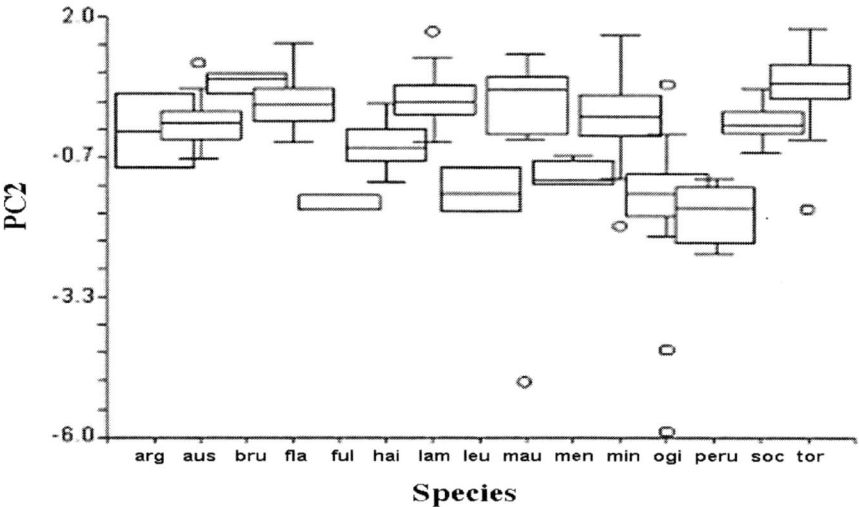

Figure 6. Box plot of PC2 scores for 15 *Ctenomys* species. For each species, the median score is shown, as are the upper and lower quartiles. Vertical lines depict minimum and maximum values falling within 1.5X quartiles of the median. Open circles denote values falling outside of 1.5X quartile values.

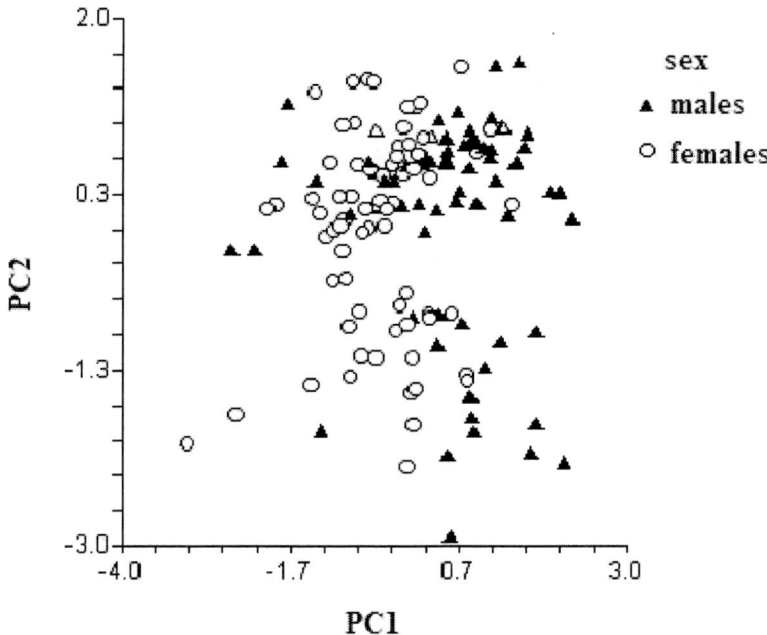

Figure 7. Distribution of 101 PCA scores for male and female *Ctenomys minutus.*

Relationships Among Skull Morphology, Geographic Distribution, Karyotype and Chromosomal Rearrangements in *Ctenomys minutus*

Because the PC2 scores in the previous analyses revealed no differences between males and females, data for both sexes were combined for subsequent analyses of skull morphology based on the variables represented by PC2. Figure 8 presents the PC1 and PC2 scores in *C. minutus*, showing that only PC2 differs between diploid numbers. These analyses reveal two groups of karyotypes, the first (a) composed of 2n = 42, 46a, 47, and 48 and the second (b) composed of 2n = 42, 46a, 46b, 49 and 50.

In Figure 9, the PC2 scores are plotted against the distance between successive collecting points. This analysis reveals that, in the North-South direction (from Jaguaruna Beach to Tavares), noticeable differences occur in skull shape. Changes in the signs of PC2 scores represent changes in skull form. From populations with 2n = 50 (locality 36) to the ones with 2n = 46a at locality 24 (Praia do Barco), the sign is negative for PC2 scores. In contrast, the sign changes to positive from population 22 (also 2n = 46a) to population 30 (2n = 48), although populations 38 and 46 (2n = 42 and 46b) again show negative scores (Figures 2 and 9).

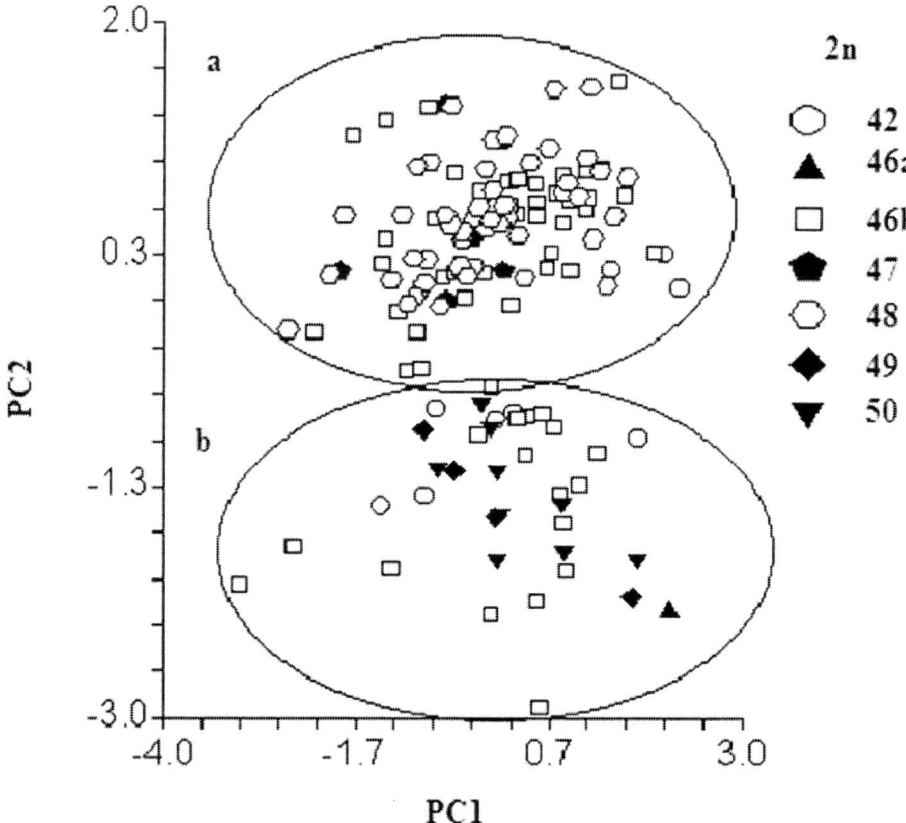

Figure 8. Distribution of 101 PCA scores for seven different karyotypes found in *Ctenomys minutus*.

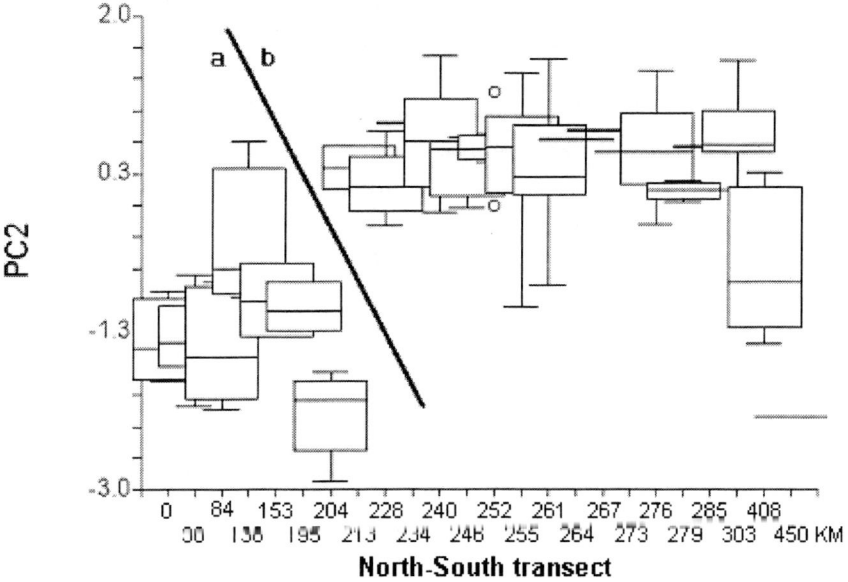

Figure 9. Box plot of PC2 scores for *Ctenomys minutus* sampled along a transect extending from Jaguaruna (O km) to Tavares (450 km). Vertical lines depict minimum and maximum values falling within 1.5X quartiles of the median. Open circles denote values falling outside of 1.5X quartile values.

Analysis of Skull Morphology and Chromosomal Rearrangements in *Ctenomys minutus*

Table 3 summarizes the chromosomal rearrangements identified by Freitas (1997) in *Ctenomys minutus*. These rearrangements were examined in relation to skull characteristics using the PC2 scores from the preceding analyses. As seen in Figure 10, the cytotypes sharing a metacentric chromosome formed by arms 20 and 17 (2n = 42, 46a, 46b, 47 and 48) show PC2 scores different from those for 2n = 50 and 49 (F = 14.93; $p < 0.0001$). The submetacentric chromosome 2 is present in 2n = 42, 46a, 46b, 49 and 50. The PC2 scores for these karyotypes are different from those for 2n = 48 and 47 (Figure 11), which are characterized by a fission of chromosome pair 2 (F = 8.29; $p < 0.001$). PC2 scores for cytotypes with chromosomes 19 and 23 (2n = 50 and 49) are similar (Figure 12), but are different from those of 2n = 42, 46a, 46b, 47 and 48 (F = 37.33; $p < 0.00001$).

PC2 scores also distinguish karyotypes with metacentric chromosomes formed by arms 16, 22 and 24 (2n = 50, 49, 48, 47 and 46a, Figure 13). The scores for these animals differ from those for animals with karyotypes 2n = 42 and 46b. The latter are characterized by a large submetacentric chromosome that results from an in tandem and centric fusion of arms 16, 22 and 24 ($F = 6.29$; $p < 0.01$).

Table 3. Collecting sites, diploid number (2n) and chromosome type for each chromosome pair involved in the polymorphism in *Ctenomys minutus*. Homozygotes for metacentric chromosomes are denoted as (+), homozygotes for acrocentric chromosomes are denoted as (-), and heterozygotes are denoted as (+/-)

Collection site number	2n	Pairs 20 and 17	Pair 2	Pairs 19 and 23	Pairs 16, 22 and 24
36	49, 50	-	+	-	-
33	49, 50	-	+	-	-
37	46a	+	+	+	-
28	46a	+	+	+	-
27	46a	+	+	+	-
26	46a	+	+	+	-
24	46a	+	+	+	-
22	46a	+	+	+	-
23	46a	+	+	+	-
71	46a	+	+	+	-
65	48	+	-	+	-
19	46a	+	+	+	-
35	46a, 47, 48	+	+/-	+	-
31	46a	+	+	+	-
67	47, 48	+	+/-	+	-
68	48	+	-	+	-
70	46a	+	+	+	-
78	46a	+	+	+	-
29	47, 48	+	+/-	+	-
34	48	+	-	+	-
80	47, 48	+	+/-	+	-
30	48	+	-	+	-
38	42	+	+	+	+
46	46b	+	+	+	+

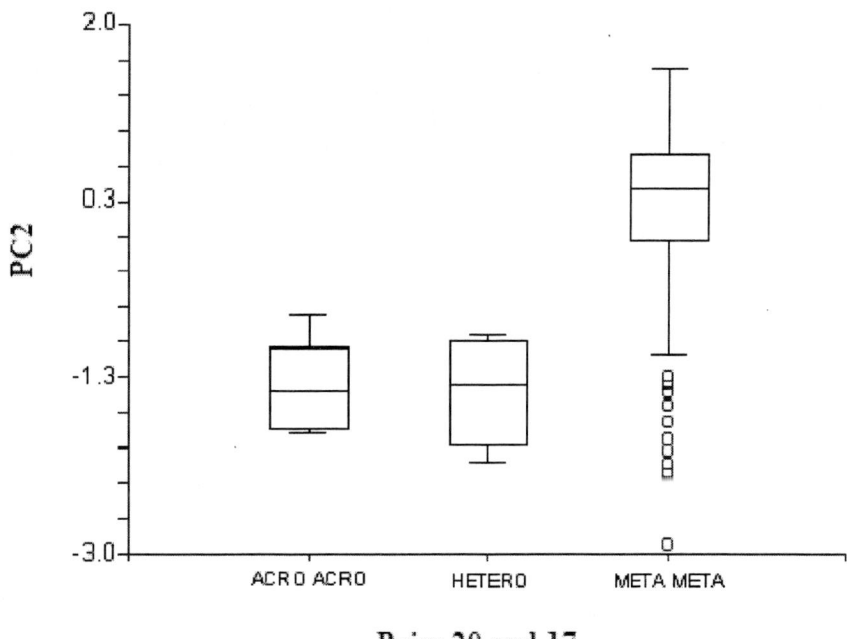

Pairs 20 and 17

Figure 10. Box plot of PC2 scores for three chromosomal forms of *C. minutus* that differ with respect to chromosomes 20 and 17.   Acro = acrocentric, meta = metacentric, and hetero = heterozygous with respect to these chromosomal morphologies.   Vertical lines depict minimum and maximum values falling within 1.5X quartiles of the median.   Open circles denote values falling outside of 1.5X quartile values.

DISCUSSION

Interspecific Analyses

The Discriminant Analysis based on morphology correctly classified only eight species.   At least some specimens from the other seven species considered were assigned to other taxa.   For example, of the 35 specimens of *C. flamarioni* examined, one was classified as *brunneus*, one as *haigi*, two as *lami*, one as *minutus*, and one as *torquatus*. Similarly, of 85 specimens of *C. minutus* considered, three were classified as *brunneus*, one as *flamarioni*, one as *haigi*, 19 as *lami*, six as *maulinus*, one as *mendocinus*, and five as *torquatus*. Finally, of the 64 specimens of *C. lami* examined, one was classified as *brunneus*, three as *flamarioni*, three as *minutus* and five as *torquatus*. This is despite work by Freitas (2001) indicating that *C. lami* and *C. minutus* differ significantly with regard to karyotype, with *minutus* including 2n = 42, 46a, 46b, 47, 48, 49 and 50 (Freitas, 1997) while *lami* is characterized by 2n = 54, 55, 56, 57 and 58. These species occur in close proximity to each other and *C. minutus* appears to have arisen from *lami* via chromosomal speciation (Freitas, in prep), although skull morphology (based on Langguth and Abella, 1970) does not differ between these species.

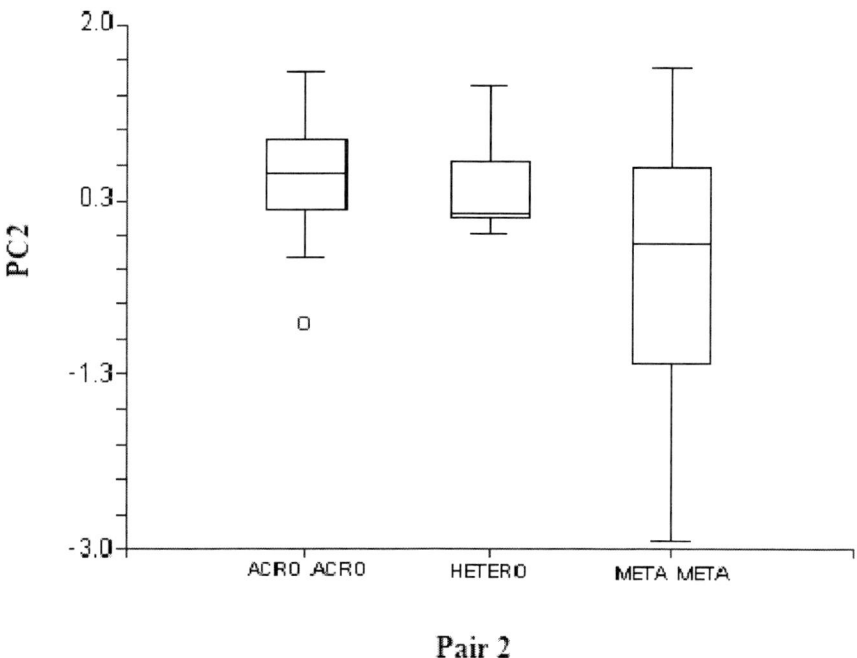

Pair 2

Figure 11. Box plot of PC2 scores for three chromosomal forms of *C. minutus* that differ with respect to chromosome 2. Acro = acrocentric, meta = metacentric, and hetero = heterozygous with respect to these chromosomal morphologies. Vertical lines depict minimum and maximum values falling within 1.5X quartiles of the median. Open circles denote values falling outside of 1.5X quartile values.

The distribution of both the canonical and PC1 scores (size) obtained from the interspecific analysis suggest that these taxa cannot be divided into two groups based on overall skull size. The large skull species identified are *australis* (2n = 48), *flamarioni* (2n = 48), *fulvus* (2n = 26), *leucodon* (2n = 36), *opimus* (2n = 26), and *peruanus*. In this group, only *flamarioni* and *australis* are related and belong to the mendocinus group (Massarini et al., 1991), displaying the same karyotype of 2n = 48 (Massarini et al., 1991; Freitas, 1994; and Massarini and Freitas, in prep.), and same sperm form (Freitas, 1995). Two other species, *fulvus* and *opimus,* also share the same karyotype of 2n=26 (Gallardo, 1979).

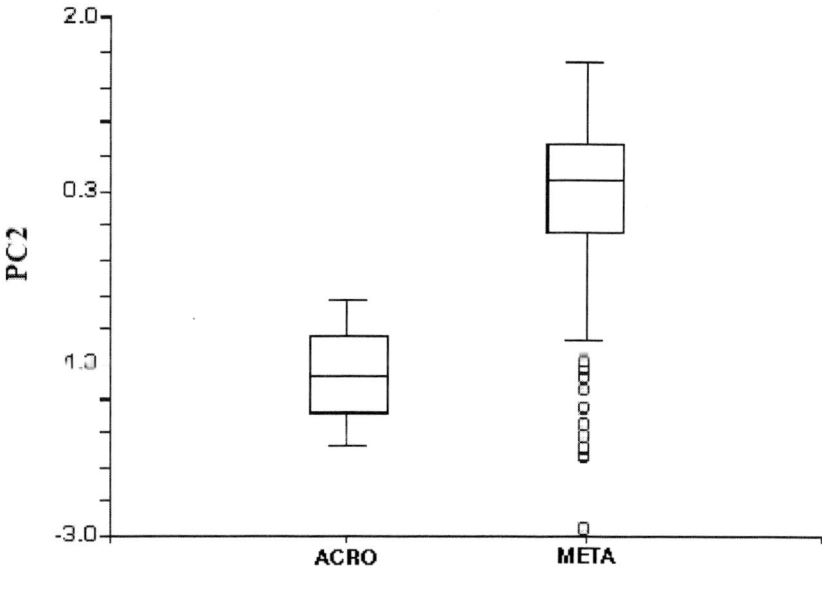

Pairs 19 and 23

Figure 12. Box plot of PC2 scores for two chromosomal forms of *C. minutus* that differ with respect to chromosomes 19 and 23. Acro = acrocentric, meta = metacentric, and hetero = heterozygous with respect to these chromosomal morphologies. Vertical lines depict minimum and maximum values falling within 1.5X quartiles of the median. Open circles denote values falling outside of 1.5X quartile values.

The small skull group of species identified by this analysis is composed of *argentinus* (2n = 44), *brunneus* (2n = 26), *haigi* (2n = 50), *lami* (2n = 54, 55, 56, 57, and 58), *maulinus* (2n = 26), *mendocinus* (2n = 48), *minutus* (2n = 42, 46a, 46b, 47, 48, 49 and 50), *sociabilis* (2n = 56) and *torquatus* (2n = 44 and 46). In this group, skull characteristics are not related to either karyotype or geographic distribution.

The second canonical axis (score 2 of Discriminant Analysis) divides the species examined into the following three groups: (1) *argentinus, haigi, minutus* and *sociabilis,* (2) *australis, fulvus, leucodon, opimus,* and *torquatus,* and (3) *brunneus, flamarioni, lami, maulinus, mendocinus* and *peruanus.* The latter group displays no consistent morphological patterns among its members. Again, these groups of species show no relationship with geographic distribution or diploid numbers. The results of the interspecific Principal Components Analysis are similar to those of the Discriminant Analysis. While PC1 scores divided the species into three groups, PC2 scores were not informative. Collectively, these results indicate that the *Ctenomys* species considered here are characterized by the same general skull form and that few characters determine skull variability among species, with overall size being the most informative aspect of skull structure.

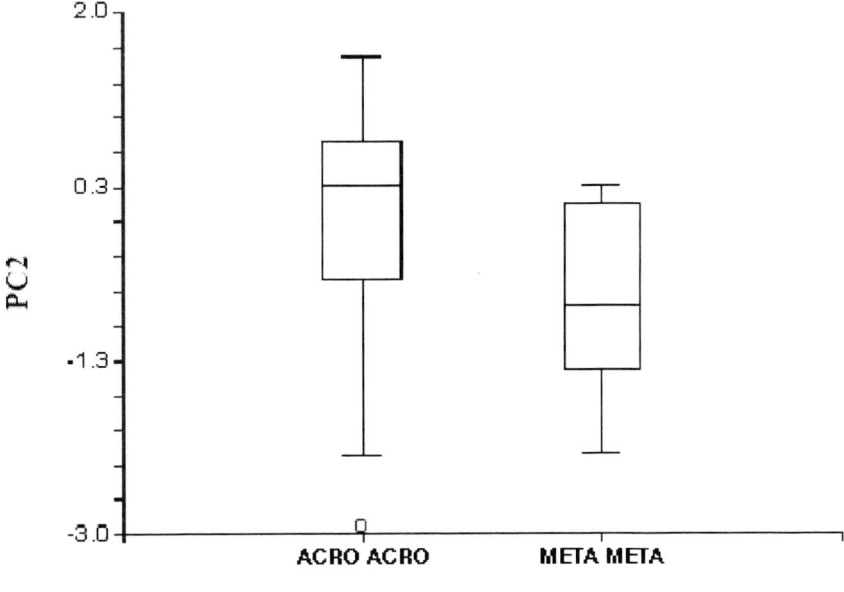

Pairs 16, 22 and 24

Figure 13. Box plot of PC2 scores of two chromosomal forms of *C. minutus* that differ with respect to chromosomes 16, 22 and 24. Acro = acrocentric, meta = metacentric, and hetero = heterozygous with respect to these chromosomal morphologies. Vertical lines depict minimum and maximum values falling within 1.5X quartiles of the median. Open circles denote values falling outside of 1.5X quartile values.

These data suggest that chromosome evolution in *Ctenomys* was very quick, allowing little opportunity for the form and size of the skull to change. Other morphological traits are also largely invariant among members of this genus. For example, all species of *Ctenomys* present the same molar morphology. As a result, other characters such as penile structure (Balbontin et al., 1996), chromosomes (Ortells, 1995), mtDNA (Lessa and Cook, 1998; Cook and Lessa, 1998; Mascheretti et al., 2000), and spermatozoid structure (Feito and Gallardo, 1982; Vitulo et al., 1988; Freitas, 1995) are more useful than skull morphology for distinguishing among species.

Intraspecific Variation in *Ctenomys minutus*

In contrast to the results of interspecific comparisons, analyses of intraspecific variation in *C. minutus* revealed that among population differences in skull morphology are related to sex and karyotype. Specifically, skull size is related to gender and skull form is related to karyotype.

Sexual dimorphism is common in many species of fossorial rodents. Reports of sexual dimorphism in the pelvic structure of *Ctenomys pearsoni* and *C. rionegrensis* (Ubilla and Altuna, 1987) are consistent with this statement. As this study indicates, sexual dimorphism is also evident in the skulls of *C. minutus*, with males clearly being distinct from females.

Skull morphology also varies with respect to geography and karyotype in this species. When PC2 scores are analyzed in relation to chromosome number, differences in skull morphology are evident among diploid numbers. Based on skull characteristics, the following two groups of karyotypes were detected: (1) a group occurring from the south to almost the center of the distribution that includes 2n = 42, 46a, 47 and 48, and (2) a group occurring from the center to the north of the distribution that includes 2n = 42, 46b, 49 and 50. This is more evident from the transect of collecting points, which reveals differences in skull shape when moving from south to north (from Tavares to Jaguaruna Beach).

Two types of skull differences were found. PC1 scores (representing skull size) revealed differences only on the basis of sex. However, the differences in PC2 scores (representing a modification in skull form) were related to diploid number, geographic distribution, and chromosomal rearrangements. These relationships were found within *C. minutus* only, and not among the different species examined. This is unusual because one would expect differences among species to be more extreme than differences within species. In *C. minutus*, skull differences were evident within and among diploid numbers: 2n = 46a included two types of populations, one with positive PC2 scores and the other with negative scores. Gava and Freitas (2002) reported a chromosome hybrid zone (HZ) between 2n = 46a and 48. For the same hybrid zone, Marinho and Freitas (2000) found differences in skull morphology between these karyotypes. In the present work, differences were also found in a 2n = 46a population beyond the location of the hybrid zone. The population with 2n = 46 in the hybrid zone has skull characteristics of 2n = 48, suggesting an introgression event from 2n = 48 to 2n = 46a. The differences in skull morphology found within 2n = 46a suggest the existence of an historical geographic barrier in this area that may have separated 2n = 46a populations into two stocks with morphological skull differences.

Chromosomal rearrangements play an important role in the evolution of species. *Ctenomys* has almost one distinct karyotype per species. Moreover, this genus contains many species with marked intraspecific chromosomal variation. In this study, we used comparisons of chromosomal rearrangements and PCA scores representing skull size and shape to examine the relationship between karyotypic and morphological variation in *C. minutus*. The chromosomal rearrangements identified by Freitas (1997) were also examined in relation to PCA scores. There were differences in the form of skull (PC2) related to the different chromosome arrangements found in 2n = 42, 46a, 46b, 47, 48, 49 and 50.

Chromosome pair 2 is found in three forms: homozygous metacentric, homozygous acrocentric, and heterozygous for these conditions. The same differences in PC2 scores are found in both sexes. Animals with 2n = 50, 46a and 42 are different from 2n = 47 and 48, indicating differential selection among chromosomal forms. An important fact is that 2n = 48 and 47 have the same skull form, this being found by Marinho and Freitas (2000) in the hybrid zone. Karyotypes 2n = 49 and 50 also share the same skull form.

The same result was found when the fusion 20/17 was analyzed, although in this case there is no hybrid zone. The Araranguá River divides the geographic distribution of *C. minutus*; animals from populations to the north of Araranguá river (2n = 49 and 50) have a different skull form than animals from south of the river (2n = 46a, 47, 48 and 42). The same pattern is observed for the fusion 23/19, in which exemplars with cytotypes 2n = 49 and 50 are morphologically different from those with 2n = 46a, 47, 48 and 42.

SUMMARY

This report showed two contrasting results related to skull morphology and species chromosomal characterizations. First, among species, there was no relationship between the form or size of any skull characters examined using the measurements described by Langguth and Abella (1970). Chromosome number, geographic distribution, sperm type, and other characteristics did not covary with the canonical or PCA scores of the 15 species considered. Second, within *C. minutus,* variation in skull morphology was evident among the seven diploid numbers examined. This suggests that the characteristics identified by Langguth and Abella (1970) show differences within species only. This was also found in two diploid numbers of *C. torquatus* (Freitas and Lessa, 1984), among different diploid numbers in *C. lami* (Freitas, 1990), and in the hybrid zone in *C. minutus* (Marinho and Freitas, 2000).

While there were differences in skull morphology in *C. minutus,* there was no direct relationship between karyotype and skull morphology. These characters are independent, although both vary with geographic distribution. The karyotypic variation found in *C. minutus* has likely resulted from reproductive isolation. As a consequence, gene flow among karyotypically distinct populations is low, leading to the evolution of differences in skull morphology that appear to be correlated with chromosomal rearrangements. A similar relationship has been observed for microsatellite analyses of this species, which reveal a concordance between karyotype, geographic distribution, and microsatellite allele frequencies (Gava and Freitas, in prep.).

ACKNOWLEDGMENTS

Special thanks are due to Jim Patton and the Museum of Vertebrate Zoology, University of California, Berkeley, for allowing measurements and analyses of *Ctenomys* specimens. I am also grateful to Eileen Lacey for the careful revision of this manuscript. Thanks are also due to the Centro de Estudos Costeiros Limnológicos e Marinhos (UFRGS). This research has been sponsored by Conselho Nacional de Desenvolvimento Cientítico e Tecnológico (CNPq), Financiadora de Estudos e Projetos (FINEP), and FAPERGS (Fundação de Amparo à Pesquisa do Rio Grande do Sul).

LITERATURE CITED

Balbontin, J., S. Reig, and S. Moreno
    1996   Evolutionary relationships in *Ctenomys* (Rodentia: Octodontidae) from Argentina, based on penis morphology. Acta Theriol. 41:237-253.

Cook, J. A., and E. P. Lessa
    1998   Are rates of diversification in subterranean South American Tuco-Tuco (Genus *Ctenomys*, Rodentia, Ocotodontidae) unusually high? Evolution 52: 1521-1527.

Feito, R., and M. H. Gallardo
    1982   Sperm morphology of chilean species of *Ctenomys* (Octodontidae). J. Mammal. 63:658-661.

Freitas, T. R. O.
    1990   Estudos citogenéticos e craniométricos em três espécies do gênero *Ctenomys*. PhD Thesis - Universidade Federal do Rio Grande do Sul, Porto Alegre.

Freitas, T. R. O.
    1994   Geographical variation of heterochromatin in *Ctenomys flamarioni* (Rodentia-Octodontidae) and its cytogenetic relationships with other species of the genus. Cyto. Cell Gen. 67:193--198.

    1995   Geographical distribution of sperm forms in the genus *Ctenomys* (Rodentia-Octodontidae). Rev. Brasileira Genét. 18:43-46.

    1997   Chromosome polymorphism in *Ctenomys minutus* (Rodentia-Octodontidae). Brazil. J. Gen. 20:1—7

    2001   Tuco-tucos (Rodentia, Octodontidae) in Southern Brazil: *Ctenomys lami* spec. nov separated from *C. minutus*. Studies on Neotropical Fauna and Environment 36:1-8.

Freitas, T. R. O., and E. P. Lessa
    1984   Cytogenetics and morphology of *Ctenomys torquatus* (Rodentia-Octodontidae). J. Mammal. 65:637-642.

Gallardo, M. H.
    1979   Las especies chilenas de *Ctenomys* (Rodentia, Octodontidae). Estabilidade cariotípica. Archivos de Biologia y Medicina Experimental, 12: 71-82.

Garcia, L.; M. Ponsá; J. Egozcue, and M. Garcia
    2000   Cytogenetic variation in *Ctenomys perrensis* (Rodentia, Octodontidae). Biol. J. Linn. Soc. 71:615-624.

Gava, A., and T. R. O. Freitas
    2002   Characterization of a hybrid zone between chromosomally divergent populations of *Ctenomys minutus* (Rodentia: Octodontidae). J. Mammal. 83:843-851.

Kiblisky, P., N. Brum-Zorrilla, G. Perez, and F. A. Saez
    1977   Variabilidade cromossómica entre diversas poblaciones uruguayas del roedor cavador del género *Ctenomys* (Rodentia-Octodontidae). Mendeliana 2:85-93.

Langguth, A., and A. Abella
  1970   Las especies uruguayas de género *Ctenomys* (Rodentia-Octodontidae). Comunicaciones Zoológicas del Museo de Historia Natural de Montevideo. 129:1-20.

Lessa, E. P., and J. A. Cook
  1998   The molecular phylogenetics of tuco-tucos (genus *Ctenomys,* Rodentia: Octodontidae) suggests an early burst of speciation. Mol. Phylog. Evol. 9:88--99.

Marinho, J. R., and T. R. O. Freitas
  2000   Intraspecific craniometric variation in chromosomal hybrid zone of *Ctenomys minutus* (Rodentia, Hystricognathi). Z. Saugertierkunde 65:226-231.

Massarini, A. I., F. J. Dyzenchauz, and S. I. Tiranti
  1998   Geographic variation of chromosomal polymorphism in nine populations of *Ctenomys azarae,* Tuco-tucos of the *Ctenomys mendocinus* group (Rodentia: Octodontidae). Hereditas 128:207--211.

Massarini, A. I., M. A. Barros, M. O. Ortells, and O. A. Reig
  1991   Evolutionary biology of fossorial Ctenomyine rodents (Caviomorph: Octodontidae). I. Chromosomal polymorphism and small karyotypic differentiation in Central Argentinian populations of tuco-tucos. Genetica 83:131-144.

Mascheretti, S., P. M. Mirol, M. D. Giménez, C. J. Bidau, J. R. Contreras, and J. B. Searle
  2000   Phylogenetics of the speciose and chromosomally variable rodent genus *Ctenomys* (Ctenomyidae, Octodontoidea) based on mitochondrial cytochrome b sequences. Biol. J. Linn. Soc. 70:361-376.

Novello, A. F., and E. P. Lessa
  1986   G-banded homology in two karyomorphs of the *Ctenomys pearsoni* complex (Rodentia: Octodontidae) of neotropical fossorial rodents. Z. Säugetierkunde 51:378-380.

Ortells M. O., J. R. Contreras, and O. A. Reig
  1990   New *Ctenomys* karyotypes (Rodentia, Octodontidae) from north-eastern Argentina and from Paraguay confirm the extreme chromosomal multiformity of the genus. Genetica 82:189-291.

Ortells, M. O.
  1995   Phylogenetic analysis of G-banded karyotypes among the South American subterranean rodents of the genus *Ctenomys* (Caviomorpha: Octodontidae), with special reference to chromosomal evolution and speciation. Biol. J. Linn. Soc. 54:43--70.

Reig, O.A., C. Busch, M. O. Ortells, and J. R. Contreras
  1990   An overview of evolution, systematics, population biology and speciation in *Ctenomys*. Pp 71-96 in Biology of Subterraneal Mammals at the Organismal and Molecular Levels. (E. Nevo and O. A. Reig, eds.). Alan R. Liss, Inc., New York.

Verzi, D. H.
  1999   The dental evidence on the differenciation of the Ctenomyine rodents (Caviomorpha, Octodontidae, Ctenomyinae). Acta Theriol. 44:263-282.

Vitulo, A. D., E. R. S. Roldan, and M. S. Merani
  1988   On the morphology of spermatozoa of tuco-tucos *Ctenomys* (Rodentia): new data and its implications for the evolution of the genus. J. Zool., London 215:675-683.

Ubilla, M., and C. A. Altuna
  1987   Morfología diferencial y dimorfismo sexual en la pelvis de *Ctenomys pearsoni* Lessa and Langguth, 1983 y *C. rionegrensis* Langguth and Abella, 1970 (Rodentia-Octodontidae). Iheringia 66-33-42.

Woods, C. A.
  1993   Suborder Hystricognathi. Pp. 771-806 in Mammals Species of the World (D. E. Wilson and D. M.Reeder, eds). Smithsonian Institution Press, Washington, D. C.

Dynamics of Genetic Differentiation in the Río Negro Tuco-tuco (*Ctenomys rionegrensis*) at the Local and Geographical Scales

Enrique P. Lessa, Gabriela Wlasiuk, and John Carlos Garza

We examined the dynamics of geographic, and to a lesser extent, local genetic differentiation in *Ctenomys rionegrensis* (the Río Negro tuco-tuco), which exhibits a remarkable variation in pelage color. Microsatellite allele frequencies and cytochrome b gene sequences were obtained and compared with previously published allozyme data to (1) examine genetic population structure, and (2) assess the likelihood of the local fixation of melanism by random genetic drift. At the geographic level, we present strong evidence for a historical range expansion by this species, as indicated by the absence of migration-drift equilibrium in all datasets and by coalescent-based estimates of demographic expansion for the cytochrome b data. Because of the former, genetic estimates of gene flow probably overestimate actual levels of gene flow. We suggest that values below $Nm = 1$ obtained from microsatellites and mitochondrial DNA are the most reliable, implying that, after expansion, the species has differentiated in near isolation, with genetic drift playing a primary role in genetic differentiation. In particular, microsatellite data suggest that the species is strongly structured geographically, with subpopulations constituting distinct genetic entities. At the local level, positive F_{IS} values in some samples indicate that *C. rionegrensis* populations are reproductively subdivided, probably pointing to some kind of assortative mating that might show an association with pelage color. Furthermore, the pattern of genetic connections among individuals is intricate and involves networks of relatives, implying that philopatry could be common in this species.

COLOR POLYMORPHISM AND GENETIC STRUCTURE IN *CTENOMYS RIONEGRENSIS*

The roles of gene flow, genetic drift and natural selection in shaping natural populations have been the subject of numerous studies, including influential research on geographic subdivision and speciation in pocket gophers (e.g., Patton and Smith, 1990) and other subterranean mammals (Nevo, 1999; Steinberg and Patton, 2000). Among subterranean mammals, *Ctenomys* is likely one of the most extreme cases of high chromosomal diversity (Ortells et al., 1990; Freitas and Lessa, 1984; Slamovits et al., 2001) and rapid speciation (Reig et al., 1990; Lessa and Cook, 1998; Cook and Lessa, 1998).

Here, we examine the dynamics of local and geographic differentiation in the Río Negro tuco-tuco (*Ctenomys rionegrensis*), using a combination of published and new data. We focus on the Río Negro tuco-tuco because of an intriguing pelage color polymorphism within its rather restricted distribution in Uruguay. The description of *C. rionegrensis* (originally as a subspecies of *C. minutus*), Langguth and Abella (1970a, b) remarked that the species was restricted to a rather narrow habitat of sandy soils but showed two distinct pelage colors: a completely melanic form and a light-colored, agouti form. Subsequent geographical sampling (Altuna et al., 1985; D' Elía et al., 1998) has revealed (1) a third pelage type, with a dark stripe along the mid-dorsal axis and grayish color fading laterally towards the ventral surface of the skin and (2) that some populations are fixed for each of the three pelage colors, while others are polymorphic with respect to pelage type (Figure 1). As in other subterranean rodents, pelage color in tuco-tucos typically matches the color of the background (e.g., Freitas and Lessa, 1984), presumably because of selective pressures imposed by predators, which are better able to detect prey against contrasting soil colors (Patton and Smith, 1990; Nevo, 1999; and references therein). In the Río Negro tuco-tuco, however, melanic individuals clearly fail to match background coloration, yet melanism is fixed in at least two populations and is present in a few others (Figure 1).

Random genetic drift can account for the maintenance of neutral genetic polymorphisms and can also lead to the local fixation of alleles. Langguth and Abella (1970b) speculated that pelage color polymorphism might be effectively neutral in *C. rionegrensis* and that local fixation of melanism could be the result of small effective population size in local subpopulations. Critical to this hypothesis is the assumption that gene flow among subpopulations is sufficiently low to allow for random local fixation of alleles (Slatkin, 1987). An allozyme study directed at testing this asssumption (D'Elía et al., 1998) yielded genetic estimates of gene flow that were too high (pairwise estimates of Nm ranging from 1.37 to 29.57) to allow for local fixation by drift alone. However, the same study indicated that the population did not match the equilibrium expectations of a model of isolation by distance. Following Slatkin (1993), D'Elía et al. (1998) suggested that the current distribution of the Río Negro tuco-tuco in Uruguay had resulted from a recent range expansion and that, consequently, genetic estimates of gene flow should be higher than current levels.

Here, we begin by providing a more detailed assessment of the geographic pattern of genetic variation in the Río Negro tuco-tuco. Specifically, we reexamine the hypothesis of a recent range expansion and obtain genetic estimates of gene flow using mitochondrial DNA sequences and microsatellite allele frequencies. We then use these microsatellite data to describe the genetic structure of a local population that is polymorphic for pelage color. From these analyses, we develop a scenario for the evolutionary dynamics of this species and suggest directions for future work.

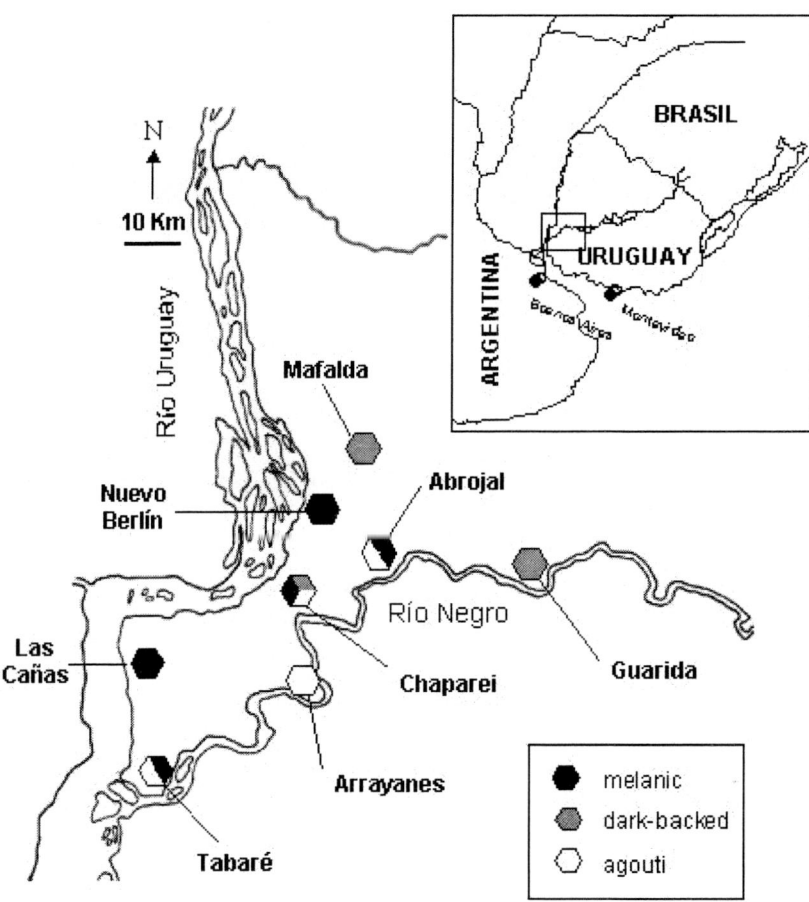

Figure 1. Geographic distribution of *Ctenomys rionegrensis* in Uruguay. Reprinted with permission from *Evolution* (Allen Press, Inc.).

MATERIALS AND METHODS

Samples

A total of 151 individuals from eight localities were analyzed to assess patterns of geographic variation and estimate levels of gene flow among subpopulations using microsatellites and mitochondrial cytochrome b (cyt b) gene sequences (Figure 1). Microsatellite genotypes from an additional 20 individuals (to complete a local sample of 57 individuals) were used to examine the local geographic structure of the La Tabaré population, which is polymorphic for pelage color.

DNA Extraction, PCR Amplification, and Genotyping

Liver tissue for DNA extraction was collected from freshly sacrificed animals. Voucher specimens were prepared and deposited in the collection of the Laboratorio de Evolución, Facultad de Ciencias, Universidad de la República, Montevideo, Uruguay. Genomic DNA was isolated following a protocol modified from Miller et al. (1988).

Two overlapping DNA fragments that, together, covered the entire 1140 bp of the cyt b gene were amplified by polymerase chain reaction (PCR) using the following primers: MVZ 05 (Smith and Patton, 1993) -TUCO 06, and TUCO 07 - TUCO 14A (Wlasiuk et al., 2003). The PCR amplifications were carried out in a reaction volume of 50 μl, using the same amplification conditions described in Lessa and Cook (1998). Details of the sequencing protocol are provided in Wlasiuk et al. (2003).

Allelic variation was examined at eleven microsatellite loci developed from *C. sociabilis* (Soci 3/4, Soci 5/6, Soci 7/8, Soci 17/18) (Lacey, 2001) and *C. haigi* (Hai 2, Hai 3, Hai 4, Hai 6, Hai 7, Hai 9, Hai 12) (Lacey et al., 1999) that were polymorphic in *Ctenomys rionegrensis*. Nine of these loci (excluding Soci 3/4 and Hai12), were used in the analysis of the sample from La Tabaré. PCR amplifications were carried out in a reaction volume of 15 μl, following the conditions in Wlasiuk et al. (2003). Products were electrophoresed through a 4% denaturing polyacrylamide gel (2000V, 1-2 hs) and visualized by Silver Nitrate staining according to Sanguinetti et al. (1994). Genotypes were scored by two independent evaluators.

Data Analysis

General Approach: The structure and dynamics of populations are reflected in their genetic makeup. At the geographical level, Wright's (1951) F_{ST} is a classical descriptor of subdivision with a known relationship to levels of gene flow (estimated as Nm, the mean number of migrants exchanged between populations per generation). Under a stable regime of gene flow, Nm values much greater than 1 effectively prevent the fixation of alternative alleles in local populations by drift. In this paper, we used analogs of F_{ST} that are better suited to sequence and microsatellite data (details below). One of the limitations of estimations of Nm from F_{ST} is that they only reflect actual levels of gene flow when there is equilibrium between migration and drift. Therefore, we used two approaches to assess equilibrium. First, we followed Slatkin (1993) and used pairwise estimates of F_{ST} analogs between populations for both sequence and microsatellite data and plotted them against geographic distances. For a model of isolation by distance, Slatkin (1993) showed that a linear inverse relationship between Nm and geographic distance on log-log plots was expected, and that the absence of such pattern might indicate recent range expansions. In turn, this would suggest that equilibrium has not been established and that current levels of gene flow would be overestimated by indirect, genetic assessments. A second way to assess equilibrium is to make use of coalescent theory, which has different expectations for the pattern of DNA polymorphism in stable and expanding populations (details below).

All the analyses outlined above were carried out separately for each locus, then averaged across loci. Multilocus microsatellite genotypes contain substantial information about population structure, both at the geographical and local levels. We made use of such information as follows: (1) at the geographical level, we used a Bayesian approach developed by Pritchard et al. (2000) to gather conclusive evidence about the distinctiveness of local populations, (2) at the local level, we used a similar approach to ask whether or not individuals with different pelage colors belonged to different subpopulations. In addition, multilocus analyses of relatedness were used to examine the pattern of genetic connections among individuals and provide preliminary assessments of philopatry. These results were supplemented by classical assessments of F_{ST} and again suitable analogs to complete an evaluation of the levels of local reproductive subdivision.

Mitochondrial Cytochrome b: Population subdivision was analyzed by assuming the Infinite Alleles Mutation Model (Kimura and Crow, 1964) and calculating γ_{ST} (Nei, 1982). Estimates of global levels of gene flow were calculated from γ_{ST} using the equation: $N_f m \approx 1/2 [(1/\gamma_{ST}) -1]$ (Wright 1951).

For isolation by distance analyses, pairwise estimates of γ_{ST} were calculated using DNASP3 (Rozas and Rozas, 1999). The log of geographic distances between pairs of populations was then plotted versus log M ($M \approx N_f m$) (Slatkin, 1993). Estimates of γ_{ST} were also used to test for isolation by distance with the Mantel permutation procedure (Mantel, 1967) using GENEPOP (version 3.2a; Raymond and Rousset, 1995). The correlation between the log of geographic distance and log M values was examined through 1000 permutations of each matrix. Relationships among cyt b haplotypes were examined by constructing a minimum spanning tree using the ARLEQUIN software package (Schneider et al., 2000).

Data were fit to a simple model that considers a single ancestral population that undergoes an instantaneous change in size (Wakeley and Hey, 1997). This model is described by three parameters, $\theta_{ancestral}$, θ_{final}, and τ (the time at which size change took place), which were estimated using the program SITES (Hey and Wakeley, 1997). Fu's F_s (Fu 1997) was also used to test for a significant excess of low frequency haplotypes, using 1000 replicates in ARLEQUIN (Schneider et al., 2000).

Microsatellites: Linkage disequilibrium between loci and deviations from Hardy-Weinberg equilibrium were examined using a Markov chain method (1000 dememorizations, 50 batches, 1000 iterations) following the algorithm of Guo and

Thompson (1992). A Bonferroni correction for multiple tests was applied to give an adjusted table-wide significance level of $\alpha = 0.05$.

Population subdivision was analyzed assuming a Stepwise Mutation Model (Ohta and Kimura, 1973) and then calculating ρ-statistics (Rousset, 1996) for the whole population. Estimates of global levels of gene flow from ρ_{ST} were calculated as $Nm \approx 1/4 \ [(1/\rho_{ST}) -1]$ (Wright, 1951) and also by the private alleles method (Slatkin, 1985), with the correction of Barton and Slatkin (1986).

Again, pairwise estimates of M based on ρ_{ST} were calculated for the isolation by distance analysis (Slatkin, 1993) and then used for the Mantel test as described above. A Mantel test was also performed to examine the correlation between pairwise estimates of gene flow from both data sets (log $M\rho_{ST}$ and log $M\gamma_{ST}$). All analyses outlined above were implemented in GENEPOP (Raymond and Rousset, 1995).

A Markov Chain Monte Carlo (MCMC) approach was used to examine the distinctiveness of subpopulations and the clustering of individual genotypes using the program STRUCTURE (Pritchard et al., 2000). First, a Bayesian test of subdivision, directed at assessing the number of subpopulations solely on the basis of the genotypes, was carried out. Second, individuals were grouped into subpopulations, again without using information concerning their origins.

At the local scale, the sample from La Tabaré was first examined for subdivision between melanic and agouti individuals using STRUCTURE (Pritchard et al., 2000), as outlined above. An assignment test was then carried out with the genotypic data from agouti and melanic individuals to characterize such presumptive subpopulations. Finally, the program RELATEDNESS (Queller and Goodnight, 1989) was used to quantify the degree of genetic similarity among individuals in La Tabaré.

RESULTS AND DISCUSSION

Assignment Tests and Local Fixation of Alleles

Multilocus genotypes contain substantial information about the ancestry of individuals. The results of our Bayesian MCMC analyses indicate that *C. rionegrensis* is highly structured into distinct, differentiated subpopulations. For example, considering a prior range of 6 to 10 subpopulations (options outside have negligible contributions to the overall likelihood), the probability that the genotypes come from 8 subpopulations is > 0.9999, whereas the corresponding probabilities for 7 or 9 subpopulations are 10^{-38} and 4.5×10^{-5}, respectively. All individuals were correctly assigned into the subpopulation in which they were captured, suggesting clear subpopulation differences.

In other words, the eight subpopulations examined have unique genetic makeups and are highly differentiated at the 11 microsatellite loci examined. Allele frequencies vary in accordance with this observation. For example, all but one locus (Soci 7/8) showed population-specific alleles. Also, local fixation of alleles was relatively common. An extreme case is Nuevo Berlín, where all 11 loci were monomorphic. Four additional subpopulations were monomorphic at one (Hai2 in El Abrojal) or two (Hai12 and Soci 3/4 in La Guarida, Hai9 and Hai12 in Las Cañas) loci. In all cases, local fixation involved alleles that were in high overall frequency across the species' distribution.

The relatively common occurrence of local fixation of microsatellite alleles contrasts with the pattern of allozyme variation reported by D' Elía et al. (1998). Nuevo Berlín does not show reduced levels of allozyme variation, and differences in allele frequencies between subpopulations are, with a few exceptions (e.g., ICD2 and AAT-2 in Los Arrayanes), relatively minor. Typically, allozyme alleles that are locally fixed are either monomorphic or are accompanied by low frequency alleles in other subpopulations. ADH is the only locus with two alleles at relatively high frequencies in all subpopulations. It is possible that, in Nuevo Berlín, natural selection has contributed to the maintenance of allozyme variation while the absence of selection on microsatellite loci has resulted in a loss of variation at these markers.

Genetic Estimates of Subdivision and Gene Flow

Our new estimates of genetic subdivision stand in sharp contrast to those obtained from allozymes by D'Elía et al. (1998). Allozyme data yielded a global value of $F_{ST} = 0.091$, whereas the analogous estimate from mtDNA cytochrome b sequences was a global $\gamma_{ST} = 0.645$, and that from microsatellites was $\rho_{ST} = 0.689$. In other words, less than 10% of the allozyme variation is partitioned among subpopulations, whereas over 60% of the variation in mtDNA and microsatellites is partitioned among subpopulations.

The corresponding differences in estimates of gene flow are $Nm = 2.49$, 0.55, and 0.45 for allozymes, mtDNA, and microsatellites, respectively. Taken at face value, allozyme data suggest that gene flow is sufficiently high to prevent the local fixation of alleles by drift alone. In contrast, mtDNA and microsatellite data yield estimates that are about five times smaller, in a range where drift can easily lead to the fixation of alternative alleles in different subpopulations.

To minimize the potential influence of positive selection and to reduce the impact of departures from the assumed microsatellite mutation model, we also generated estimates of gene flow based on private alleles (Slatkin, 1985; Barton and Slatkin, 1986; see also Hedrick, 1999). For all three datasets, these estimates are smaller than the corresponding values based on F_{ST} -like estimates. However, as seen in Figure 2a, the contrast between estimates of Nm for allozymes versus the other two datasets is evident for both methods of estimating gene flow.

Evidence of Recent Demographic Expansion

Following an approach suggested by Slatkin (1993), D' Elía et al. (1998) showed that the allozyme data were inconsistent with the relationship expected between pairwise geographic distances and corresponding estimates of gene flow under an isolation by distance model. Specifically, no significant correlation ($p > 0.05$) was found between pairwise estimates of gene flow based on allozymes and geographic distances. Our cytochrome b and microsatellite data are consistent with these findings (Figure 2b). Mantel tests show no significant correlations between either set of Nm estimates and geographic distances, or between the two sets of Nm values ($p > 0.05$). However, like global estimates of Nm, pairwise estimates of gene flow tend to be reduced as one moves from allozymes ($1.37 < Nm < 29.57$; D' Elía et al., 1998) to cytochrome b ($0.18 < Nm < 4.0$) to microsatellites ($0.07 < Nm < 2.69$). Once again, there are consistent differences in gene flow estimates among markers, regardless of the method (F_{ST} or rare alleles) or the level (global or pairwise) of analysis considered.

Using the coalescent, it is possible to examine further the possibility of a recent demographic expansion of the Río Negro tuco-tuco. We found the following evidence to support such a scenario: (1) $\theta_w = 4.232$ larger than $\pi = 2.237$, (2) negative and highly significant Fu's test ($F_s = -26.85$, $p < 0.00001$), and (3) $\theta_{ancestral} = 0$ smaller than $\theta_{final} = 11.06$ using the model of Wakeley and Hey (1997), with an estimated time of expansion of $\tau = 2.513$ (in units of N_f generations). The fit of the latter model was not good, suggesting that the history of the population differed from the assumptions of the model (namely a panmictic population that underwent a single, instantaneous change in size sometime in the past). Our data indicate that the population is not panmictic, but is instead highly subdivided. This observation suffices to account for the poor fit of the model. Thus, both historical expansion and subsequent isolation appear to have played a role in forming the genetic structure of the Río Negro tuco-tuco.

Dynamics of Geographical Differentiation in *C. rionegrensis*: A Possible Scenario

The analyses discussed above provide several concordant, and a few seemingly discordant, clues about the history of this species of tuco-tuco. Here, we attempt to integrate this information to provide a coherent historical scenario, as follows:

Phase 1: Expansion: The current range of the species in Uruguay is the result of an expansion from an even more restricted area. This is supported by the (a) failure of all three datasets to conform to a model of isolation by distance, (b) coalescent-based tests of expansion, and (c) estimates of $\theta_{ancestral} \ll \theta_{final}$. All data are consistent with a recent (in population genetic terms) expansion. Additionally, the numerous locally fixed microsatellite alleles have high overall frequencies, suggesting that they were present in the common ancestral population.

A

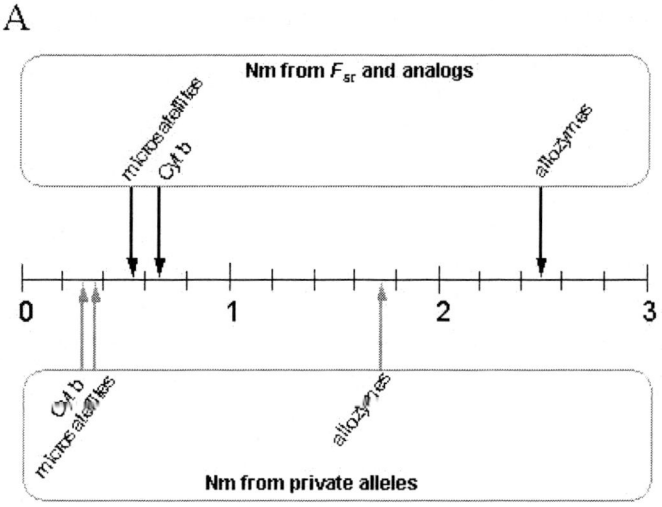

B

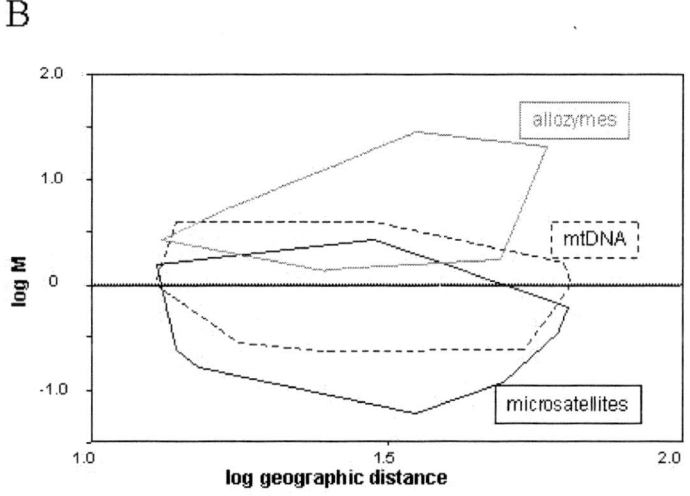

Figure 2. Genetic estimates of gene flow among subpopulations of *Ctenomys rionegrensis*. (a) Global estimates based on Fst, rst, and gst for allozymes, microsatellites, and mtDNA, respectively. (b) Pairwise estimates of M plotted against geographic distances for the three datasets.   Figure 2b modified from Wlasiuk et al. (2003).

Phase 2: Isolation: Substantial isolation is required for the different subpopulations to have acquired the unique profiles suggested by assignment tests and by the relatively numerous instances of local fixation of alleles. The greatest apparent discordance among datasets occurs among estimates of gene flow, which reflect the degree of isolation of subpopulations since their establishment. It is important to recall, however, that all of the genetic estimates of gene flow we have considered are based on an assumption of equilibrium between local drift and migration, yet we have conclusive evidence that such equilibrium has not been reached. Therefore, all of our estimates of gene flow include a substantial historical component. This is not surprising given that the proposed expansion of range constitutes historical gene flow.

Despite the apparent violation of the assumption of equilibrium, it is worth asking which estimates of gene flow may more reliably reflect current levels of genetic exchange among subpopulations. Figure 2 indicates that estimates based on private alleles, which are less dependent upon assumptions concerning strict neutrality and specific mutational models, are consistently lower than those based on F_{ST} -like measurements. An additional reason to favor the former is that, under neutrality, private alleles likely include those recently arisen by mutation, and are therefore less likely to be influenced by more ancient demographic events. A rooted allele tree may be used to distinguish ancestral and derived alleles. We have made use of this possibility with our mtDNA sequences, using sequences from two geographically isolated populations of *C. rionegrensis* in Argentina as outgroups (D' Elía et al., 1999). These analyses indicate that the most frequent haplotype is ancestral to more localized alleles (Figure 3). Finally, the potential for positive selection is reduced as one moves from allozymes (for which only some replacement changes are detected) to cytochrome b sequences (in which a sizeable fraction of the variation is silent) to microsatellites (generally assumed to be selectively neutral). A growing number of comparative analyses of these different genetic markers are uncovering similar contrasts (e.g., Dufresne et al., 2002). Thus, current levels of gene flow in *C. rionegrensis* are likely as low as those suggested by the private alleles method for both mtDNA and microsatellites, averaging $Nm \cong 0.3$, if not simply zero. Naturally, this conclusion holds for the scale of geographic sampling depicted in Figure 1, where subpopulations are separated by at least 13 km.

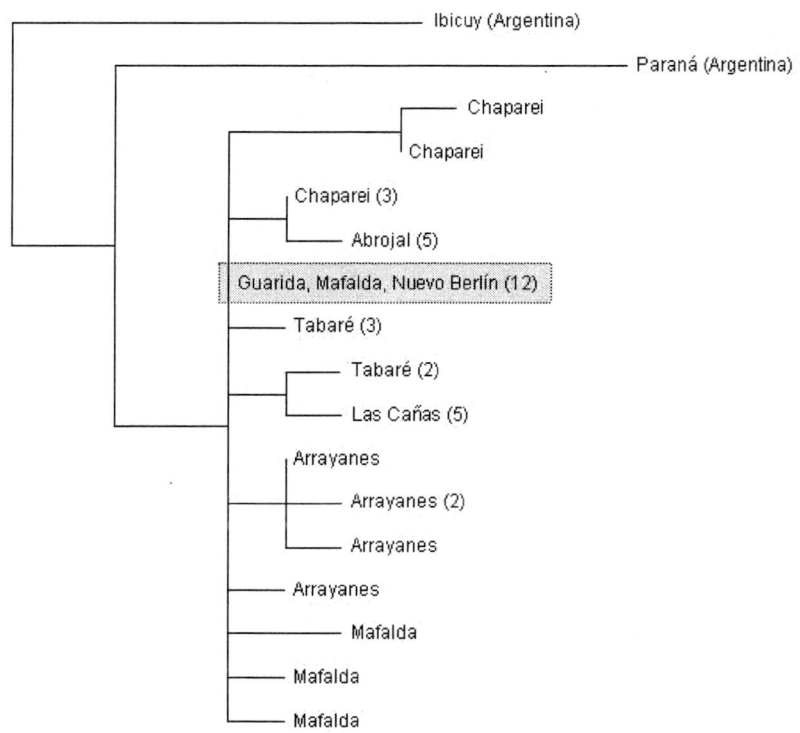

Figure 3. Maximum parsimony tree of mtDNA cytochrome b haplotypes from Uruguayan *Ctenomys rionegrensis*, rooted using two samples from Argentina. The most common haplotype (highlighted) is inferred to be basal to more localized alleles in Uruguayan subpopulations.

Local Genetic Structure

Local Subdivision--F_{IS} Estimates: Using allozymes, D'Elía et al. (1998) detected positive and significant F_{IS} across subpopulations at five loci. In that study, positive F_{IS} was also detected across loci in two out of seven subpopulations. Microsatellites concur in yielding generally positive, and occasionally significant, values of F_{IS}. An overall test favors a positive F_{IS} over all loci ($p < 0.001$). Additionally, three loci have positive, significant F_{IS} across subpopulations. In La Tabaré, significant positive F_{IS} values were found across loci and several loci were marginally above the 0.05 significance level in various other subpopulations. In general, however, the average values obtained from microsatellites are less extreme than those reported by D'Elía et al. (1998) for allozymes.

Not surprisingly, the best data on local genetic subdivision in a subterranean mammal are those from the pocket gopher, *Thomomys bottae*, in which F_{IS} tends to be positive but generally not significant (summarized in Patton and Smith, 1990, Table 3.4). Thus, it seems that local subdivision is of greater significance in the Río Negro tuco-tuco than in *T. bottae*. As shown in Figure 4, there is a trend in *C. rionegrensis* towards positive values of F_{IS} across loci at the local level and a similar trend for average F_{IS} estimates across geography.

Population Structure in La Tabaré: Multilocus genotypes carry information on population structure not revealed by F-statistics. We have initiated detailed analyses of genetic structure at the local level at La Tabaré, a subpopulation that is polymorphic for pelage color (Figure 1). Here, we report microsatellite analyses of 57 individuals collected in 1999 to generate a preliminary view of local genetic structure in this population.

Using a Bayesian approach similar to that employed at the geographical scale, we examined the extent of genetic subdivision within La Tabaré. The most likely source of the genotypes is a single, rather than two or more subpopulations. This is not surprising given the significant but modest evidence of subdivision provided by F_{IS}.

Using the program Relatedness (Queller and Goodnight, 1989), we generated pairwise estimates of the genetic similarity among individuals relative to the background genetic makeup of the subpopulation. Using a conservative relatedness of $r \geq 0.40$ as an indication of close kinship, 5.8% of the comparisons are among close relatives. The figure does not vary much if we restrict our attention to specific subsets of the La Tabaré sample. For example, 37 individuals, including 11 juveniles, were captured in the early fall of 1999. At that time of the year, one finds no pregnant females or newborns, and the population consists of adults plus juveniles born in the preceding spring that have had ample opportunities to disperse. In this subset of animals, 5.9% of the comparisons were between close relatives ($r = 0.4$-0.82). Of the 11 juveniles, only two had no identifiable close relatives, whereas eight had between two and five identifiable close relatives.

These initial estimates and the intricate patterns of genetic similarity that they reveal suggest that La Tabaré includes numerous cases of coexistence by close relatives of different ages. Field observations suggest that this may be, at least in part, related to burrow sharing among adults and subadults (E. Lessa and G. Wlasiuk, unpublished observation). In the spring of 1999, we recorded several cases of captures of multiple adults (most often females, but occasionally including one male) from the same burrow opening, in addition to the expected capture of juveniles with a lactating female. Most species of tuco-tucos are thought to be solitary, but various levels of burrow sharing and sociality are known (reviewed in

Lacey, 2000) and J. R. Contreras has reported burrow sharing in an isolate of the Río Negro tuco-tuco in Victoria, Argentina (cf. Reig et al., 1990).

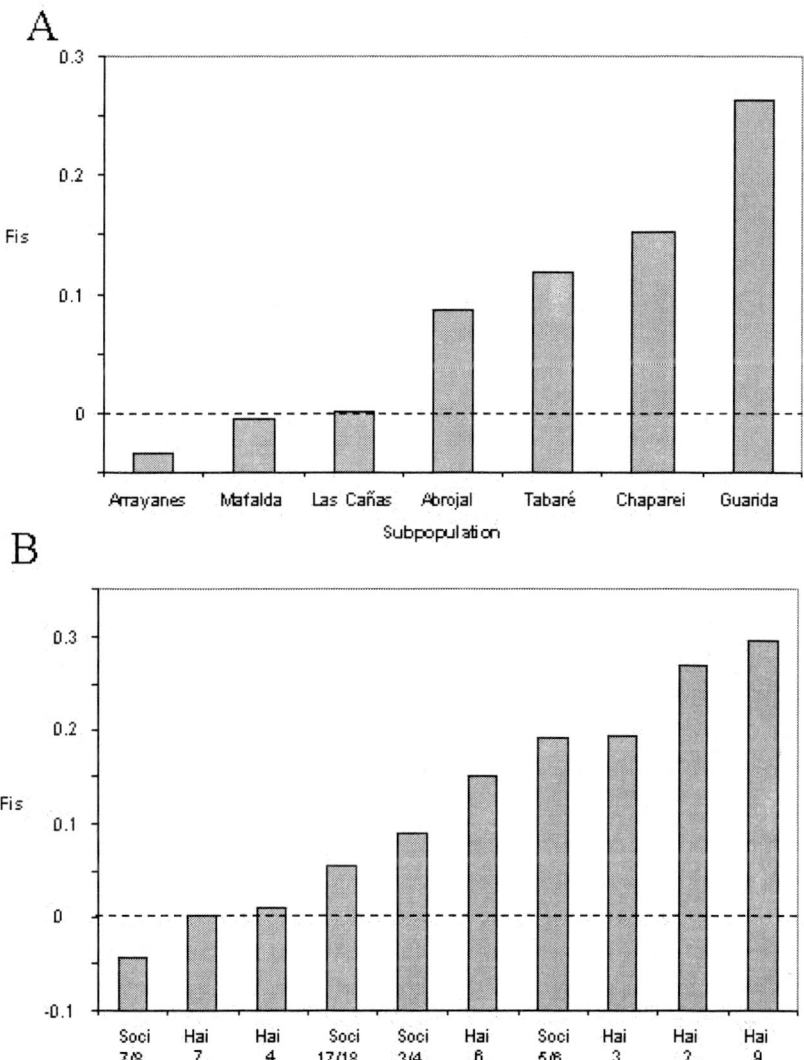

Figure 4. F_{IS} estimates for microsatellite loci in *Ctenomys rionegrensis*. (a) Averages across loci for each of seven subpopulations (Nuevo Berlín is excluded because it is monomorphic). (b) Estimates per locus in the subpopulation of La Tabaré.

CONCLUSIONS AND PROSPECTS

Our examination of genetic population structure in *C. rionegrensis* has several implications for the evolutionary dynamics of this and, possibly, other species of tuco-tucos. These implications include the following:

(1) There is ample evidence for a historical demographic expansion of the species from a more restricted distribution to its current range in Uruguay. Far from showing equilibrium between local genetic drift and gene flow, our analyses suggest that the footprints of such an expansion are substantial.

(2) One consequence of this historical expansion is that genetic estimates of gene flow that assume that drift and migration have reached an equilibrium almost certainly overestimate current levels of gene flow. Methods that are less dependent on such an assumption are likely to provide more reliable estimates of Nm.

(3) Regardless of the method employed, there are consistent differences between the relatively high estimates of gene flow obtained from allozymes and the relatively low values suggested by mtDNA and microsatellites. We suggest that the latter are more reliable and that current gene flow probably is very low ($Nm < 0.3$, perhaps even $Nm = 0$).

(4) Estimations of F_{IS} are consistent in suggesting that local subpopulations of *C. rionegrensis* are reproductively subdivided. Additionally, estimates of genetic relatedness and field observations indicate that subpopulations include networks of related individuals, implying that philopatry is common in this species. Initial observations of burrow sharing suggest that the possibility of sociality should be carefully examined in the species.

(5) One of the motivations for the research presented here was to assess the likelihood of the local fixation of melanism by random genetic drift. This hypothesis was set forth by Langguth and Abella (1970b) to account for the discordance of pelage and background color in the melanic form of the Río Negro tuco-tuco. The allozyme data of D'Elía et al. (1998) suggested that levels of gene flow were too high to allow for such a possibility. The new microsatellite and mtDNA data, in contrast, suggest otherwise and include many instances of local fixation of alleles. Pelage color may have followed a similar path. However, only direct analysis of the genetic basis and biological consequences of pelage color polymorphism will tell whether or not positive selection affects this trait (see Hoekstra and Nachman, this volume). Our current efforts include the direct study of loci that are good candidates for controlling pelage color polymorphism, such as Agouti (Bultman et al., 1992) and MCR-1 (Mountjoy et al., 1992).

ACKNOWLEDGMENTS

We dedicate this paper to James L. Patton as a small measure of our appreciation for years of inspiration and support. Enrique P. Lessa and John Carlos Garza were postdoctoral fellows in the laboratory of James L. Patton. Gabriela Wlasiuk was a masters' student in the laboratory of Enrique P. Lessa. We are grateful to CSIC-Universidad de la República for continuing funding of our research on tuco-tucos. EPL acknowledges the support of a Guggenheim Fellowship and the expert guidance of John Wakeley (Harvard University) in the study of population genetics.

LITERATURE CITED

Altuna, C., M. Ubilla, and E. P. Lessa
     1985   Estado actual del conocimiento de *Ctenomys rionegrensis* Langguth y
             Abella, 1970 (Rodentia: Octodontidae). Actas Jor. Zool. Uruguay 1:8-9.

Barton, N. H., and M. Slatkin
     1986   A quasi-equilibrium theory of the distribution of rare alleles in a
             subdivided population. Heredity 56:409-415.

Bultman S. J., E. J. Michaud, and R. P. Woychik
     1992   Molecular characterization of the mouse agouti locus. Cell 71:1195-204.

Cook, J. A., and E. P. Lessa
     1998   Are rates of diversification in subterranean South American tuco-tucos
             (genus *Ctenomys*, Rodentia: Octodontidae) unusually high? Evolution
             52:1521-1527.

D'Elía, G., E. P. Lessa, and J. A. Cook
     1998   Goegraphic structure, gene flow and maintenance of melanism in
             Ctenomys rionegrensis (Rodentia: Octodontidae). Z. Säugetierk. 63:285-
             296.

     1999   Molecular phylogeny of tuco-tucos, genus *Ctenomys* (Rodentia:
             Octodontidae): evaluation of the mendocinus species group and the
             evolution of asymmetric sperm. J. Mamm. Evol. 6:19-37.

Dufresne, F., E. Bourget, and L. Bernatchez
     2002   Differential patterns of spatial divergence in microsatellite and
             allozyme alleles: further evidence for locus-specific selection in the
             acorn barnacle, *Semibalanus balanoides*? Mol. Ecol. 11:113-123.

Freitas, T. R. O., and E. P. Lessa
    1984   Cytogenetics and morphology of *Ctenomys torquatus* (Rodentia: Octodontidae). J. Mammalogy 65:637-642.

Fu, Y. X.
    1997   Statistical test of neutrality of mutations against population growth, hitchhiking and background selection. Genetics 147:915-925.

Guo, S. W., and E. A. Thompson
    1992   Performing the exact test for Hardy-Weinberg proportion for multiple alleles. Biometrics 48:361-372.

Hedrick, P. W.
    1999   Perspective: highly variable loci and their interpretation in evolution and conservation. Evolution 53:313-318.

Hey, J., and J. Wakeley
    1997   A coalescent estimator of the population recombination rate. Genetics 145:833-846.

Kimura, M., and J. F. Crow
    1964   The number of alleles that can be maintained in a finite population. Genetics 49:725-738.

Lacey, E. A.
    2000   Spatial and social systems of subterranean rodents. Pp. 257-296 in Life Underground: the Biology of Subterranean Rodents (E. A. Lacey, J. L. Patton, and G. N. Cameron, eds.). University of Chicago Press, Chicago, Ill.

    2001   Microsatellite variation in solitary and social tuco-tucos: molecular properties and population dynamics. Heredity 86:628-37.

Lacey, E. A., J. E. Maldonado, J. P. Clabaugh, and M. D. Matocq
    1999   Interspecific variation in microsatellites isolated from tuco-tucos (Rodentia: Ctenomyidae). Mol. Ecol. 8:1754-1756.

Langguth, A., and A. Abella
    1970 a   Las especies uruguayas del género *Ctenomys*. Com. Zool. Mus. Hist. Nat. Montevideo 10:1-27.

Langguth, A., and A. Abella
  1970 b Sobre una población de tuco-tucos melánicos (Rodentia: Octodontidae).
          Acta Zool. Lilloana 28:101-108.

Lessa. E. P., and J. A. Cook
  1998   The molecular phylogenetics of tuco-tucos (genus *Ctenomys*, Rodentia:
         Octodontidae) suggests an early burst of speciation. Mol. Phylogenet.
         Evol. 9:88-99.

Mantel, N.
  1967   The detection of disease clustering and a generalized regression
         approach. Cancer Res. 27:209-220.

Miller, S. A., D. D. Dikes, and H. F. Polesky
  1988   A simple salting procedure for extracting DNA for human nucleated
         cells. Nucl. Acids Res. 16:215.

Mountjoy, K. G., L. S. Robbins, M. T. Mortrud, and R. D. Cone
  1992   The cloning of a family of genes that encode the melanocortin receptors.
         Science 257:1248-1251.

Nei, M.
  1982   Evolution of human races at the gene level. Pp. 167-181 in Human
         Genetics, part A: the Unfolding Genome (B. Bonne-Tamir, T. Cohen and
         R. M. Goodman, eds.). Alan R. Liss, New York.

Nevo, E.
  1999   Mosaic Evolution of Subterranean Mammals. Regression, Progression,
         and Global Convergence. Oxford University Press, New York, 413 pp.

Ortells, M. O., J. R. Contreras, and O. A. Reig
  1990   New *Ctenomys* karyotypes (Rodentia, Octodontidae) from north-eastern
         Argentina and from Paraguay confirm the extreme chromosomal
         multiformity of the genus. Genetica 82:189-201.

Ohta, T., and M. Kimura
  1973   A model of mutation appropiate to estimate the number of
         electrophoretically detectable alleles in a finite population. Genet. Res.
         22:201-204.

Patton, J. L., and M. F. Smith
     1990   The evolutionary dynamics of the pocket gopher, *Thomomys bottae*, with emphasis on California populations.   Univ. California Publ. Zoology 123:1-161.

Pritchard, J. K., M. Stephens, and P. Donelly
     2000   Inference of population structure using multilocus genotype data. Genetics 155:945-959.

Queller, D. C., and K. F. Goodnight
     1989   Estimating reletedness using genetic markers.   Evolution 43:258-275

Raymond, M., and F. Rousset
     1995   Genepop: Population Genetic Software for Exact Tests and Ecumenicism, version 1.2. J. Hered. 86:248-249.

Reig, O. A., C. Bush, M. O. Ortells, and J. R. Contreras
     1990   An overview of evolution, systematics, population biology cytogenetics, molecular biology and speciation in *Ctenomys*.   Pp. 71-96 in Evolution of Subterranean Mammals at the Organismal and Molecular Levels (E. Nevo and O. A. Reig, eds.).   Wiley-Liss, New York.

Rousset, F.
     1996   Equilibrium values of measures of population subdivision for stepwise mutation processes.   Genetics 142:1357-1362.

Rousset, F., and M. Raymond
     1995   Testing heterozygote excess and deficiency.   Genetics 140:1413-1419.

Rozas, J., and R. Rozas
     1999   DnaSP: an Integrated Program for Molecular Population Genetics and Molecular Evolution Analysis, version 3.   Bioinformatics 15:174-175.

Sanguinetti, C. J., E. D. Neto, and A. J. G. Simpson
     1994   Rapid silver staining and recovery of PCR products separated on polyacrylamide gels.   Biotechniques 17:915-918.

Schneider, S., D. Roessli, and L. Excoffier
     2000   Arlequin: A Software for Population Genetic Data Analysis, version 2,000.   Genetics and Biometry Laboratory, University of Geneva, Geneva.

Slamovits, C. H., J. A. Cook, E. P. Lessa, and M. S. Rossi

    2001   Recurrent amplifications and deletions of satellite DNA accompanied chromosomal diversification in South American tuco-tucos (genus *Ctenomys*, Rodentia: Octodontidae): a phylogenetic approach. Mol. Biol. Evol. 18:1708-1719.

Slatkin, M.

    1985   Rare alleles as indicators of gene flow. Evolution 39:53-65.

    1987   Gene flow and the geographic structure of natural populations. Science 236:787-792.

    1993   Isolation by distance in equilibrium and non-equilibrium populations. Evolution 47:264-279.

Smith, M. F., and J. L. Patton

    1993   The diversification of South American murid rodents: evidence from mitochondrial DNA sequence data for the akodontine tribe. Biol. J. Linn. Soc. 50:149-177.

Steinberg, E. K, and J. L. Patton

    2000   Genetic structure and the geography of speciation in subterranean rodents: oportunities and constraints for evolutionary diversification. Pp. 301-331 in Life Underground: the Biology of Subterranean Rodents (E. A. Lacey, J. L. Patton, and G. N. Cameron, eds.). University of Chicago Press, Chicago, Ill.

Wakeley, J., and J. Hey

    1997   Estimating ancestral population parameters. Genetics 145:847-855.

Wlasiuk, G, J. C. Garza, and E. P. Lessa

    2003   Genetic and geographic differentiation in the Río Negro tuco-tuco (*Ctenomys rionegrensis*): inferring the roles of migration and drift from multiple genetic markers. Evolution 57:913-926.

Wright, S

    1950   The genetical structure of populations. Ann. Eugenics 15:323-354.

Morphological and Cytogenetic Analyses of *Bibimys labiosus* (Winge, 1887) (Rodentia, Sigmodontinae): Implications for its Affinities with the Scapteromyine Group

Pablo Rodrigues Gonçalves, João Alves de Oliveira,
Margaret Oliveira Corrêa, and Leila Maria Pessôa

A recently obtained series referable to *Bibimys labiosus* (Winge, 1887) from Viçosa, state of Minas Gerais, represents the first documented record for this species from a Brazilian locality since the original specimen was collected at nearby Lagoa Santa in 1837. We designate a lectotype and redescribe the species, introducing additional morphological (external, cranial, gastric and phallic) characters; we also describe its karyotype, including C-, G-, and NOR banding. The diploid (2n=70) and autosomal arm numbers (AN = 80) of *Bibimys labiosus* depart from the previously described karyotype of the Argentinean *B. torresi* (2n = 70, AN = 76). Morphological and cytogenetic comparisons between *Bibimys* and two other genera grouped in the tribe Scapteromyini (*Kunsia* and *Scapteromys*) were also performed. In addition to previously postulated cranial and dental homologies, the present analyses revealed phallic and cytogenetic (heterochromatin distribution) similarities between *Bibimys* and the other two scapteromyine genera. Nonetheless, more inclusive sampling of external, cranial and gastric characters, as well as conventional and NOR staining of chromosomes of available representative species, do not support the grouping of these three genera and suggest that *Bibimys* is a very divergent genus.

INTRODUCTION

Of the sigmodontine rodents gathered by the Danish palaeontologist Peter W. Lund at Lagoa Santa, Minas Gerais, Brazil, during the first half of the nineteenth century, a number of forms are still known from very few reports subsequent to the original descriptions. Herluf Winge (1887) was the first to revise Lund's work, redefining many of Lund's original names and proposing new ones to account for the notable diversity revealed among the sub-fossil and extant taxa recovered.

Scapteromys labiosus, one of the species originally described by Winge (1887), was later transferred to the genus *Bibimys* by Massoia (1980a). Also including *B. torresi* and *B. chacoensis*, *Bibimys* is a well-differentiated genus characterized by an unusually enlarged labial region covered with very short, velvety pelage - a remarkable trait that seems unique among New World muroids (Figure 1).

Figure 1. Live specimen of *Bibimys labiosus* captured in Viçosa, Minas Gerais (MZUFV 752).

Bibimys labiosus and its congeners are currently grouped in the tribe Scapteromyini together with the genera *Scapteromys* and *Kunsia* (Massoia, 1979; Musser and Carleton, 1993; Pardiñas, 1996), although certain authors have questioned this arrangement (Cabrera, 1961; Gyldenstolpe, 1932; Hershkovitz, 1966). This grouping has been supported mainly by craniodental similarities (Massoia, 1979; Pardiñas, 1996), although evidence from other character systems is lacking. In fact, comprehensive analyses of the variation within *Bibimys* and of its suprageneric affinities have been hindered by its scarcity in collections. Very few specimens have been reported and studied since the original descriptions of each species, most of them consisting only of skull and mandible fragments from owl pellets. In the case of *B. labiosus*, besides Lund's material in the Copenhagen Museum, only a single bibliographical record exists in the literature for a Brazilian locality (Paglia et al., 1995). As a consequence, more than a century after the original description, Winge's account (Winge, 1887) remains the most complete description of this species.

The present study was motivated by the collection of five specimens during a one-year mark-recapture study in Viçosa, Minas Gerais, at the same locality as the single previous Brazilian record referred to above. In view of the fact that *Bibimys labiosus* was described on the basis of incomplete specimens, and because the original description was based almost exclusively on comparisons with "*Hesperomys" expulsus* (currently allocated to the genus *Calomys*), we begin by redescribing the species. Our redescription is based on the original material collected by Lund and the complete specimens obtained by us at Viçosa. It includes a wider variety of external, craniodental and phallic characters than the original description, as well as the karyotype. Finally, we discuss differentiation within *Bibimys* and its affinities with *Scapteromys* and *Kunsia* in light of the new morphological and cytogenetic information.

MATERIALS AND METHODS

Two males and three females were collected with Sherman and pitfall traps at Mata do Paraíso (20° 45′ S, 42° 53′ W, alt. 650 m), a reserve managed by the Universidade Federal de Viçosa Forestry Research Centre at Viçosa, Minas Gerais, Brazil, during a year-long mark-recapture study conducted in 1999-2000. The specimens were kept under observation while alive, and one female and two males were karyotyped. The specimens were prepared as skins and skulls or preserved in fluid, and were deposited in the collections of the Museu Nacional, Universidade Federal do Rio de Janeiro (MN 62061; MN 62062; MN 62063) and Museu de Zoologia João Moojen de Oliveira, Universidade Federal de Viçosa (MZUFV 752 and MZUFV 753).

In order to determine the specific identity of these specimens, the skull and skin of one (MN 62061) were compared with the original specimen of *Bibimys labiosus* in the Universitets Zoologiske Museum, Copenhagen (ZMC). The descriptions and comparisons were later expanded with information obtained from the other four specimens collected at Viçosa.

Cranial and Pelage Morphology

The descriptions were made following the anatomical terminology used by Carleton (1980), Carleton and Musser (1989) and Voss (1993) for cranial characters, and by Reig (1977) and Hershkovitz (1993) for dental and mandibular characters. Intergeneric comparisons were carried out using 25 specimens of *Kunsia fronto planaltensis* (MN 21300; 21826; 21828-21850) from Planaltina, Goiás State, Brazil, and nine specimens of *Scapteromys tumidus* (MN 62304–MN 62312) from Ensenada, Buenos Aires Province, Argentina.

Stomach Morphology

The stomachs of the fluid preserved specimens were observed, and one of them was removed and dissected. Both specimens were also examined for the presence of a gall bladder. Stomach dissection and gastric terminology follows Carleton (1973).

Penile and Bacular Morphology

The glans penis was removed from specimens MN 62061, 62062 and 62063 before taxidermy and the prepuce cut off. After fixation in 4% formalin, the penises were cleared in 4% KOH and stained with Alizarin Red and Alcian Blue to visualize osseous and cartilaginous tissues (Hooper and Hart, 1962; Lidicker, 1968). Sketches were done under a binocular dissecting microscope with a camera lucida. A lateral incision was made to visualize the internal structures of the glans without damaging them, following Spotorno (1992). The terminology of the anatomical descriptions follows Hooper and Musser (1964).

Cytogenetic Analyses

All cytogenetic analyses were based on mitotic metaphase chromosomes from the bone marrow of two males (MN 62061 and 62062) and one female (MZUFV 753) previously injected with 0.05% colchicine solution. The procedure used was modified from Ford and Harmerton (1956) and involves the extraction of the femur bone marrow with KCl 0.075M solution, incubation for 30 min at 37°C, pre-fixation with Carnoy solution for 5 min, followed by three rounds of alternating centrifugation (5 min at 1000 rpm) and fixation in Carnoy solution. G-banding was performed as described by Seabright (1972). Nucleolar organizer regions were detected by means of silver nitrate staining (Ag-NOR; Howell and Black, 1980). The distribution of constitutive heterochromatin (C-bands) was determined by Sumner's (1972) method. Chromosomes were classified according to Levan et al. (1964). Metacentric and submetacentric chromosomes were considered to be biarmed, and subtelocentric and acrocentric chromosomes were considered uniarmed.

RESULTS

Bibimys labiosus (Winge, 1887)

   Scapteromys labiosus Winge, 1887

   Scapteromys labiosus: Trouessart (1898: 534) (distribution)

   Scapteromys labiosus: Gyldenstolpe (1932: 151) (incertae sedis)

   Scapteromys labiosus: Moojen (1952: 80) (original measurements)

   Scapteromys labiosus: Cabrera (1961: 475) (incertae sedis)

   Scapteromys labiosus: Hershkovitz (1966: 96) (incertae sedis)

   Bibimys labiosus Massoia, 1980*a*: 281 (new name combination)

Type Material

The most complete specimen of *B. labiosus* described by Winge (1887) is represented by a partial skull and skin (ZMC 259) obtained by Lund in Lagoa Santa on May 25, 1837. This individual, the skull of which was used in the original description of this species, is here designated the lectotype of *Bibimys labiosus* (Figure 2a).

   Lectotype: ZMC 259, skin and skull, young male (gender determined by Winge, 1887). Skin unstuffed and torn on the back, missing both forelimbs and part of the ventral surface. The right ear, left hind limb and a fragment of tail have been torn off and are currently kept in a separate envelope, which is labelled in Winge's handwriting. Two additional fragments, the soft palate and lips, are kept in the cardboard box containing the skull. Skull incomplete: base of braincase fragmented; all basal bones posterior to the left alisphenoid and to the basisphenoid missing; left bulla loose. Left mandibular ramus broken just anterior to the molar series: anterior part attached to right ramus at symphysis, posterior part separate. Dentition complete.

   Paralectotypes: Winge (1887) also studied five uncataloged skull fragments, two of them composed of rostral parts (including both molar rows, palatal area, and premaxillary and frontal bones) and the other three composed of isolated maxillary fragments with incomplete molar rows. In addition to these, 44 fragments of mandibles and a complete mandibular ramus (all from a set currently labelled ZMC 6529) recovered from cave floor sediments from Lapa da Escrivania No. 5, were also available to Winge, who regarded this material as belonging to *Scapteromys labiosus*. All of these specimens were also used in the following comparisons with the recently collected material.

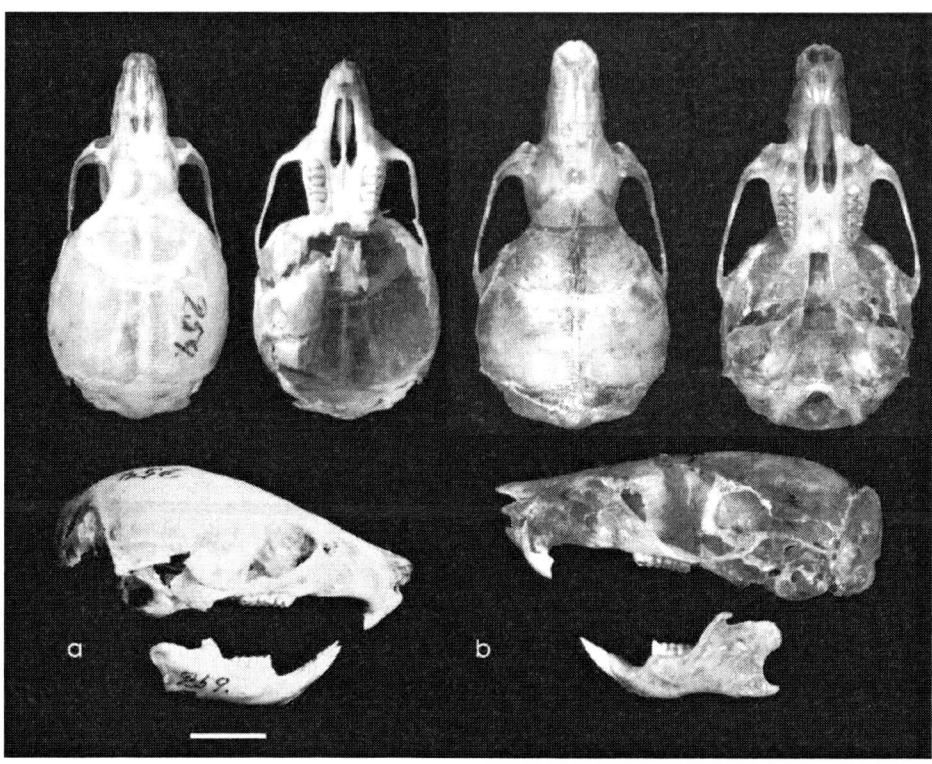

Figure 2. Dorsal, ventral and lateral views of the skulls of (a) the lectotype (ZMC 259), and (b) a recently obtained specimen of *Bibimys labiosus* (MZUFV 753). Scale bar equals 10 mm.

External Characters

The pelage of the lectotype is soft, long and dense. The dorsal surface is a homogeneous brownish tone, with slightly paler fur on the sides of the body. Some of the guard hairs have a yellow-orange subapical band, giving the dorsum a black-lined appearance. The subapical band becomes wider and paler laterally on the body. The underparts were damaged in taxidermy but a well-defined limit to the sides is marked by the prevalence of whitish distal parts to the ventral hairs, which, like the dorsal hairs, are dark grey at the base. The recently obtained specimens display some degree of variability in the dark dorsal tone and some are paler than the lectotype. The pelage of specimen MN 62061 is dark brown over the rump. The sides of head and body are paler, with the pale yellowish bands more apparent than in the lectotype. In one of the specimens (MZUFV 753), the dorsum is notably paler, apparently because the banded yellowish hairs predominate over the entirely dark ones.

The ears of the lectotype as well as those of the Viçosa specimens are covered on their proximal half by hairs originating on the head or on the proximal upper end of the helix; the pinnae are well furred internally in the auricular region with short banded hairs extending nearly to the outer edge. The external surface of the ear has relatively fewer hairs, which are mainly a uniform dark brown and concentrated along the upper edge. The sides of the head are colored like the flanks (dark yellow in the recently collected individuals, approaching a pale reddish brown in the lectotype), and become darker like the dorsum on the frontal region of the head. The labial area between the nostrils and oral margins (including the lower lip) is covered with densely distributed, short, white hairs, forming a round, white, velvety area on the tip of the muzzle of preserved skins (Figure 3). This region is very swollen and pinkish in color in live specimens suggesting a high degree of vascularization and a possible sensory function. Live specimens observed in captivity used this structure to explore the substrate by touch. In fluid-preserved specimens the color is lost, and the swollen velvet area spreads to a circle nearly 8 mm in diameter that includes the internarial and infranarial portions of the rhinarium and the upper lip. The mystacial vibrissae are 2–2.5 cm long, dark brown proximally and whitish distally. Hairs in the submental region are longer than those on the swollen muzzle, less dense, and entirely white, as opposed to the remaining ventral hairs, which are a dark slate color at the base.

The forelimbs of the lectotype are missing. The manus in recently collected specimens is well covered dorsally with hairs, which are either proximally grey and distally unpigmented (whitish) or entirely whitish, allowing the underlying skin color to show through. A greatly reduced claw-bearing pollex and five palmar pads - three reduced interdigital and two larger metacarpal pads - are present. Carpal vibrissae are not distinguishable. Like the manus, the pes of the lectotype and recently obtained specimens are also skin colored. The ungual tufts are mainly whitish but also have some dark brown hairs (their tips always white), the longest extending beyond the tips of the claws. The claws on the hind feet are long, those on digits 2-4 extending to more than 1/3 of the length of the digits, and are paler in color than the skin of the pes. The hallux has a claw similar in size to that of the 5th toe and about half the length of the remaining claws. The plantar surface has 2 metatarsal (1 reduced hypothenar and 1 thenar) and 4 interdigital pads in both the lectotype and the fluid preserved specimens (Figure 4).

The fragmentary tail of the lectotype (about 3 cm near the proximal end) and the tails of the other specimens are densely haired; the hairs, which are about 2 rows of epidermal scales in length, are entirely dark brown on the upper surface of the tail and dark brown proximally, with whitish distal halves on the ventral surface. External measurements of recently collected specimens are presented in Table 1.

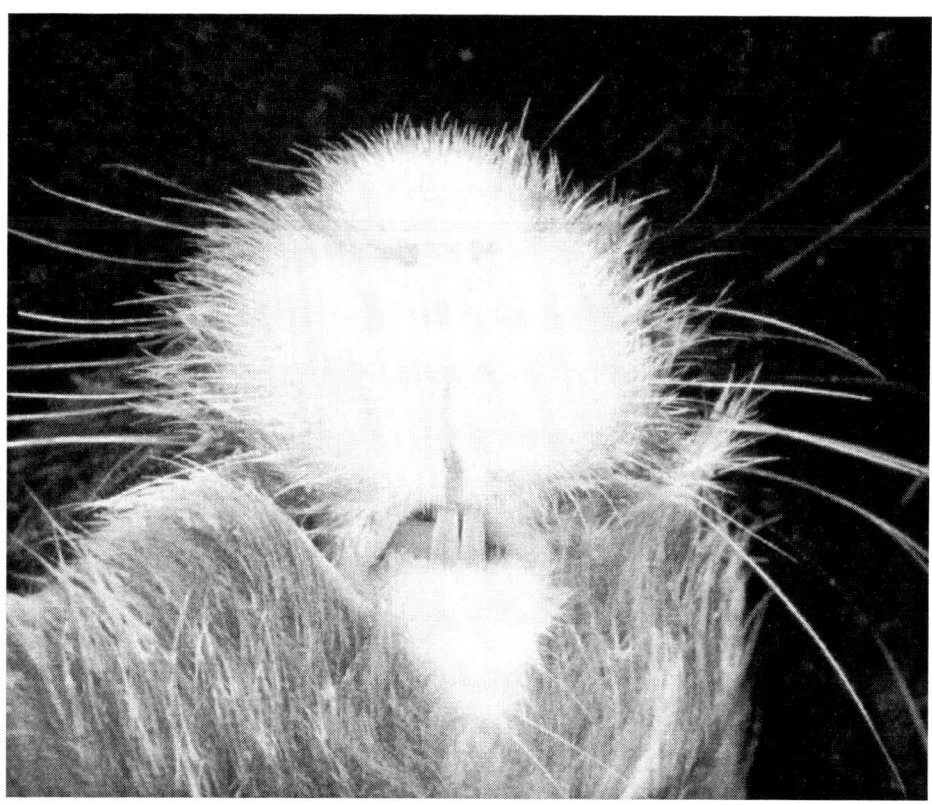

Figure 3. Ventral view of the labial region of *Bibimys labiosus* (MN 62062).

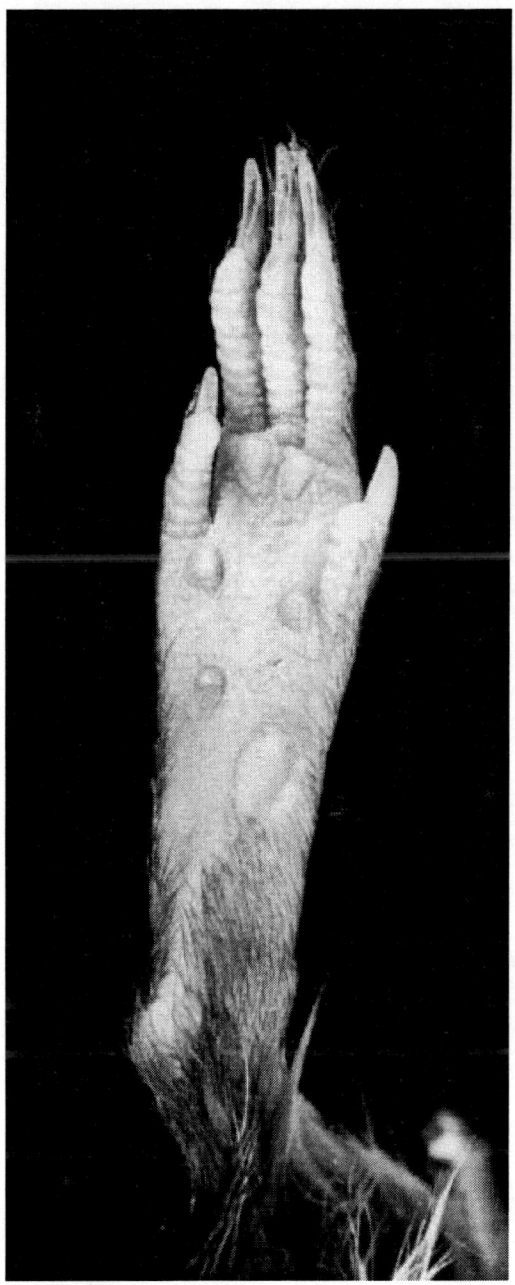

Figure 4. Ventral view of the hind foot of *Bibimys labiosus* (MN62062).

Table 1. External and cranial measurements of recently collected specimens of *Bibimys labiosus* from Viçosa, Minas Gerais, Brazil (MZUFV 752, MZUFV 753 and MN 62061). All measurements are in mm. Means and ranges for measurements are given in the right column.

Character	MZUFV 752	MZUFV 753	MN 62061	Mean [max.-min.]
HBL (head and body length)[1]	76	90	89	85 [76-90]
TL (tail length)[1]	69	--	--	69
HF (hind foot length without claw/with claw)[2]	19/21	20/22	20/23	20/22 [19/21-20/23]
EAR[2]	15	18	14	16 [14-18]
ONL (occipitonasal length)[1]	25.69	26.36	--	26.02 [25.69-26.36]
COL (condyloincisive length)[2]	23.80	24.41	24.54	24.25 [23.8-24.54]
ZB (zygomatic breadth)[2]	13.20	13.44	13.60	13.41 [13.2-13.6]
BB (braincase breadth)[1]	12.91	12.98	13.51	13.13 [12.91-13.51]

1 – Measurements defined by Myers *et al.* (1990).
2 – Measurements defined by Carleton and Musser (1993).

Table 1. Continued.

Character	MZUFV 752	MZUFV 753	MN 62061	Mean [max.-min.]
IOB (interorbital breadth)[1]	4.76	4.96	4.53	4.75 [4.53-4.96]
NL (nasal length)[1]	9.36	9.06	--	9.21 [9.06-9.36]
BR (rostrum breadth)[1]	4.41	4.25	4.40	4.35 [4.25-4.41]
LD (diastema length)[1]	5.31	5.66	5.36	5.44 [5.31-5.66]
LIF (incisive foramina length)[1]	5.62	5.90	5.45	5.65 [5.45-5.90]
MTRL (maxillary toothrow length)[1]	3.73	3.75	3.80	3.76 [3.73-3.80]
ITRL (inferior toothrow length)	3.86	3.91	4.00	3.93 [3.86-4.00]
AW (alveolar width)[1]	5.48	5.42	5.53	5.48 [5.42-5.53]
OCW (occipital condyle width)[1]	6.48	6.56	6.80	6.61 [6.48-6.80]

1 – Measurements defined by Myers *et al.* (1990).
2 – Measurements defined by Carleton and Musser (1993).

Table 1. Continued.

Character	MZUFV 752	MZUFV 753	MN 62061	Mean [max.-min.]
BOL (basioccipital length)[1]	3.91	3.75	4.03	3.89 [3.75-4.03]
ZP (zygomatic plate)[2]	2.51	2.38	2.31	2.40 [2.31-2.51]
BD (braincase depth)[2]	7.88	7.99	8.20	8.02 [7.88-8.2]

1 – Measurements defined by Myers *et al.* (1990).
2 – Measurements defined by Carleton and Musser (1993).

Skull Morphology

The skulls of the lectotype and remaining specimens are slender, with a rounded, laterally inflated and dorso-ventrally compressed braincase and a short, narrow rostrum (Figures 2a, 2b). The dorsal profile is faintly convex in lateral view, especially in the lectotype, which is a younger individual. Older specimens are less curved, mainly in the rostral and interorbital region.

A well-developed gnathic process is present in the holotype and in the remaining series, projecting forward as far as the anteriormost extension of the premaxillary bones. The zygomatic plate is not projected anteriorly, forming only a moderately pronounced zygomatic notch when the skull is observed from above. The anterior edge of the zygomatic plate is straight in the lectotype, but varies in the remaining specimens from slightly concave (MN 62061) to slightly convex (MZUFV 752). The maxillary ramus of the zygomatic arch is thick and connected to the thin squamosal ramus by a large jugal bone. The lacrimals are incomplete (broken) in the lectotype and laterally expanded in the recently obtained specimens. The nasals are depressed along their longitudinal suture and do not extend behind the suture between the pre-maxillary and frontal bones. The frontal bones do not form crests, but the margin of the interorbital region is barely rounded, almost forming a right angle at the level of the supraorbital foramen. The fronto-parietal suture is U-shaped; temporal crests and ridges are absent in the lectotype, but there is a thickening along the anterior part of the parieto-squamosal suture in the remaining specimens. The parietals are large, encompassing part of the lateral walls of the braincase. The interparietal is wide and antero-posteriorly compressed.

The incisive foramina extend to a line joining the protoflexi of the upper first molars. The premaxillary portion of the septum between foramina is expanded, occupying 2/3 of the length of the septum. The maxillary portion of the septum is superimposed on the premaxillary, with the vomer sheath intermediary in the contact region. Three pairs of palatine foramina are present. The bony palate is long, extending posteriorly beyond the line of the hypoflexus of the upper third molar.

Because the skull of the lectotype is badly damaged ventrally, the following ventral traits of the braincase were observed only in the recently obtained skeletal material. The parapterygoid fossae are shallow and nearly 50% wider than the mesopterygoid fossa at mid-point. The sides of the mesopterygoid fossa are straight, producing inconspicuous alisphenoid wings; the anterior limit is usually biconcave due to the presence of a short, pointed median process. The sphenopalatine vacuities are large, extending posteriorly beyond the sphenopalatine suture, thus delineating a reduced mesopterygoid roof. An alisphenoid strut is present on both sides in all three cleaned skulls. A large posterior opening of the alisphenoid canal, a conspicuous stapedial foramen, and a large petrotympanic fissure accommodating the stapedial artery are present in the contact between the bulla and the petrosal. These characters, together with the lack of an internal groove in the lateral wall of the alisphenoid and of a sphenofrontal foramen in the posterior wall of the orbit, constitute evidence for a pattern 2 cephalic blood supply (Voss, 1988). A slender extension of the periotic (tegmen tympani) connects it anteriorly to the alisphenoid and a long, thin hamular process of the squamosal connects posteriorly to the periotic, defining a large postglenoid fenestra and a reduced subsquamosal foramen. The foramen magnum is oriented more posteriorly than ventrally as a result of the weak basicranial flexion. The supraoccipital is trilobate, without lambdoidal crests.

The mandibles (Figures 2a, 2b) are deep and robust, with shallow sigmoid and lunate notches. The ventral outline is slightly inflected, forming a weakly pronounced angular process. The symphysis is elongated posteriorly by enlargement of the antero-ventral surfaces of the rami. The capsular process in both rami is weakly developed and does not form prominent incisor capsules in any of the specimens observed, including the lectotype. The mental foramen is positioned anteromedially to the root of m1. The superior and inferior masseteric crests are weakly developed, and identifiable in the lectotype only by convergent scars that meet in a plane perpendicular to the procingulum of m1. The coronoid process projects a little further upwards than the condylar process and is notably curved posteriorly in the two specimens in which it is complete. The condylar process is relatively short, not extending posteriorly beyond the level of the angular process. See Table 1 for cranial measurements of recently collected specimens.

Dentition

The incisors are orthodont, with anterior surfaces that are yellow-orange, smoothly rounded and not grooved. The molars of the lectotype (Figures 5a, 5b) show little wear; the third molars are apparently recently erupted and the molar roots are completely enclosed in the alveoli (not apparent externally). The molar rows are parallel and worn in a terraced pattern (Hershkovitz, 1962) in the additional specimens examined. M1 is longer than the combined length of M2 and M3 in all specimens. The procingulum on M1 has a slight depression on its anterior face that separates unequal labial and lingual anteroconules, but lacks an indentation (anteromedian flexus) dividing the anterior cingulum. The smaller anterolingual conule projects lingually, whereas the anterolabial conule is directed anteriorly. The paraflexus and metaflexus are posteriorly elongated and occupy a wide area on the occlusal surface of M1. Both M1 and M2 display labial enamelled projections of the paracone and metacone in the lectotype (M2 drawn in original description) and other specimens. The anteriormost projection is here interpreted as the paralophule due to its apparent origin from the paracone and not from the median mure, whereas the second is regarded as a posteroloph, following Hershkovitz (1993). A true mesoloph is absent in the lectotype and in the remaining specimens. The protoflexus and hypoflexus are diagonally oriented, delineating elongated, diagonal anterior and median mures. An anteroloph is present on M2 in the lectotype and in the other young specimen available (MN 62061). M3 is greatly reduced and cylindrical, its unworn occlusal surface showing two flexi. The more anterior (paraflexus) is roughly perpendicular to the antero-posterior axis and separates a slender anteroloph from the paracone. Unlike the paraflexus, the medial part of the metaflexus is oriented parallel to the antero-posterior axis of the tooth. These two flexi have worn to become enamelled islands in the other specimens examined.

The lower incisors are chisel-shaped. Their wear surfaces extend almost the entire lingual length of the tooth, which is relatively short in the specimens studied, including the lectotype. The three molars are of similar size, m1 being the largest (Figure 5b). The anterior face of the procingulum is slightly depressed but without a conspicuous indentation, as in M1. The metaflexid, entoflexid and posteroflexid are well defined and diagonally oriented, the second being the largest in all the lower molars. A distinct protoflexid is present in m1 in all specimens, with the young specimen in the recently obtained series presenting an enamel island at the medial end of this flexid. A conspicuous posterolophid is present in m1 and m2 in the lectotype and persists even in the oldest specimens examined. The metalophid, entolophid and posterolophid are long and narrow, giving m2 an E-shaped trilophodont (*sensu* Hershkovitz, 1966) appearance. A reduced mesolophid is present in m1 and m2 in the lectotype, separated from the entoloph by a tenuous mesoflexid that slightly penetrates the occlusal surface of m2. This small projection

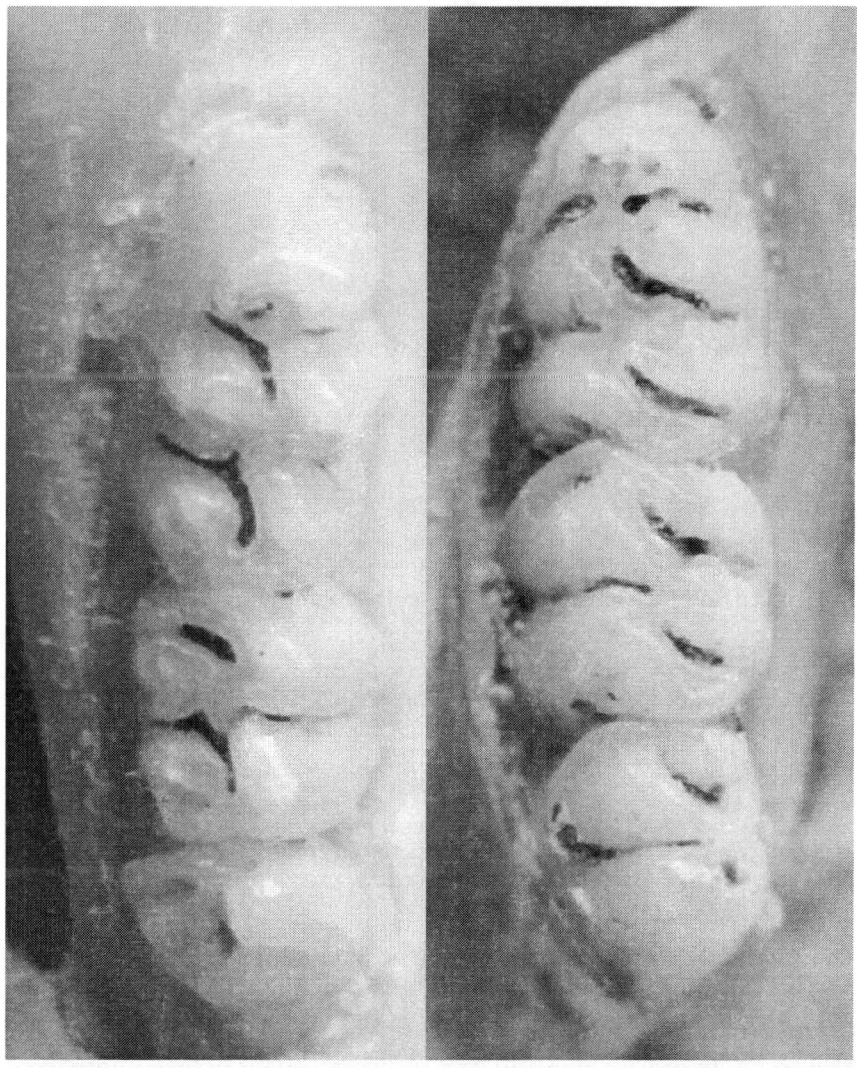

Figure 5. (a) Right upper and (b) left lower molar rows of the lectotype of *Bibimys labiosus* (ZMC 259). Original magnification was 20x; the photo was resized for publication.

is lost in the older specimens available. A reduced entolophulid is present in m1 and m2 in the lectotype, but it also is lost in the older specimens available. In m3, a reduced posterolophid is distinct only in the lectotype and in the youngest specimen of the additional series. Three flexids are evident on the occlusal surface of m3: a large hypoflexid, an entoflexid, and a reduced posteroflexid, which is absent in older specimens. Two main cusps, the protoconid and hypoconid, remain evident in all specimens, whereas a smaller third cusp, the metaconid, is absent in the oldest individual studied.

Digestive Tract

A gall bladder is present. The stomach is unilocular-hemiglandular, with an incisura angularis that does not extend beyond the level of the esophageal opening, and an antrum covered by glandular epithelium that does not reach the limit of the incisura glandularis - a condition that is very similar to that shown in Figure 1A of Carleton (1973).

Glans and Baculum

The glans of *Bibimys labiosus* is an elongated and laterally compressed rod-like structure. The main body is uniformly covered with spinous epidermis, except on the rim of the crater where there is a subapical collar of non-spinous epidermis. The body wall is marked by a major median groove running the length of its ventral surface (Figure 6a). In lateral view, the main body broadens at the level of the crater rim in the younger specimen examined (MN 62061). This enlargement is not so prominent in the older specimen analysed (MN 62063). The lateral and dorsal general outlines of the body wall are straight (Figures 6a, 6b).

The bacular mound is a flame-like structure covered by spinous epidermis with three distally directed lobules on its extremity. Alcian Blue staining revealed a reduced distal cartilaginous baculum, with no traces of lateral digits (Figures 6d, 6e). Alizarin staining showed a proximal osseous baculum composed of a straight shaft with its proximal and distal ends laterally expanded. The proximal end has a slight depression in the ventral surface giving a concave appearance; the proximal end of the bacular base is paddle-like without a conspicuous notch. A dorso-median keel is present on the proximal half of the shaft.

The dorsal papilla is a finger-like structure, flattened dorso-ventrally in most specimens, but in one (MN 62061) the tip of the papilla is slightly divided into three lobes, two small lateral ones and a more developed median one (Figure 6f). The urethral process is a bifurcated projection that extends beyond the ventral wall. Each lobe of this process has a distally directed spine fixed in a blunt base in the apical surface (Figure 6g). Minute spines are present throughout the ventral surface.

The urethral process emerges from the ventral wall of the glans at approximately the same level as the dorsal papilla, constituting part of the urethral aperture (the *meatus urinarius*).

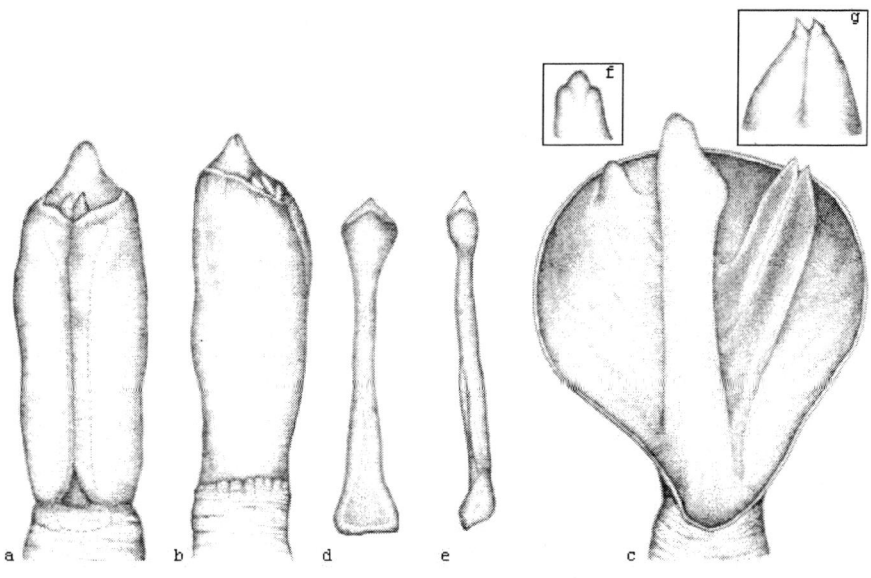

Figure 6. Views of the glans and baculum of *Bibimys labiosus*: (a) ventral, (b) lateral and (c) lateral internal aspects of glans, showing the simple dorsal papilla and the dissected dorsal processes exposing the urethral lumen; (d) ventral and (e) lateral aspects of dissected baculum of *Bibimys labiosus* (MN 62063). In the inserts: (f) dorsal papilla displaying the complex configuration and (g) ventral view of the urethral processes (MN 62061).

Cytogenetic Analyses

The karyotype of three specimens of *Bibimys labiosus* was obtained from Giemsa staining. It revealed a diploid number (2n) of 70 and an autosomal arm number (AN) of 80 (Figure 7a). This karyotype comprises one pair of large submetacentric, five pairs of small metacentric and 28 pairs of subtelocentric-acrocentric autosomes, which vary from large to very small. The X chromosome is a large subtelocentric (between pairs 7 and 8), that corresponds to 4.74% of the total arm length, and the Y is a small acrocentric chromosome (between pairs 23 and 24), corresponding to 2.56% of the genome (Figure 7a). Staining for G-bands, C-bands and NORs was performed on one male. G-banding identified all the chromosome pairs (Figure 8a). A conspicuous heterochromatic block was observed only in the centromeric region of the X chromosome, whereas C-bands in the autosomes were absent (Figure 8b). Silver nitrate staining revealed eight subtelo-acrocentric chromosomes (pairs 14, 17, 23 and 30) bearing telomeric NOR sites on the short arms (Figure 7b).

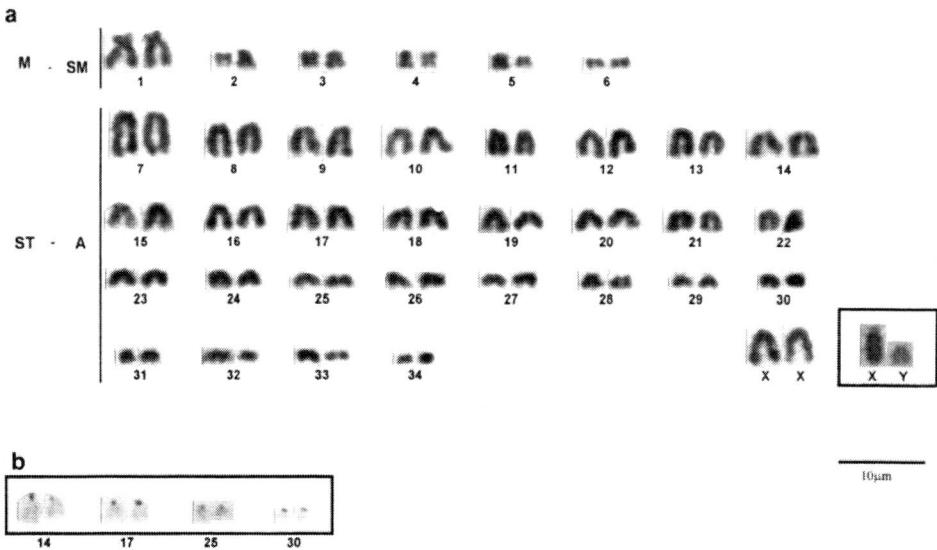

Figure 7. (a) Karyotype of *Bibimys labiosus* (2n = 70; AN = 80). Conventional staining in a female, and sex chromosomes of a male (insert). (b) The four Ag-NOR stained pairs of *Bibimys labiosus*.

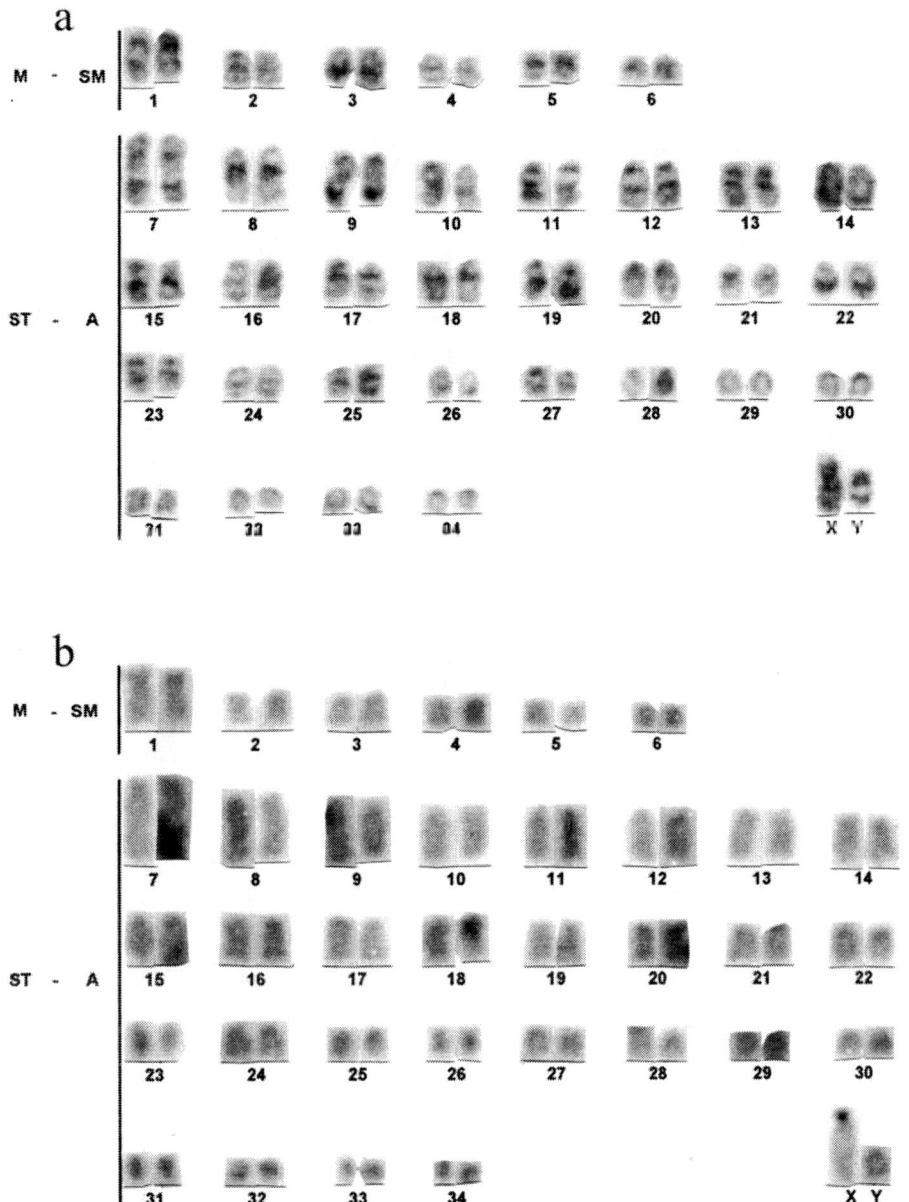

Figure 8. (a) G-banding karyotype of a male of *Bibimys labiosus*; (b) C-banding karyotype of a male of *Bibimys labiosus*.

DISCUSSION

Extant status of B. labiosus and variability in the genus Bibimys

Although Winge (1887) based his description of *Scapteromys labiosus* on the skin and skeleton of a specimen captured by Lund, recent authors (Eisenberg et al., 1999; Nowak, 1999) considered the species to have been described from fossil remains (but see Voss and Myers, 1991). This mistake was probably due to the fact that much of the material studied by Winge had been recovered from the limestone caves of Lagoa Santa. The determination of the Viçosa specimens as *Bibimys labiosus* confirms the extant status of this species, as originally stated by Winge (1887).

The first records of *B. labiosus* from Argentina were provided by Massoia (1980b; 1983; 1988a; 1988b), Massoia et al. (1989a; 1989b; 1989c; 1989d) and Massoia et al. (1989) from seven localities in Misiones province, based mainly on material recovered from owl pellets.

Pardiñas (1996) compared the Argentinean populations referred by Massoia to *Bibimys torresi* and *B. labiosus*. This comparison led him to suggest subspecific rather than specific differentiation for *labiosus*. The characters emphasized were tooth measurements and alveolar composition, with *B. labiosus* showing relatively smaller tooth-rows (Table 2) and three alveolar loci, versus the variable two-three loci found in specimens from the type locality of *B. torresi*.

In spite of these analyses, no comprehensive morphological distinction based on larger samples of complete specimens has been proposed for the nominal forms referred to the genus *Bibimys*. As a consequence and as stated by Pardiñas (1996), the craniodental differences among *Bibimys* species are far from clear. This fact raises a question about the precise identity of the Argentinean *B. labiosus* mentioned by Massoia (*op. cit.*) and Pardiñas (1995, 1996). The northernmost Argentinean locality cited by Massoia for *B. labiosus* is Río Victoria, Guaraní Department, Misiones Province, which lies nearly 1500 km from Lagoa Santa and 300 km from Las Palmas, El Chaco province, the type locality of *B. chacoensis* (Figure 9). The distances from the Misiones localities cited by Massoia (*op. cit.*) to the type localities of *B. labiosus* and *B. chacoensis* suggest that Massoia's samples may be closer to the latter, which is the least known form of the genus.

Additional comparisons between the two *Bibimys* species were undertaken using the bacular and cytogenetic data provided here. The osseous bacula of young and old individuals of *Bibimys torresi* that were examined by Massoia (1979) barely differ from those reported for *B. labiosus* in the present study. An older individual of *B. torresi* presents a laterally expanded proximal base, which is more pronounced than in the samples of *B. labiosus* analysed here. This trait is subject to the same sampling problem as the previously mentioned characters, and it is possible that the differences revealed are of an intrapopulational rather than geographical origin.

Table 2. Mean and minimum-maximum values (in brackets) of undamaged structures available for comparison among samples of *Bibimys labiosus* (type series and recently obtained specimens from Brazil and Argentina) and *Bibimys torresi*. Sample sizes are given in parentheses.

Species	*Bibimys labiosus*				*Bibimys torresi* [2]
Locality	Lagoa Santa, Minas Gerais, Brazil [1]		Viçosa, Minas Gerais, Brazil	Misiones, Argentina [2]	Buenos Aires, Argentina
	Lectotype	Paralectotypes			
M1-M3 length (mm)	3.75	3.92 (n = 3) [3.5 – 4.25]	3.76 (n=3) [3.73 – 3.8]	3.98 (n=6) [3.72 – 4.13]	4.36 (n=11) [4.04 – 4.46]
m1-m3 length (mm)	3.75	4.29 (n = 13) [4.00 – 4.50]	3.93 (n=3) [3.86 – 4.00]	4.28 (n=6) [4.00 – 4.44]	4.27 (n=11) [4.04 – 4.70]

1 – Measurements provided by Winge (1887).
2 – Measurements provided by Pardiñas (1995)

The karyotype of *Bibimys labiosus* presented here is very similar to that of *B. torresi* as recently described from Otamendi, Argentina (Dyzenchauz and Massarini, 1999). Both species have a diploid number of 70, but differ in the number of autosomal arms; AN = 80 for *B. labiosus*, while AN = 76 for *B. torresi*. Pericentric inversions and deletions/additions of chromosomal segments can explain such differences in the autosomal number. Analysis of C-bands also reveals similarities between *Bibimys torresi* and *B. labiosus*. In both species, there is a very low amount of constitutive heterochromatin. C-positive bands are present in the X chromosome and in pair 3 of *B. torresi* (Dyzenchauz and Massarini, 1999), but are present only in the X chromosome of *B. labiosus* (present study).

The difference in number of autosomal arms as well as the geographic discontinuities discussed here suggest that *B. labiosus* and *B. torresi* should be regarded as distinct species. Analysis of G-bands in *B. torresi* could provide greater detail as to the kinds of rearrangements that might be involved in the karyotypic diversification of the genus.

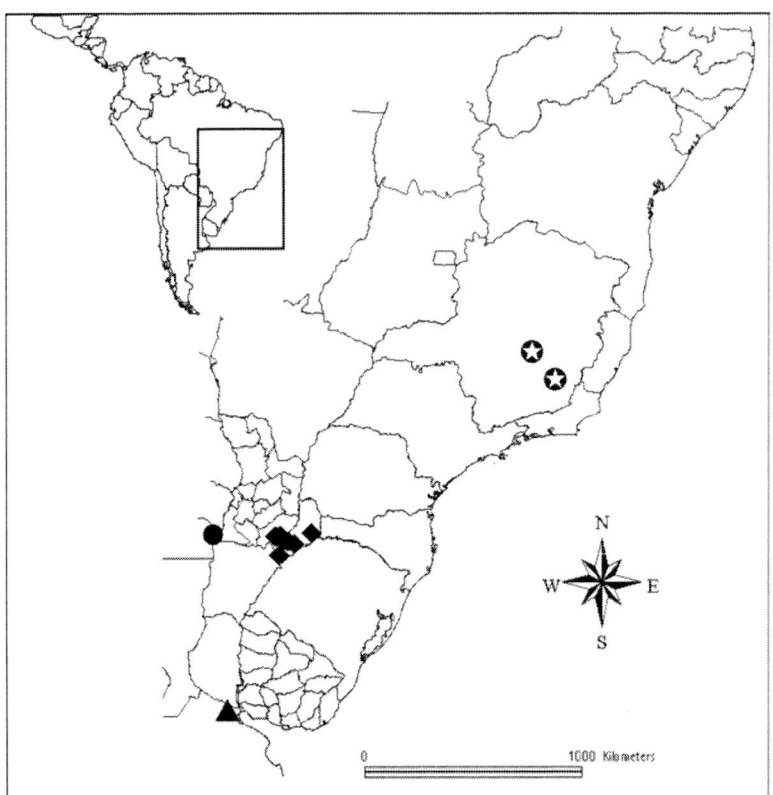

Figure 9. Geographic records of *Bibimys labiosus* from Lagoa Santa (left star) and Viçosa (right star) in Minas Gerais State, Brazil, and from seven localities in Misiones Province (diamonds), Argentina. The type locality for *Bibimys torresi* in Buenos Aires Province is indicated (triangle), as is the type locality for *B. chacoensis* in El Chaco Province (circle), Argentina.

Affinities of *Bibimys* and the Scapteromyini

Concept of the Scapteromyini: The concept of a scapteromyine group dates back to Winge (1887), who was the first to add three species (*labiosus, fronto* and *principalis,* currently allocated to the genera *Bibimys* and *Kunsia*) to the genus *Scapteromys* Waterhouse, 1837. Winge grouped the three forms mainly on the basis of "a thickening of the outer wall of the lacrimal canal" (i.e., an enlargement of the maxillary portion of the zygomatic arch). In spite of this assignment, Winge (1887) noted that *labiosus* departed considerably from the other two forms (*fronto* and *principalis*), which, in his opinion, was more compatible with the diagnosis of *Scapteromys* provided by Thomas (1884). A closer inspection of the zygomatic arch among representatives of the three genera in question reveals that while the maxillary part is enlarged in *Bibimys labiosus* and *Kunsia fronto*, *Scapteromys tumidus* displays a slender maxillary portion that does not differ in width from the jugal. Therefore, this character groups *Bibimys* and *Kunsia*, but does not relate *Bibimys* or *Kunsia* to *Scapteromys*. Consequently, the criteria used by Winge to group the three genera remain unclear.

Bibimys labiosus is nevertheless considerably divergent from *Scapteromys* and *Kunsia*, and this morphological variation led later authors to consider this species to be *incertae sedis* (Cabrera, 1961; Gyldenstolpe, 1932; Hershkovitz, 1966). Hershkovitz (1966) expressed his doubts regarding the placement of *labiosus*, noting its similarity to *Akodon azarae* and concluding that, "there is absolutely nothing about *labiosus* that suggests more than remote relationship to scapteromyines."

Finally, Massoia (1979) created the genus *Bibimys* and assigned it to the Scapteromyini based on the same zygomatic condition mentioned by Winge (1887) as well as on additional dental similarities, among them the presence of a long and narrow mesoloph. By later inclusion of *labiosus* in *Bibimys*, Massoia (1980) consequently reassigned this form to the scapteromyines.

Pardiñas (1996) reinforced Massoia's hypothesis regarding generic relationships, citing additional putative dental and alveolar homologies between *Bibimys* and *Scapteromys*, as well as mandibular homologies between *Bibimys* and *Kunsia*. He also postulated that the median enamel projections in the first and second upper and lower molars of *Bibimys*, which he believed constituted the mesoloph and mesolophid, were homologous to those revealed for *Scapteromys*. Based on these findings, *Bibimys* was hypothesized to represent an intermediate state between the complete pentalophodont and the tetralophodont molar stages, a condition referred to by Pardiñas (1996) as "complex tetralophodont."

We confirm the presence of a reduced mesolophid in the lower molars of the lectotype of *B. labiosus* (Figure 5b), but we disagree that the name mesoloph applies to the labial projections of the upper molars. Reig (1977) proposed that the paralophule and entolophulid constituted remnants of the mesoloph/id, an untested evolutionary hypothesis implicit in the terminology followed by Massoia (1979) and Pardiñas (1996). The adoption of different nomenclatures to describe the enamelled projections of the molar cusps is the basis of the divergence between these authors' descriptions and ours. Hershkovitz (1993) provided a more precise definition of the mesoloph/id and distinguished it from the paralophule and entolophulid; we therefore follow his terminology to describe the enamelled projections of the upper molars as paralophules, which originate from the paracone. It is interesting to note that an entolophulid and mesolophid are both present on the molars of *B. torresi* examined by Pardiñas (1996), providing evidence for the independent origin of these structures.

Table 3: Character state distributions for selected specimens of *Bibimys, Scapteromys* and *Kunsia*. Italicized character states are considered derived among Neotropical muroids.

Character	*Bibimys labiosus*	*Scapteromys tumidus*	*Kunsia fronto planaltensis*
1- Mesolophid	*reduced*	present	*absent*
2- Zygomatic arch	enlarged maxillary part	slender maxillary part	enlarged maxillary part
3- Zygomatic notch	moderate	*pronounced*	*pronounced*
4- Interorbital region	hourglass shape	hourglass shape	hourglass shape
5- Alisphenoid strut	present	absent	present
6- Parapterygoid fossae	wider than mesopterygoid fossa	narrower than mesopterygoid fossa	narrower than mesopterygoid fossa
7- Palate	*long, extending beyond the level of m3*	relatively short, not extending beyond the level of m3	relatively short, not extending beyond the level of of m3
8- Braincase	rounded	oblong	oblong
9- Tegmen tympani	contacts the squamosal	contacts the squamosal	contacts the squamosal

Table 3. Continued.

Character	*Bibimys labiosus*	*Scapteromys tumidus*	*Kunsia fronto planaltensis*
10- Carotid circulation	*pattern 2*	pattern 1	pattern 1
11- Plantar pads	six, with reduced hypothenar	six, with reduced hypothenar	five ** or six ***
12- Gall bladder	present	*absent*	unknown
13- Stomach	hemiglandular	*discoglandular*	unknown
14- Subapical collar of nonspinous epidermis	present	present	unknown
15- Dorsal papilla configuration	compound and simple	compound	simple
16- Urethral processes configuration	two-lobed	four-lobed	four-lobed
17- Cartilaginous bacular configuration	reduced, without lateral digits	reduced, with lateral digits	reduced, with lateral digits
18- Osseous baculum base	paddle-shaped	spoon-shaped	spoon-shaped
19- Diploid number	70	24, 34 and 36	44 *

*   Karyotype of *Kunsia tomentosus* (reported by Andrades-Miranda et al., 1999)
**   Values for *Kunsia fronto* (reported by Hershkovitz, 1966)
***   Values for *Kunsia tomentosus* (reported by Hershkovitz, 1966)

Much of the discussion about the scapteromyine affinities of *Bibimys* has, until now, focused on a small set of dental and cranial characters, a consequence of the scarcity of available specimens. Therefore, a comparative analysis of the new morphological and cytogenetic data provided in this study is essential to re-evaluating the affinities of *Bibimys* with *Scapteromys* and *Kunsia* (Table 3).

Phallic morphology: Few data are available regarding the phallic morphology of scapteromyines, the most complete descriptions being those of *Scapteromys aquaticus* and *S. tumidus* (Hershkovitz, 1966; Hooper and Musser, 1964). We therefore compared *Bibimys labiosus* with *Scapteromys tumidus* on the basis of data given by Hooper and Musser (1964). In the absence of descriptions for *Kunsia* species, we report on the conditions observed in one specimen of the original series of *K. f. planaltensis* (MN34500).

Bibimys labiosus has a well-differentiated glans. Nevertheless, some character states are shared between *B. labiosus* and *S. tumidus* and to a lesser extent, between *B. labiosus* and *K. f. planaltensis*. The multilobulate dorsal papilla, the presence of a subapical collar of non-spinous tissue and the reduction of the cartilaginous digits are features that help group *Bibimys* and *Scapteromys*. The reduction of the bacular mounds and distal parts of the baculum are the only phallic character shared by the three genera. Some of these similarities, however, are not absolute and the polarities for these characters deserve special attention in evaluating this grouping.

The compound dorsal papilla was present in only one of the three specimens examined, suggesting individual variation in the occurrence of this structure. It is interesting to note that samples of *Scapteromys* also showed variation in the configuration of this papilla (Hershkovitz, 1966; Hooper and Musser, 1964). The only specimen of *Kunsia f. planaltensis* examined had a simple papilla. Hooper and Hart (1962) documented the presence of compound dorsal papillae in microtine rodents, and its retention in *Scapteromys* and *Bibimys* could represent a plesiomorphic state. Alternatively, this condition may have been independently acquired by these two genera. It has not been reported for any other South American sigmodontine so far.

A subapical collar of non-spinous epidermis, present in *Scapteromys* and *Bibimys*, also occurs in other Sigmodontinae (e.g., *Oryzomys*, *Sigmodon*, *Oxymycterus*, *Akodon*), as well as in some representatives of the North American Neotomini-Peromyscini. Thus, it does not constitute a unique condition shared by *Scapteromys* and *Bibimys*.

Other striking phallic features shared by *Bibimys*, *Scapteromys* and *Kunsia* are the reduced cartilaginous baculum and bacular mounds. Based on ontogenetic evidence obtained from species of the genus *Abrothrix*, Spotorno (1992) postulated that reduction of cartilaginous digits and bacular mounds leading towards simplification of the glans architecture is a derived condition among the South American Sigmodontinae. While reduction in the cartilaginous baculum contributes to grouping of the three taxa, its configuration in the three forms is quite different. For example, in *B. labiosus,* the lateral digits are absent in all specimens examined, while in *S. tumidus* the lateral digits are present although reduced. Thus, although the three taxa share reductions of the distal baculum, this trend is accentuated in *Bibimys labiosus*.

Karyological considerations: Cytogenetic analyses of Brazilian specimens of *Scapteromys* revealed diploid numbers of 36, 34 and 24 (Freitas et al., 1984). Samples collected in Argentina had a diploid number of 32 (Fronza et al., 1976), and those from Uruguay a diploid number of 24 (Brum-Zorrilla et al., 1972). All karyomorphs displayed AN = 40. These karyotypes apparently show a gradient of centric fusions in populations from southern Brazil to northern Argentina (Andrades-Miranda et al., 1999). The karyotype of *Kunsia tomentosus* analysed by Andrades-Miranda et al. (1999) had 2n = 44 and AN = 42. G- and C-band and NOR analyses, fluorescence *in situ* hybridization (FISH) with telomeric probes, and restriction endonuclease bands all demonstrated an almost complete lack of homologies between the karyotypes of *B. labiosus* and *Scapteromys tumidus*. The diploid number of 70 described for both species of *Bibimys* analysed thus far differs strikingly from that reported for *Scapteromys* and *Kunsia*.

With regard to heterochromatin distribution, *Kunsia* revealed a pattern widespread in mammals, with C-bands located in the centromeric region of all autosomes in addition to the sexual pair. In *Scapteromys*, however, the 2n = 36 and 2n = 34 karyomorphs had pericentromeric C-bands restricted to one pair of autosomes and the X chromosome, while the Y chromosome was entirely heterochromatic. In the 2n = 24 karyomorph of *Scapteromys*, heterochromatin was restricted to the sex chromosomes (Freitas et al., 1984). *Bibimys* exhibited a C-banding pattern somewhat similar to that of *Scapteromys*, with heterochromatin almost restricted to the X chromosome. Although Freitas et al. (1984) considered this pattern of heterochromatin reduction to be unusual among mammals, recent reports show low amounts of heterochromatin in C-bands from other South American rodents (Reig et al., 1992; Rossi et al., 1995). It is difficult to postulate homologies between the karyotypes of *B. labiosus* and *S. tumidus* because the diploid and autosomal arms numbers are widely divergent and this pattern of heterochromatin distribution may not be unique to *Scapteromys* and *Bibimys labiosus*. The NOR staining pattern in *B. labiosus* departs from the pattern reported for *Scapteromys*, in which stained sites are present in the X chromosome in addition to three autosomal pairs (Andrades-Miranda et al., 1999).

Bibimys as a divergent genus: Complete cytogenetic and phallic information is still lacking for many sigmodontine taxa, preventing more comprehensive comparisons on the basis of these character systems. It may be relevant to consider additional cranial characters commonly emphasized in comparative studies of sigmodontines

(Carleton, 1980; Voss, 1993; Voss and Carleton, 1993). *Bibimys, Scapteromys* and *Kunsia* share only two of eight additional cranial characters compared (Table 3), namely the presence of a slender tegmen tympani and an hourglass-shaped interorbital region. Both conditions are widely distributed among sigmodontines and are considered to be plesiomorphic (Carleton, 1980; Voss, 1993). Therefore, similarities in bulla attachment and morphology of the interorbital region do not unambiguously support a grouping of the three taxa.

Moreover, a further subset of cranial characters reveals trenchant differences between *Bibimys* and the other two genera. *Bibimys* diverges from *Scapteromys* and *Kunsia* by its shallow zygomatic notch, derived carotid circulation pattern, relatively long palate, wider parapterygoid fossae, and rounded braincase. Most of these conditions probably constitute derived traits (Carleton, 1980), suggesting that *Bibimys* is a strongly differentiated genus.

The morphological and cytogenetic evidence provided in this study is ambiguous in evaluating the affinities of *Bibimys* with *Scapteromys* and *Kunsia*, given that the polarities of some phallic characters are dubious and karyological homologies are difficult to trace. Phallic similarities between *Bibimys labiosus* and *Scapteromys tumidus* could be interpreted as homologies that evolved in parallel with the conditions seen in microtines (Hooper and Hart, 1962), some North American neotomines, and in *Abrothrix* species. At any rate, information from the phallus and karyotype is not decisive in determining the integrity and relationships among these forms. Further, the additional cranial and stomach characters surveyed do not support the natural grouping of *Bibimys* with *Scapteromys* or *Kunsia*, but, rather, indicate that the former genus is considerably divergent.

Recent sequence analyses of the cytochrome-*b* gene (Smith and Patton, 1999), which did not include *Bibimys*, place *Scapteromys* and *Kunsia* within an expanded Akodontini tribe, despite low confidence intervals. The placement of *Bibimys* within this group is not supported by phallic and karyological characters. A further evaluation of the integrity and relationships of the Scapteromyini will require additional evidence drawn from a multidisciplinary data analysis based on a broader sample of the three genera in question as well as members of the Akodontini. Such an approach will help to define closely related outgroups that will eventually allow the formulation of appropriate hypotheses of morphological evolution for the three genera in question.

ACKNOWLEDGEMENTS

The authors would like to dedicate this contribution to Dr. James L. Patton, as an acknowledgment of his outstanding career, which has represented an inspiration to recent generations of Brazilian mammalogists. We thank Dr. H. Baagøe and Prof. M. Andersen for permission to study Lund's specimens and for their hospitality during the visits of J. A. Oliveira and L. M. Pessôa to the ZMC. Dr. G. A. Ribeiro

(Departamento de Engenharia Florestal, UFV) authorized fieldwork in the Mata do Paraíso. J. Dergam, R. Feio and G. Lessa gently allowed access to the facilities of the MZUFV, and J. L. Pontes and J. B. dos Santos provided valuable assistance during trapping sessions. We are grateful to Dr. C. J. Tribe for reading an earlier version of the manuscript and for providing valuable comments, to M. Pessôa for the ink-drawing in Figure 3, and to Geert Brovad (ZMC) for providing slides used to compose figures of type specimens. Our license for collecting was provided by Instituto Brasileiro do Meio Ambiente e dos Recursos Naturais Renováveis (IBAMA). This research was partially funded by resources from the Vice-Reitoria para Ensino de Pós-graduação e Pesquisa (SR2-UFRJ). Work by P. R. Gonçalves and L. M. Pessôa were supported by graduate and research fellowships from Conselho Nacional de Desenvolvimento Científico e Tecnológico (CNPq).

LITERATURE CITED

Andrades-Miranda, J., A. P. Nunes, L. F. B. Oliveira, and M. S. Matevi
   1999   The karyotype of the South American rodent *Kunsia tomentosus* (Lichtenstein, 1830). Cytobios 98:137-147.

Brum-Zorrilla, N., N. Lafuente, and P. Kiblisky
   1972   Cytogenetic studies in the cricetid rodent *Scapteromys tumidus* (Rodentia - Cricetidae). Experientia 28:1373-1373.

Cabrera, A.
   1961   Catalogo de los mamíferos de América del Sur. Revista del Museo Argentino de Ciencias Naturales "Bernardino Rivadavia", Ciencias Zoológicas 4:309-732.

Carleton, M. D.
   1973   Survey of gross stomach morphology in New World cricetinae (Rodentia, Muroidea), with comments on functional interpretations. Miscellaneous Publications, Museum of Zoology, University of Michigan 146:1-43.

   1980   Phylogenetic relationships in neotomine-peromyscine rodents (Muroidea) and a reappraisal of the dichotomy within New World Cricetinae. Miscellaneous Publications, Museum of Zoology, University of Michigan 157:1-145.

Carleton, M. D., and G. G. Musser
    1989   Systematic studies of oryzomyine rodents (Muridae, Sigmodontinae): a synopsis of *Microryzomys*. Bulletin of the American Museum of Natural History 191:1-83.

Dyzenchauz, F. J., and A. I. Massarini
    1999   First cytogenetic analysis of the genus *Bibimys* (Rodentia, Cricetidae). Zeitschrift für Säugetierkunde 64:59-62.

Eisenberg, J. F., and K. H. Redford
    1999   Mammals of the Neotropics. Vol. 3. The University of Chicago Press, Chicago and London, 609 pp.

Ford, C. E., and J. L. Hamerton
    1956   A colchicine hypotonic citrate squash sequences for mammalian chromosomes. Stain Technology 31:247-251.

Freitas, T. R. O., M. S. Mattevi, and L. F. B. Oliveira
    1984   Unusual C-band patterns in three karyotypically rearranged forms of *Scapteromys* (Rodentia, Cricetidae) from Brazil. Cytogenetics and Cell Genetics 38:39-44.

Fronza, T. G., R. L. Wainberg, and B. E. Llorente
    1976   Polimorfismo del cromosoma X y significación filogenética del cariotipo de la ``Rata aquatica'' *Scapteromys aquaticus* (Rodentia, Cricetidae) de la ribera de Punta Lara (Argentina). Mendeliana 1:41-48.

Gyldenstolpe, N. C. G.
    1932   A manual of Neotropical sigmodont rodents. Küngliche Svenska Vetenskapsakad, Handlinger Series 3.B 3:1-164.

Hershkovitz, P.
    1962   Evolution of Neotropical cricetine rodents (Muridae) with special reference to the phyllotine group. Fieldiana: Zoology 46:1-524.

    1966   South American swamp and fossorial rats of the scapteromyine group (Cricetinae, Muridae) with comments on the glans penis in murid taxonomy. Zeitschrift für Säugetierkunde 31:81-149.

Hershkovitz, P.
    1993   A new central Brazilian genus and species of sigmodontine rodent (Sigmodontinae) transitional between akodonts and oryzomyines, with a discussion of muroid molar morphology and evolution. Fieldiana: Zoology 75:1-18.

Hooper, E. T., and B. S. Hart
    1962   A synopsis of Recent North American microtine rodents. Miscellaneous Publications Museum of Zoology, University of Michigan 120:1-68.

Hooper, E. T., and G. G. Musser
    1964   The glans penis in neotropical cricetines (Family Muridae) with comments on classification of muroid rodents. Miscellaneous Publications Museum of Zoology, University of Michigan 123:1-57.

Howell, W. M., and D. A. Black
    1980   Controlled silver staining of nucleolus organizer regions with a protective colloidal developer: a 1-step method. Experientia 36:1014-1015.

Levan, A., K. Fredga, and A. A. Sandberg
    1964   Nomenclature for centromeric position on chromosomes. Hereditas 52:201-220.

Lidicker, W. Z., Jr.
    1968   A phylogeny of New Guinea rodent genera based on phallic morphology. Journal of Mammalogy 49:609-643.

Massoia, E.
    1979   Descripción de un gênero y especie nuevos: *Bibimys torresi* (Mammalia - Rodentia - Cricetidae - Sigmodontinae - Scapteromyni). Physis 38:1-7.

    1980a   El estado sistemático de cuatro especies de cricetidos sudamericanos y comentarios sobre otras especies congenericas. Ameghiniana 17:280-287.

    1980b   Mammalia de Argentina. I - Los mamíferos silvestres de la provincia de Misiones. Iguazú 1:15-43.

Massoia, E.
    1983   La alimentación de algumas aves del Orden Strigiformes en la
            Argentina. El Hornero, Numero Extraordinário:125-148.

    1988a   Análisis de regurgitados de *Tyto alba* de Ituzaingó, partido de Morón,
            provincia de Buenos Aires. Boletím Científico, Asociación para la
            Protección de la Naturaleza 2:13-20.

    1988b   Presas de *Tyto alba* en Campo Ramón, Departamento Oberá, Provincia
            de Misiones. Boletím Científico, Asociación para la Protección de la
            Naturaleza 7:4-16.

Massoia, E., J. C. Chebez, and S. Heinonen Fortabat
    1989a   Segundo análisis de egagrópilas de *Tyto alba tuindara* en el
            Departamento de Apóstoles, Província de Misiones. Boletím Científico,
            Asociación para la Protección de la Naturaleza 13:3-8.

    1989b   Análisis de regurgitados de *Tyto alba tuindara* de Los Helechos,
            Departamento Oberá, Provincia de Misiones. Boletím Científico,
            Asociación para la Protección de la Naturaleza 14:16-22.

    1989c   Mamíferos e aves depredados por *Tyto alba tuindara* en Bonpland,
            Departamento Candelária, Provincia de Misiones. Boletím Científico,
            Asociación para la Protección de la Naturaleza 15:19-24.

    1989d   Mamíferos e aves depredados por *Tyto alba* en el arroyo Yabebyrí,
            Departamento de Candelária, Provincia de Misiones. Boletím
            Científico, Asociación para la Protección de la Naturaleza 15:8-13.

Massoia, E., S. I. Tiranti, and M. P. Torres
    1989   La depredación de pequenos mamíferos por *Tyto alba* en Canal 6,
            Partido de Campana, Província de Buenos Aires. Boletím Científico,
            Asociación para la Protección de la Naturaleza 13:14-19.

Moojen, J.
    1952   Os roedores do Brasil. Instituto Nacional do Livro. Rio de Janeiro, 214
            pp.

Musser, G., and M. D. Carleton
  1993   Family Muridae. Pp. 501-755 in Mammal Species of the World: a
          taxonomic and   geographic reference (D.E. Wilson and D.M. Reeder,
          eds.). Smithsonian Institution Press, Washington, 1207 pp.

Myers, P., J. L. Patton, and M. F. Smith
  1990   A review of the *boliviensis* group of *Akodon* (Muridae: Sigmodontinae),
          with emphasis on Peru and Bolivia. Miscellaneous Publications
          Museum of Zoology, University of Michigan 177: 1-104.

Nowak, R. M.
  1999   Walker's mammals of the World.   6th edition. The Johns Hopkins
          University Press, Baltimore and London, 1936 pp.

Paglia, A. P., P. De Marco Junior, F. M. Costa, R. F. Pereira, and G. Lessa
  1995   Heterogeneidade estrutural e diversidade de pequenos mamíferos em
          um fragmento de mata secundária de Minas Gerais, Brasil. Revista
          Brasileira de Zoologia 12:67-79.

Pardiñas, U. F. J.
  1995   Novedosos cricétidos (Mammalia, Rodentia) en el Holoceno de la
          region pampeana, Argentina. Ameghiniana 32:197-203.

  1996   El registro fosil de *Bibimys* Massoia, 1979 (Rodentia) en la Argentina.
          Consideraciones sobre los Scapteromyini (Cricetidae, Sigmodontinae) y
          su distribuicion durante el Plioceno-Holoceno en la region Pampeana.
          Mastozoología Neotropical 3:15-38.

Reig, O. A.
  1977   A proposed unified nomenclature for the enamelled components of the
          molar teeth of the Cricetidae (Rodentia). Journal of Zoology, London
          181:227-241.

Reig, O. A., A. I. Massarini, M. O. Ortells, M. A. Barros, S. I. Tiranti, and F. J.
Dyzenchauz
  1992   New karyotypes and C-banding patterns of the subterranean rodents of
          the genus *Ctenomys* (Caviomorpha, Octodontidae) from Argentina.
          Mammalia 56:603-623.

Rossi, M. S., C. A. Redi, G. Viale, A. I. Massarini, and E. Capanna
    1995   Chromosomal distribution of the major satellite DNA of South American rodents of the genus *Ctenomys*. Cytogenetics and Cell Genetics 69:179-184.

Seabright, M. A.
    1972   A rapid banding technique for human chromosomes. Lancet 2:971-972.

Smith, M. F., and J. L. Patton
    1999   Phylogenetic relationships and the radiation of sigmodontine rodents in South America: Evidence from cytochrome *b*. Journal of Mammalian Evolution 6:89-128.

Spotorno, A. E.
    1992   Parallel evolution and ontogeny of simple penis among New World cricetid rodents. Journal of Mammalogy 74:504-514.

Sumner, A. T.
    1972   A simple technique for demostrating centromeric heterochromatin. Experimental Cell Reseach 75:304-306.

Thomas, O.
    1884   On a collection of Muridae from central Peru. Proceedings of the Zoological Society of London 447-458.

Trouessart, E.-L.
    1898   Catalogus Mammalium tam viventium quam fossilium. Tomus I. R. Friedländer & Son, 664 pp.

Voss, R. S.
    1988   Systematics and ecology of ichthyomyine rodents (Muroidea): patterns of morphological evolution in a small adaptive radiation. Bulletin of the American Museum of Natural History 188:259-493.

    1993   A revision of the Brazilian muroid rodent genus *Delomys* with remarks on "thomasomyine" characters. American Museum Novitates 3073:1-44.

Voss, R. S., and M. D. Carleton
  1993   A new genus for *Hesperomys molitor* Winge and *Holochilus magnus* Hershkovitz (Mammalia, Muridae) with an analysis of its phylogenetic relationships. American Museum Novitates 3085:1-39.

Voss, R. S., and P. Myers
  1991   *Pseudoryzomys simplex* (Rodentia: Muridae) and the significance of Lund's collections from the caves of Lagoa Santa, Brazil. Bulletin of the American Museum of Natural History 206:414-432.

Winge, H.
  1887   Jordfundne og nulevende Gnavere (Rodentia) fra Lagoa Santa, Minas Geraes, Brasilien. E Museo Lundii 1:1-200.

An Introduction to the Genus *Bibimys* (Rodentia: Sigmodontinae): Phylogenetic Position and Alpha Taxonomy

Guillermo D'Elía, Ulyses F. J. Pardiñas, and Philip Myers

Bibimys is an exceptionally poorly understood genus of sigmodontine rodents. It is known from both fossil and recent specimens. Currently, *Bibimys*, with three species, is placed with the genera *Kunsia* and *Scapteromys* in the tribe Scapteromyini. This study is the first to examine the phylogenetic position of *Bibimys* within the Sigmodontinae and, in particular, to examine its purported close relationship with *Kunsia* and *Scapteromys*. It is the first study to include samples of all three of the currently recognized species of *Bibimys*. We found that *Bibimys*, *Kunsia* and *Scapteromys* do not form a monophyletic group and that all three genera seem to be part of the akodont radiation. We present an amended diagnosis of the genus. Our analyses cast doubt on the distinctiveness of the three forms currently included within *Bibimys*. A gazetteer of *Bibimys* localities is presented. Finally, we suggest future research directions that should shed new light on the evolutionary history of this little-known genus.

SYSTEMATICS OF *BIBIMYS* AND THE SCAPTERMYINI

Sigmodontine rodents traditionally have been arranged in groups of genera based on morphological similarity. Some of these groups have received the formal rank of tribes (e.g., McKenna and Bell, 1997). Few, however, have been examined from a formal phylogenetic perspective. One of the proposed groups is the tribe Scapteromyini, which was informally described by Hershkovitz (1966) and later formally introduced by Massoia (1979) although without a diagnosis. The scapteromyines are one of the smallest of the sigmodontine tribes, consisting of only three genera: *Bibimys* (3 recent species currently recognized), *Kunsia* (2 recent species) and *Scapteromys* (2 recent species). For convenience, we shall informally refer to these three genera as the scapteromyine genera, but testing the phylogenetic validity of the tribe Scapteromyini and determining its evolutionary position within the Sigmodontinae is one of our goals in this paper.

Few studies deal with the scapteromyine genera. In part, this is because scapteromyine rodents, with the exception of *Scapteromys*, are known from very few specimens. Some of the type specimens are badly preserved (e.g., *Bibimys labiosus*, *Kunsia fronto chacoensis*) or were described from fossil material (e.g., *Kunsia fronto fronto*). In addition, these rodents are mainly distributed in the Río de la Plata basin, which traditionally has not been the main focus of authorities working on sigmodontine evolutionary history. As a result of the limited number of studies dealing with these genera, our knowledge of scapteromyine systematics is exceptionally poor even in such basic aspects as alpha taxonomy, patterns of diversification, and place within the sigmodontine radiation. Only after these fundamental issues are clarified will we be able to investigate the processes by which the diversity of the group arose, its biogeographic history, and its conservation status.

The present contribution is part of a revisionary study of the scapteromyine rodents. It is based on both morphological and molecular evidence. New material from the genus *Bibimys* has recently become available as the result of fieldwork in Argentina, Brazil, and Paraguay. Here, we examine the genus *Bibimys* at two hierarchical levels. First, we analyze the phylogenetic position of *Bibimys* within the sigmodontine radiation and, in particular, its relationships with the other scapteromyine genera. Second, we provide an emended diagnosis of the genus *Bibimys* and present new evidence concerning the validity of its three species.

PHYLOGENETIC ANALYSIS

Mice and rats of the subfamily Sigmodontinae (*sensu* Reig, 1980) constitute one of the most diverse components of the Neotropical mammal fauna. Further, current and past sigmodontine diversity is still being characterized with the continuing description of both extant and fossil genera and species (e.g., González et al., 1998; Ortiz et al., 2000). Recently, the widespread adoption of a cladistic approach (e.g., Engel et al., 1998; Smith and Patton, 1999; Steppan, 1995) has lead to substantial progress in clarifying the phylogenetic relationships of the numerous sigmodontine taxa. However, our understanding of sigmodontine evolutionary history is far from complete (D'Elía, 2000), in part because large and/or relevant components of the sigmodontine radiation have not yet been included in any phylogenetic analysis.

The taxonomic history of the scapteromyine group is brief and relatively simple. A scapteromyine concept can be traced back to Peters (1860), who placed *Mus tomentosus* together with *Hesperomys* (*Scapteromys*) *tumidus* in *Scapteromys*. Thirty years after its original description, Fitzinger (1867) elevated *Scapteromys* to generic status, although some authors continued to use *Scapteromys* as a subgenus of *Hesperomys* (e.g., Thomas, 1884). In the following years, more fossil and recent species were added to *Scapteromys* (Gyldenstolpe, 1932; Miranda-Ribeiro, 1914; Winge, 1887). Hershkovitz (1966) removed taxa previously assigned to *Scapteromys* (e.g., *S. tomentosus*) to create the genus *Kunsia* and suggested that both genera be combined into an informal group known as the scapteromyines. He further suggested that the scapteromyines were part of the akodont radiation. At the same time, he called into question the scapteromyine condition of three other taxa, including *B. labiosus* (at that time *Scapteromys labiosus*). Avila-Pires (1972) described a new subspecies of *Kunsia fronto*. Later, Massoia (1979), the first author to use the name Scapteromyini, expanded the group with the description of a new species and genus, *Bibimys torresi*. One year later, Massoia (1980a) allocated two species - *Akodon chacoensis* and *Scapteromys labiosus* – to *Bibimys*. More recently, as result of a broad phylogenetic study, Smith and Patton (1999) found that the scapteromyine genera *Kunsia* and *Scapteromys* formed a clade that fell within their akodont lineage, and therefore they did not recognize the Scapteromyini as a tribe separate from the Akodontini. These authors suggested that analysis of *Bibimys* DNA sequences would shed new light on the relationships of their expanded Akodontini.

Materials and Methods

Voucher specimens for the individuals sequenced in this study are or will be deposited in the following collections: Argentina - Centro Nacional Patagónico Colección Mamíferos (CNP) and Museo de Mar del Plata Colección Mamíferos (MMP-Ma); Brazil - Museu Nacional do Rio do Janeiro (MN); Paraguay - Museo Nacional de Historia Natural del Paraguay (MNHNP; GD: field number of Guillermo D'Elía); and United States of America - Museum of Vertebrate Zoology (MVZ) and The University of Michigan Museum of Zoology (UMMZ). Complete cytochrome *b* gene sequences from specimens belonging to the three recognized species of *Bibimys* were studied as follows. *B. chacoensis* ($n = 2$): CNP 756 (Argentina, Province of Chaco, Department of Bermejo, Cancha Larga, S 27° 04 W 58° 43) and GD 153 (Paraguay, Myers et al., *in prep.*). *B. labiosus* ($n = 2$): MN 62062 and MN 62063 (Brazil, State of Minas Gerais, Viçosa, S 20º 45' W 42º 53'). *B. torresi* ($n = 1$): MMP-Ma 3620 (Argentina, Province of Buenos Aires, Canal 6, S 34º 09' W 58º 57'). The specimens from *B. chacoensis* and *B. labiosus* are from localities close to the type localities for each form and agree morphologically with the descriptions of the corresponding holotypes (pers. obs. and Gonçalves et al., this volume). The specimen of *B. torresi* comes from the type locality of the species. In addition we sequenced one specimen of each of the following species: *Akodon montensis* (MNHNP 2910: Paraguay, Department of Itapua, Estancia San Isidro, S 26º 29' W 55º 54'), *A. azarae* (GD 264: Paraguay, Department of Paraguari, Cost of the Tebicuary River, S 26º 24' W 57º 02'), *Scapteromys aquaticus* (UMMZ 17499: Paraguay, Department of Paraguari, Cost of the Tebicuary River, S 26º 24' W 57º 02') and *S. tumidus* (MVZ 183269: Uruguay, Department of Maldonado, Las Flores, S 34º 49' W 55º 19'). We completed our sigmodontine dataset with sequences available through GENBANK as of May 2001, and with new sequences kindly provided by J. L. Patton and M. F. Smith (Museum of Vertebrate Zoology).

Although sigmodontine (*sensu* Reig, 1980) monophyly is well corroborated (Engel et al., 1998; but see also Steppan, 1995) the identity of the sister group of sigmodontines is not clear. The subfamily Sigmodontinae falls in a large cricetid clade composed of other main branches of the muroid radiation (see D'Elía, 2000). Currently, the relationships among those groups are not understood. Therefore, to polarize sigmodontine character states, we have included as outgroups (Nixon and Carpenter, 1993) representatives of the other primary clades that make up the large cricetid clade. These are the arvicolines, the cricetines, the neotomines, the peromyscines and the tylomyines. Outgroup sequences were retrieved from GENBANK, with the exception of that for *Tylomys*, which was generated in this study. A total of 115 taxa were included in our phylogenetic analysis, of which 93 belong to the ingroup and 22 belong to the outgroup. All taxa analyzed as well as the sources of their cytochrome *b* sequences are listed in Appendix 1.

The cytochrome *b* gene sequences reported in this study were amplified and sequenced in two fragments using primers located both internally and in the flanking regions of the gene (MVZ 05–MVZ 16 and MVZ 103-MVZ 14, Smith and Patton, 1993; Smith *pers. com.*). Negative controls were included in all experiments. Dye-labeled PCR products were sequenced using an ABI 377 automated sequencer. In all cases both heavy and light DNA strands were sequenced and compared.

Sequence alignment was done with the program Clustal X (Thompson et al., 1997) using the default values for all alignment parameters. Cytochrome *b* sequences of sigmodontine taxa reported in GENBANK vary from 1140 to 1144 base pairs. From the alignment it was clear that the position of the indel responsible for the difference is at the very end of the sequence, but it was impossible to determine unambiguously its exact position (i.e, if it corresponds to the codon number 380 or 381). To avoid this problem, we chose to work with the first 1134 bases of the sequences.

Aligned sequences were subjected to maximum parsimony analysis (MP; Kluge and Farris, 1969; Farris, 1982). In all cases, characters were treated as unordered and equally weighted. We employed two strategies to search for the most parsimonious tree(s). First, PAUP* 4 (Swofford, 2000) was used to perform 200 replicates of traditional heuristic searches with random addition of sequences and tree bisection-reconnection branch swapping. Second, two batches of parsimony ratchet (Nixon, 1999) were performed in PAUP* using command files written in PAUPRat (Sikes and Lewis, 2001). The difference between the two ratchet batches was the percentage of perturbed characters, which were 15% and 25%, respectively. Each ratchet batch consisted of 20 series of 200 iterations. For each series a different command file was used. Two measures of clade support were calculated. First, 500 bootstrap replications with 3 addition sequence replicates each were executed. Second, we performed 500 parsimony jackknife replications with 3 addition sequence replicates each and the deletion of one-third of the character data. In both bootstrap and jackknife searches, the branches with less than 50 % of support were allowed to collapse.

Results and Discussion

Four different cytochrome *b* haplotypes were found among the five specimens of *Bibimys* that were sequenced. Both individuals from Viçosa (Minas Gerais, Brazil) shared the same haplotype. Therefore, only one specimen from that locality was considered in the phylogenetic analyses. The cytochrome *b* gene of *Bibimys* had a length of 1141 base pairs. *Bibimys* cytochrome *b* haplotypes showed a strong base compositional bias (Table 1), with a marked deficit of guanine (12.27 % of all positions, 3.68 % in third positions). Similar compositional biases are common in sigmodontine rodents (e.g., our complete dataset; Myers et al., 1995; Smith and Patton, 1993, 1999) and in mammals in general (Irwin et al., 1991).

Our total dataset had 677 variable characters, of which 570 were parsimony-informative. The 200 replicates of heuristic search found 48 shortest trees, while the parsimony ratchets recovered those 48 trees plus 107 additional shortest trees, totaling 155 shortest trees. A final TBR swapping of those 155 trees performed in PAUP* 4 (Swofford, 2000) found an additional shortest tree not originally recovered by either the replicated heuristic searches or the parsimony ratchets. In total, we recovered 156 most parsimonious trees of 10280 steps (CI = 0.120, RI = 0.474). The strict consensus of those trees is shown in Figure 1.

Table 1. Base composition of *Bibimys* cytochrome *b* haplotypes (1141 bases). Haplotype data from four specimens are presented; each specimen is identified by its museum record number (e.g., MN 62062).

	A	C	G	T
MN 62062[a]	0.29448	0.30149	0.12445	0.27958
CNP 756[a]	0.29886	0.30061	0.12182	0.27870
GD 153[a]	0.29886	0.29886	0.12182	0.28046
MMP 3620[a]	0.29798	0.30149	0.12270	0.27783
Mean across haplotypes[b]	0.29755	0.30061	0.12270	0.27914
1st position[c]	0.30577	0.23950	0.20997	0.24475
2nd position[c]	0.21053	0.25263	0.12105	0.41579
3rd poition[c]	0.37632	0.40987	0.03684	0.17697

[a] The mean frequency of each base, averaged across all codon positions.
[b] The mean frequency of each base, averaged across all haplotypes.
[c] The mean frequency of each base, averaged across all haplotypes and listed by codon position.

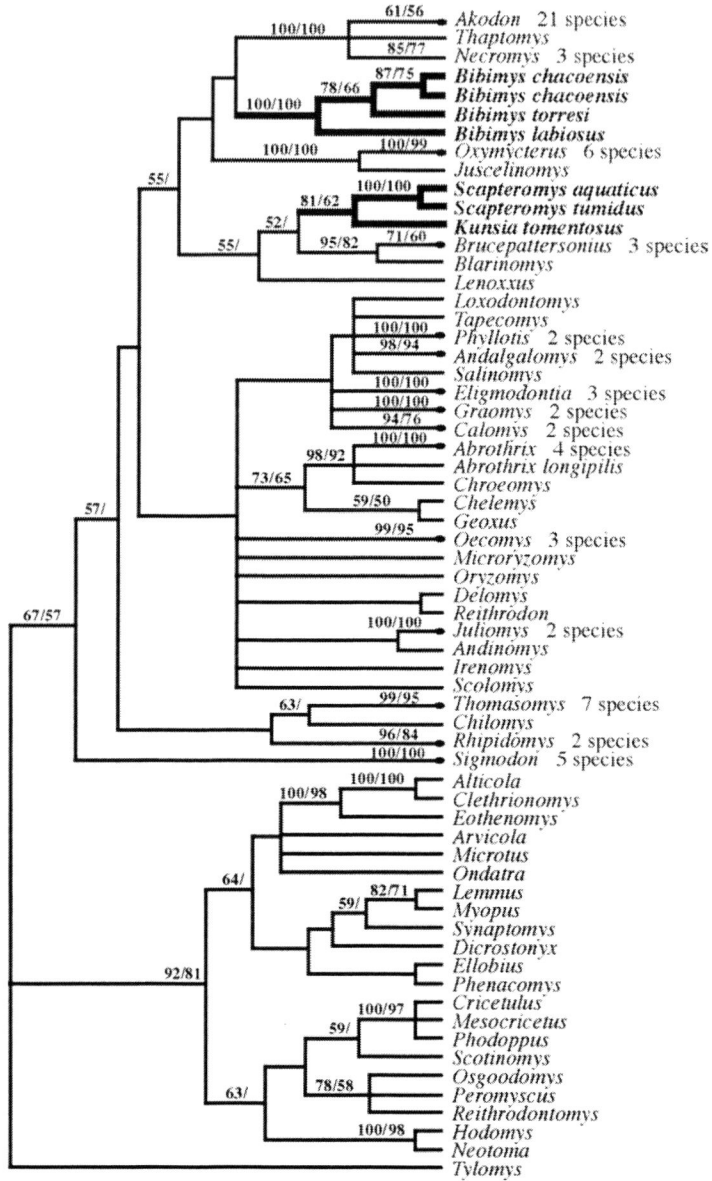

Figure 1. Strict consensus of the 156 most-parsimonious trees (length 10280, CI = 0.120, RI = 0.474) obtained by maximum parsimony analyses (i.e., heuristic searches and parsimony ratchets) of 93 sigmodontine and 22 outgroup taxa. More than one species of several non-scapteromyine genera were included in the analyses (see Appendix 1). To simplify the figure, only genera are shown as terminal taxa. Exceptions are the scapteromyine taxa and the genus *Abrothrix*, which was not monophyletic. Numbers above branches indicate parsimony jackknife (right of the diagonal) and bootstrap (left of the diagonal). Only values above 50 % are shown. For technical details of the analyses, refer to the text.

In the most recent treatise of mammal classification (McKenna and Bell, 1997) the Scapteromyini is listed as a sigmodontine tribe, implying that these rodents constitute a monophyletic group. However, the results obtained here (Figure 1) do not support the grouping of *Bibimys*, *Kunsia* and *Scapteromys* as a natural group. *Bibimys* is the sister group of a clade containing *Akodon*, *Necromys* and *Thaptomys*, whereas *Kunsia* and *Scapteromys* are sister to each other in a clade also containing *Brucepattersonius*, *Blarinomys* and *Lenoxus*. Both clades are part of the tribe Akodontini as defined by Smith and Patton (1999). However, the two clades containing the scapteromyine genera, as well as the whole akodont clade, have low levels of support (<50%) as measured by bootstrap values (Figure 1). When only those clades with bootstrap values ≥ 50 % are retained, *Bibimys* and the clade *Kunsia-Scapteromys* fall into a basal polytomy involving several sigmodontine lineages. Although parsimony jackknife values are in general somewhat higher than bootstrap values, they are also low. This lack of support for most sigmodontine suprageneric groups (in analyses based on cytochrome *b* gene sequences) was earlier documented by Smith and Patton (1999). Currently, it is unclear if the lack of resolution of suprageneric sigmodontine groups is the result of the limited resolving power of the cytochrome *b* gene at that taxonomic level or the result of an explosive radiation of the group following its entry into the South American continent.

These results indicate that one must be cautious with the interpretation of scapteromyine relationships. Clearly, phylogenetic analyses of other data sets are needed to evaluate hypotheses about the phylogenetic relationships of *Bibimys* with respect to *Kunsia* and *Scapteromys* in particular, and with the rest of the sigmodontines in general. Of those possible data sets, the study of more slowly evolving nuclear DNA sequences, as well as a broad morphological analyses not limited to craniodental features, are promising. We are currently working in both directions.

Given the lack of support for a scapteromyine clade at the molecular (cytochrome *b*) level, it is interesting to explore the reasons why *Bibimys* has traditionally been considered related to *Scapteromys* and *Kunsia*. As noted above, few authors have analyzed the phylogenetic relationships of *Bibimys*. Massoia (1979), when including *Bibimys* in the tribe Scapteromyini, implied a close phyletic relationship of this genus with *Kunsia* and *Scapteromys*. This evolutionary hypothesis was based exclusively on similarities in molar morphology, including the presence of a slender mesoloph in M1 and M2, a cylindromorphic M3, and reduced posterolophids on m1 and m2. Pardiñas (1996), who studied dental and jaw morphology of both extant and extinct specimens, later reinforced Massoia's hypothesis. However, Hershkovitz (1966) explicitly excluded *Bibimys* (*Scapteromys labiosus* at that time) from his concept of a scapteromyine group, stating that it morphologically resembled *Akodon azarae*. His conclusions were based mainly on the cranial and molar morphology of *B. labiosus* as portrayed in Winge's (1887) description of that species. Thus, different authors reached strikingly different conclusions from the study of the same characters. This situation, common in rodent systematics, is due in part to the difficulty of establishing homologies among dental features such as cusps, lophs, etc., as well as the relatively small number of characters provided by molar morphology. Recent progress in the study of molar development (see Jernvall and Jung, 2000; Jernvall. at al., 2000; Peters and Balling, 1999) should improve our ability to determine the homologies of these traits.

In addition to the small set of craniodental characters mentioned above and the DNA sequence data presented here, two other types of data have been used to explore the phylogenetic affinities of *Bibimys*. Gonçalves et al. (this volume), based on a phenetic comparison of phallic and karyotypic data, cast doubt on the affinity of *Bibimys* to *Kunsia* and *Scapteromys*. Unfortunately, as those authors indicate, their results are difficult to interpret because these characters were not analyzed in a cladistic framework and the taxonomic coverage of the study was limited to scapteromyine genera. As a result, it is not possible to determine if suggested shared character states are synapomorphies or plesiomorphies, or if they evolved independently.

Interestingly, a new species of sucking lice has recently been recorded from *Bibimys* specimens (Castro and Gonzalez, 2003). This parasite seems to belong to the *aitkeni* species group of the genus *Hoplopleura*. The other species of this group are parasites of mice of the genera *Akodon* and *Necromys*. In contrast, *H. scapteromydis*, recorded from *Scapteromys*, does not belong to the *aitkeni* group (Castro and Gonzalez, 2003). The relationships among these parasites have not been analyzed phylogenetically, however, and this evidence should be interpreted with caution.

Our analysis supports the inclusion of *Bibimys*, *Kunsia* and *Scapteromys* in the akodontine tribe as defined by Smith and Patton (1999). This result is an extension of analyses presented by Smith and Patton (1999), who found evidence for the placement of the clade *Kunsia*-*Scapteromys* within the akodont radiation. This finding is important because it implies that the akodont rodents are both morphologically and ecologically more diverse than traditionally recognized, including cursorial akodon-like mice and long nosed mice, as well as the crimson nosed *Bibimys*, the semiaquatic *Scapteromys* and the largest sigmodontine rat, the fossorial *Kunsia*. Hershkovitz (1966), who suggested an akodont origin for his scapteromyine group (*Kunsia* plus *Scapteromys*), was the first to point out the akodont condition of the scapteromyines. Nevertheless, one must be cautious in expanding the akodont tribe to include the scapteromyines due to the low levels of support found for this hypothesis (Figure 1). In addition, it is worth noting that the akodont tribe, as currently defined, lacks a formal diagnosis. Overall, in the light of these results and other sigmodontine phylogenetic studies, it is clear that the sigmodontine tree is far from resolved. Future cladistic analyses of morphological characters along with new DNA sequences are needed to provide a set of derived character states useful for defining the limits and contents of the akodonts and other related groups.

BIBIMYS ALPHA TAXONOMY

Currently, three species of *Bibimys* are recognized. The genus was erected by Massoia in 1979, with the type species *B. torresi* Massoia, 1979. Later, Massoia (1980a) referred *Scapteromys labiosus* Winge, 1887, and *Akodon chacoensis* Shamel, 1931, to *Bibimys* but did not critically review the genus or provide diagnostic characters. *Bibimys* has a large distribution in tropical and subtropical lowlands of eastern Argentina, east-central Brazil, and eastern Paraguay (Myers et al., *in prep.*), although records are scattered and include very few known localities. Interestingly, there are specimens of *Bibimys* in Argentinean and Brazilian Late Pleistocene-Holocene deposits (Pardiñas, 1996; Pardiñas et al., 2004; Winge, 1887).

The existence of three species in the genus *Bibimys* has never been questioned. The genus, however, has never been critically revised. Trapping members of the genus is difficult and, as a result, few specimens are available (see notes in Dyzenchauz and Massarini, 1999). The type localities of *Bibimys* are separated by considerable distances (Figure 2), which we suspect has supported the common assumption of the existence of significant differences among these three forms. Pardiñas (1996), based on craniodental characters, suggested that *B. chacoensis* (there referred to as *B. labiosus* following Massoia, 1988) and *B. torresi* were conspecific. In contrast, Gonçalves et al. (this volume), based on geographic distances between known populations and differences in the number of autosomal arms of *B. torresi* and *B. labiosus*, argued for the validity of both species. The present contribution is the first attempt to evaluate the alpha taxonomy of *Bibimys* that includes material referred to all recognized forms.

Materials and methods

Morphological analyses of *Bibimys* were based on fossil and recent specimens (Appendix 2). The recent specimens included here were live trapped by us, recovered from owl pellets, or borrowed from museum collections. All specimens are or will be housed in the following collections: Argentina - Museo de La Plata Colección Mamíferos (MLP), Centro Nacional Patagónico Colección Mamíferos (CNP), Colección Elio Massoia (CEM), Museo Argentino de Ciencias Naturales "Bernardino Rivadavia" (MACN Zoología), Museo de Ciencias Naturales de Mar del Plata "Lorenzo Scaglia" (MMP-Ma), and Colección Felix de

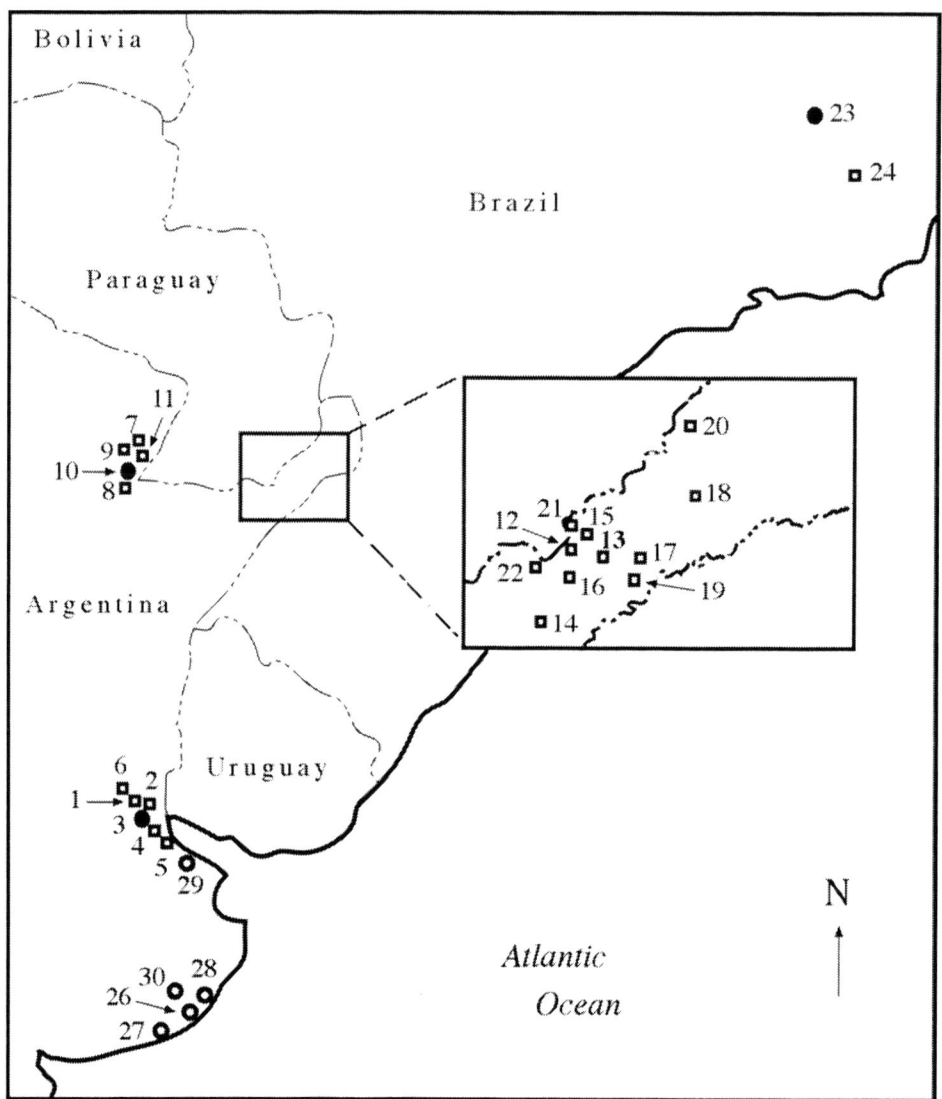

Figure 2. Map showing all recorded localities for *Bibimys*. Black circles indicate type localities. Open squares indicate recent localities. Open circles indicate fossil localities. Both fossil and recent specimens have been taken from locality 23, the type locality of *B. labiosus*. Localities 10 and 3 (recent) are the type localities of *B. chacoensis* and *B. torresi*, respectively. The Brazilian (Minas Gerais) locality number 31 is not mapped because it is not specified in its museum record. For details on locality names, reference sources, as well the specimens analyzed in this contribution, see Appendix 2. For details on the Paraguayan locality number 25 see Myers et al. (in prep.).

Azara (CFA); Denmark - Universitets Zoologiske Museum (ZMC); Paraguay, Museo Nacional de Historia Natural del Paraguay (GD, field number of Guillermo D'Elía); United Kingdom, Natural History Museum (BMNH); United States of America - National Museum of Natural History (USNM), and University of Michigan Museum of Zoology (UMMZ). Craniodental measurements were taken with digital calipers following Myers et al. (1990). Terminology for molar structures follows Reig (1977). External measurements were obtained from the specimen labels and field catalogues. In addition, information about *B. labiosus* was obtained from Gonçalves et al. (this volume). Genetic variation in *Bibimys* was assessed from the specimens reported in the phylogenetic analysis (see above).

Results and discussion

We have put together by far the largest series of *Bibimys* specimens to date, and the only one that includes specimens from all three recognized species. Although this series is not adequate to address some issues (see below), it allows us to evaluate the taxonomic utility of some characters mentioned in the literature and to evaluate for the first time other characters in the context of *Bibimys* taxonomy. In the light of this new information we present the following emended diagnosis of the genus *Bibimys*.

Bibimys Massoia, 1979

Type species: *Bibimys torresi* Massoia, 1979

Species included:
  Bibimys labiosus (Winge, 1887)
  Bibimys chacoensis (Shamel, 1931)
  Bibimys torresi Massoia, 1979.

Emended diagnosis: Sigmodontine rodents (*sensu* Reig 1980) with the following combination of characters (Figure 3): small size (head-body length < 130 mm); gnathic process of the premaxilla well developed and forming a sharply projecting plate anterior to the incisors; anterior part of nasals inflected dorsally; sides of interorbital region squared, not beaded or ledged; interorbital region hour-glass shaped; zygomatic notches moderately deep and broad; anterior margin of zygomatic plate approximately vertical; anterior half of the zygomatic arch flattened and expanded dorsal-ventrally; braincase inflated and slightly narrower than the zygomatic breadth; upper incisors strongly opisthodont; molars terraced and moderately hypsodont; M1/m1 with anteromedian flexus/xid present; M1-2/m1-2 with mesolophs/phids fused with paralophule/entolophulid; M3 with cylindrical outline and two persistent enamel islands but no evidence of a hypoflexus; m3 rectangular with large oblique hypoflexid; length of m3 slightly shorter than that of m2; mesopterygoid fossae narrower than parapterygoid plates; anterior margin of mesopterygoid fossa squared and lacking a median spine; alisphenoid strut present; squamosal root of the zygomatic arch connected to a slender hamular process by a crest; small subsquamosal fenestra above large postglenoid foramen; tegmen tympani anteriorly enlarged and reaching the squamosal; stapedial foramen and posterior opening of alisphenoid canal large; squamosal-alisphenoid groove and sphenofrontal foramen absent; gall bladder present; stomach unilocular and hemiglandular; anterior part of snout, from just below the external nares to the opening of the mouth, strikingly swollen, appearing bulbous; entire area from just dorsal to the rhinarium to the lower lips covered with remarkable short, bristle-like, all-white hairs that run stiffly outward from (i.e., perpendicular to the surface of) the lips, skin underlying the hairs distinctively reddish in living specimens (color is lost in both skins and formalin-preserved specimens).

Distribution: Late Pleistocene-Late Holocene of Argentina: Buenos Aires Province. Late Pleistocene-Holocene of Brazil: Minas Gerais State. Recent Argentina: Buenos Aires, Chaco, Entre Ríos, and Misiones Provinces; Brazil: Minas Gerais State, and Paraguay (Myers et al., *in prep.*). See Figure 2 and Appendix 2 for details.

Species boundaries: As stated above, the validity of the three species currently assigned to *Bibimys* has never been critically assessed. Previous studies such as Pardiñas (1996) and Gonçalves et al. (this volume) limited their comparisons to two forms. The only qualitative morphologic character proposed to date to distinguish among *Bibimys* species was the number of roots of M3. Pardiñas (1996) reported that the M3 of *B. chacoensis* (referred to as *B. labiosus* in that paper) had three roots whereas that of *B. torresi* had only two. Increased sample sizes reveal that this character is polymorphic in *B. chacoensis*. Of a sample of 14 specimens from Vedia, Chaco (Argentina) nine individuals had three roots while the rest had two. Some characters do appear to vary slightly among populations referred to different forms. For instance, the basioccipital of *B. chacoensis* is more excavated than that of the other forms, a feature also noted by Shamel (1931) in his description of the type specimen. The premaxillary bones of *B. torresi* appear to project more strongly anteriorly than in the other forms. However, these characters show great intrapopulation variability (Figure 4) and, further, this variation seems to be related to age. Analyses of larger sample sizes, especially of *B. torresi* and *B. labiosus*, are needed to clarify the taxonomic utility of these cranial character states.

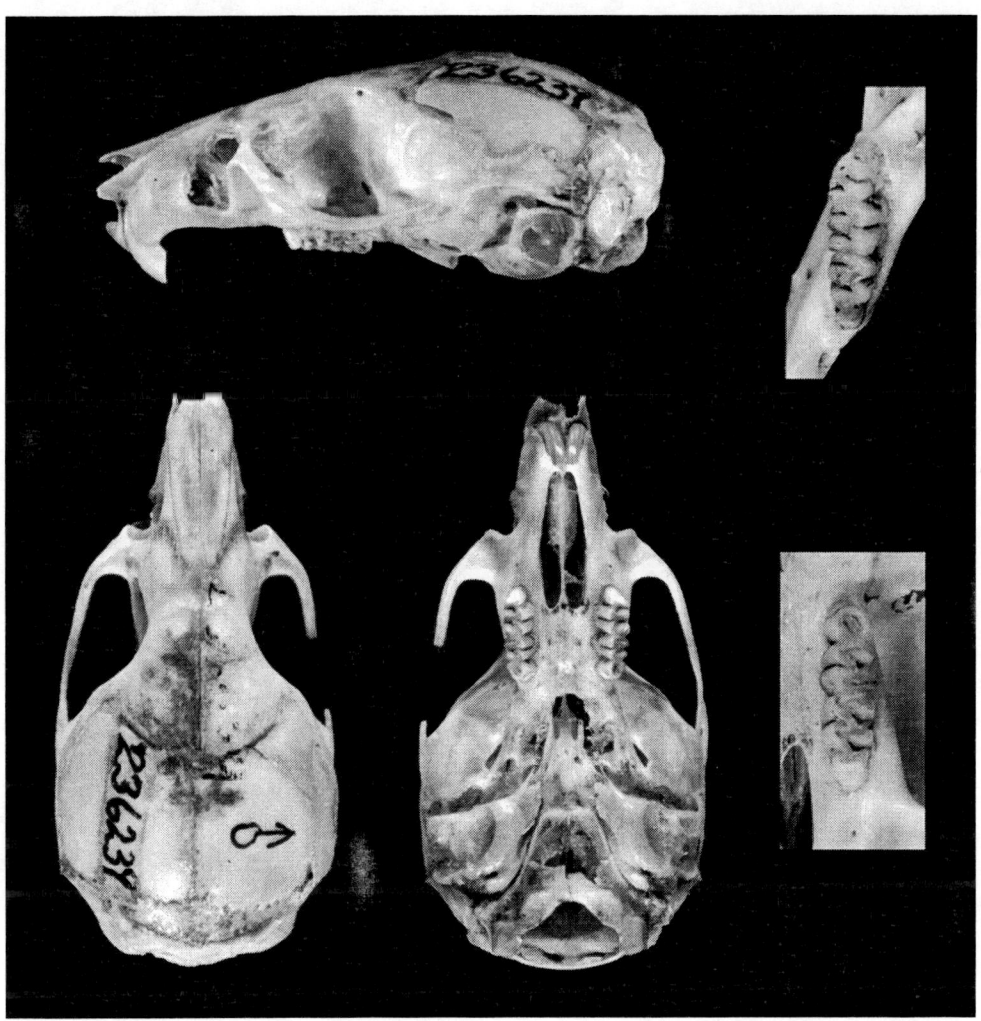

Figure 3. Skull of and molars of *Bibimys chacoensis* (holotype, USNM 236234).

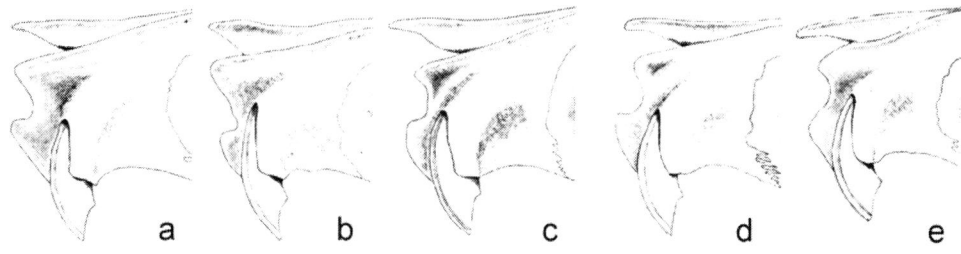

Figure 4. Lateral views of the anterior skull of 5 specimens of *Bibimys*. (a) MMP-Ma 3705, (b) MMP-Ma 3620 (c) MMP-Ma 2502 (d) C-0911, and (e) C-1580. Specimens (a) and (b) are referred to *B. torresi* from Estación Experimental INTA Delta del Parana, Argentina; specimens (c), (d), and (e) are referred to *B. chacoensis* from Paso Mono, Argentina. Note the large intrapopulation variation in the shape of the gnathic process and anterior part of the nasals.

Small sample sizes precluded our testing the statistical significance of differences in measurements among populations of the same versus different forms of *Bibimys*. The skulls of three specimens of *B. torresi* appeared to be slightly larger than specimens of the two other forms (Table 2). For most variables, however, sample sizes are too small to permit any estimate of variation within or among populations. Larger samples, from owl pellets, are available for two variables (upper and lower toothrow length). Analyses of these samples indicate that measurements for the three forms overlap (Figure 5). We note that these measurements were taken by at least five different people; the possibility of systematic error in recording these data cannot be ignored.

In summary, in a critical examination of cranial and dental morphology of both fossil and extant specimens referred to the three species, including the type specimens, we have not discovered any quantitative or qualitative dental or cranial character that unambiguously differentiates *B. chacoensis*, *B. labiosus*, and *B. torresi*. Sample sizes are still much too small, however, for much weight to be given to conclusions based on quantitative traits.

Table 2. External and cranial measurements (in mm) of *Bibimys* specimens. For measurement definitions see Myers et al. (1990).

Species	*chacoensis*	*labiosus*[1]	*torresi*
Number	USNM 236239	MZUFV 752	MMP-Ma 3620
	CNP 756	MZUFV 753	MMP-Ma 3705
	MMP-Ma 2502	MN 62061	MACN 20337
	CAF-0911		
	CAF-01580		
Locality	Las Palmas	Viçosa	INTA Delta del Paraná
	Cancha Larga	Viçosa	INTA Delta del Paraná
	Paso Mono	Viçosa	INTA Delta del Paraná
	Paso Mono		
	Paso Mono		
Sex	male	-	female
	male	female	male
	male	male	female
	female		
	male		
Age	adult	-	adult
	subadult	-	old adult
	subadult	-	old adult
	adult		
	adult		
Head and body	94	76	106
length	88	90	127
	107	89	-
	-		
	-		

Table 2. Continued.

Species	chacoensis	labiosus[1]	torresi
Hind foot length (with claw)	22.5	21	21.8
	21.5	22	22.9
	20.8	23	-
	-		
	-		
Ear length	-	15	17.4
	14	18	17
	17.7	14	-
	-		
	-		
Weight (g)	-	-	35
	23	-	42
	32	-	-
	-		
	-		
Condylobasal length	22.25	23.80	24.41
	22.21	24.41	26.00
	23.26	24.54	26.00
	23.80		
	23.51		
Zygomatic breadth	12.63	13.20	13.97
	12.24	13.44	13.93
	13.34	13.60	13.50
	12.64		
	13.33		
Interorbital constriction	4.39	4.76	4.61
	4.41	4.96	4.76
	4.48	4.53	4.45
	4.33		
	4.45		

Table 2. Continued.

Species	chacoensis	labiosus[1]	torresi
Rostral length	-	-	9.08
	7.65	-	9.70
	8.43	-	8.70
	8.35		
	8.26		
Nasal length	-	9.36	9.35
	8.32	9.06	10.04
	8.77	-	9.49
	9.33		
	9.05		
Rostral width	4.28	4.41	4.70
	4.40	4.25	4.62
	4.17	4.40	4.45
	4.53		
	4.18		
Length frontal along midline	7.88	-	9.37
	8.22	-	9.76
	8.75	-	8.99
	8.5		
	8.56		
Length interparietal along midline	2.24	-	2.72
	2.34	-	2.97
	1.92	-	-
	2.39		
	1.96		
Length of orbit	8.50	-	9.18
	8.23	-	9.78
	9.01	-	-
	8.58		
	8.71		

Table 2. Continued.

Species	*chacoensis*	*labiosus*[1]	*torresi*
Diastema length	5.55	5.31	5.91
	5.46	5.66	6.56
	5.54	5.36	6.05
	5.40		
	5.52		
Maxillary toothrow length	3.79	3.73	4.16
	3.69	3.75	4.22
	3.70	3.80	4.46
	3.75		
	3.83		
Length of incisive foramen	5.63	5.62	5.88
	5.07	5.90	6.77
	5.28	5.45	6.19
	5.73		
	5.30		
Length of palatal bridge	3.58	-	4.17
	3.81	-	3.94
	4.12	-	-
	3.82		
	4.17		
Width of toothrow	5.21	5.48	5.98
	4.86	5.42	5.70
	5.45	5.53	-
	5.33		
	5.38		
Width across occipital condyles	6.79	6.48	6.68
	6.61	6.56	7.04
	6.61	6.80	-
	6.56		
	6.56		

Table 2. Continued.

Species	chacoensis	labiosus[1]	torresi
Breadth across mastoid region	12.24	-	12.73
	12.10	-	13.89
	12.19	-	12.62
	12.12		
	12.40		
Length of basioccipital at midline	3.67	3.91	4.75
	3.94	3.75	4.94
	4.62	4.03	5.12
	4.77		
	4.66		
Breadth of mesopterygoid fossa	1.27	-	1.34
	0.93	-	1.65
	1.23	-	1.16
	1.28		
	1.48		
Breadth of zygomatic plate	2.09	2.51	2.58
	2.19	2.38	2.86
	2.58	2.31	2.65
	2.50		
	2.71		
Cranial depth	9.61	-	10.11
	9.70	-	9.88
	9.50	-	9.90
	9.38		
	9.66		

[1] Data from Gonçalves et al. (this volume).

Gonçalves et al. (this volume) studied the penile and bacular morphology of three individuals of *B. labiosus* and Massoia (1979) provided a description of the baculum of two specimens of *B. torresi*. Minor differences can be seen in these descriptions. The two bacula shown by Massoia (1979: Figure 1), however, differ in a number of aspects; the extent of intrapopulation variability in bacular structure has not been documented for either *B. labiosus* or *B. torresi*.

The karyotype of two of the three *Bibimys* forms is known. Dyzenchauz and Massarini (1999) reported the karyotype of three *B. torresi* specimens from Otamendi, while Gonçalves et al. (this volume) reported the karyotype of three specimens of *B. labiosus* from Viçosa. All six individuals were characterized by $2n = 70$. Two differences, however, are evident among these karyotypes. First, *B. torresi* specimens have 76 arms - four fewer than individuals of *B. labiosus*. In addition, *B. torresi* exhibits C-bands in the X chromosome and in pair 3, while *B. labiosus* has C-bands in the X chromosome only. The biological significance of these differences is unknown. Cytogenetic data, usually differences in diploid number, have been useful for discriminating among morphologically similar sigmodontine species (e.g., Geise et al., 2001). However, analyses of additional individuals from these species, as well as from *B. chacoensis*, are essential given the well-documented chromosomal polytypism and polymorphism reported for other sigmodontine species (e.g., Bianchi et al., 1971; Nachman and Myers, 1989; Sbalqueiro and Nascimento, 1996).

In the phylogenetic analysis described above (Figure 1), all specimens of *Bibimys* constituted a monophyletic group. The level of variation of the cytochrome *b* gene among these putative species was extremely low (Table 3). Only 22 variable sites were found among the four *Bibimys* haplotypes recovered in this study. Of those, two are in first codon positions, and the remaining 20 are in third codon positions. Among those 22 changes, only two are non-synonymous. Of the four haplotypes recovered, the one corresponding to the Viçosa specimens is the most divergent, differing in 16 positions (1.40 %) from the other three haplotypes. The most similar haplotypes are those recovered from individuals from Cancha Larga and the Paraguayan locality (4 substitutions, 0.35 %). These values are much lower than those reported for other akodont species- pairs and are even lower than among-population variation in some akodont species (D'Elía unpublished data; Smith and Patton 1991, 1993). The extremely low levels of genetic variation reported here are remarkable considering the geographic distance among populations (Figure 2).

Despite the lack of critical morphological differences among *Bibimys* taxa and the similarity of their cytochrome *b* sequences, we do not at this time recommend any formal changes in the nomenclature for *Bibimys* species. While the

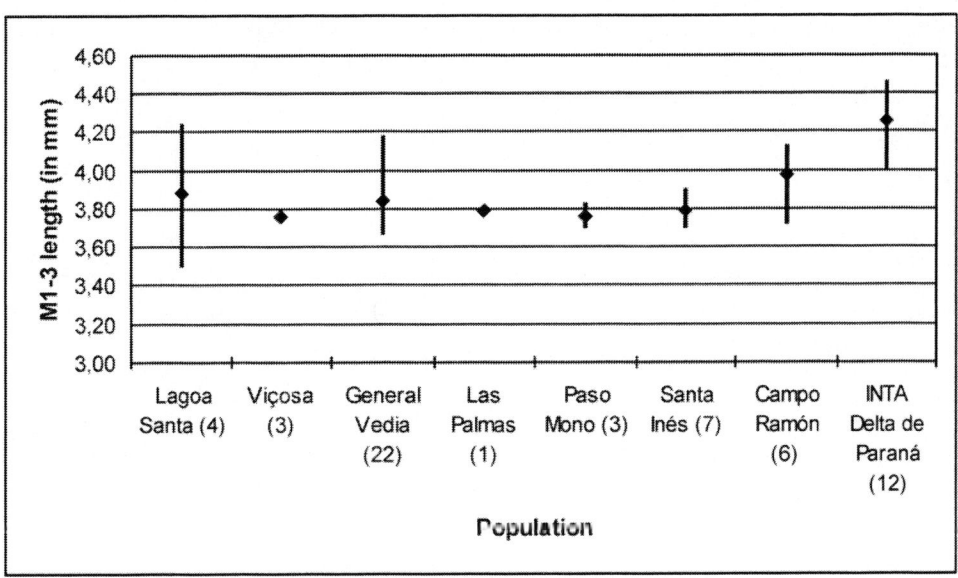

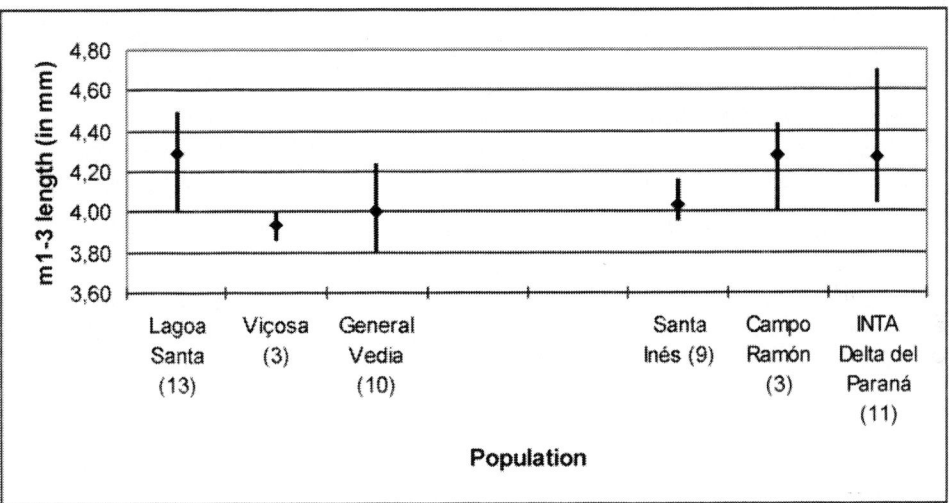

Figure 5. Variation in the length of upper (M1-3) and lower (m1-3) molar series across populations of *Bibimys*. The mean and range of values for each population are indicated by a dot and a bar, respectively. The number next to each population name indicates the sample size. Data from specimens from Lagoa Santa and Viçosa were taken from Gonçalves et al. (this volume). Data from specimens from Campo Ramón were taken from Pardiñas (1996).

available evidence tends to indicate that the genus is composed of a single species, widely distributed in SE Brazil, E Paraguay, and NE Argentina, sample sizes from each of these populations are still very small. If future studies reach the same conclusion, *B. chacoensis* and *B. torresi* would be regarded as junior synonyms of *B. labiosus*.

Table 3. Genetic differences among cytochrome *b* haplotypes recovered from *Bibimys* specimens (see text for details). Values above the diagonal correspond to the observed percentage of divergence. Values below the diagonal are observed number of substitutions.

	MN 62062	CNP 756	GD 153	MMP 3620
MN 62062		1.40	1.40	1.40
CNP 756	16		0.35	0.70
GD 153	16	4		0.70
MMP 3620	16	8	8	

Independent of the validity of the species discussed above, the extremely low levels of geographic variation found are remarkable. The apparent lack of populations located between those represented here is probably not due entirely to lack of sampling. Interestingly, the fossil record indicates that the distribution of *Bibimys* has fluctuated greatly over the last 120,000 years (Pardiñas, 1996, 1999a). Fossil remains of *Bibimys* were recovered from deposits as recent as 300 years before present at Mar del Plata, Argentina, more than 500 km south of the southernmost extant population known (Figure 2). This distributional change is parallel to that of other sigmodontine genera such as *Pseudoryzomys*, *Eligmodontia*, and *Phyllotis*, and has also taken place in other mammals (e.g., *Desmodus*, see Pardiñas and Tonni, 2000). The available evidence suggests that the current fragmentary distribution of *Bibimys* populations may be a recent event. If so, this may explain the low levels of variation found among extant populations. If *Bibimys* populations were interconnected until the recent past, then perhaps not enough time has elapsed for these populations to accumulate significant differences. This hypothesis seems to be supported by the molecular data. Although the number of *Bibimys* specimens sequenced by us ($n = 5$) is far from adequate to evaluate thoroughly the geographic structure of genetic variation in this genus, it is worth noting that the geographic patter of variation among the four haplotypes detected seems to differ from the expected under a pattern of isolation by distance. This scenario of a recent contraction of the geographic distribution of *Bibimys* can be tested further with additional studies of new fossil and recent specimens, as well as additional and more variable DNA markers.

CONCLUSIONS

Results of the first phylogenetic analysis that include the sigmodontine genus *Bibimys* indicate that this genus does not form a monophyletic group with *Kunsia* and *Scapteromys*. If this is the case, the Tribe Scapteromyini (*sensu* Massoia, 1979) as currently envisioned (e.g., McKenna and Bell, 1997) is not a natural group. Our results support the conclusion of Smith and Patton (1999) that *Bibimys*, *Kunsia* and *Scapteromys* are part of the akodontine radiation. However, our tree is far from robust. These hypotheses should be tested further using phylogenetic analyses of nuclear DNA sequences and morphological characters.

Results of the first analysis of geographic variation that included individuals assigned to the three currently recognized species of *Bibimys* revealed low levels of morphologic and genetic variation. This homogeneity leads us to question the distinctiveness of the three forms currently recognized at the species level. Fossil evidence shows that *Bibimys* has suffered a recent contraction of its geographic range. This may explain the low levels of variation detected among populations that are several hundred kilometers distant from one another. Analysis of additional specimens is required to test further this scenario.

Note: The material of this paper was presented at Pattonfest in 2001 and formed the basis for subsequent broader phylogenetic analyses published in 2003 (D'Elía, 2003; D'Elía et al., 2003). Regarding *Bibimys*, both studies corroborate the phylogenetic hypothesis advanced here.

ACKNOWLEDGMENTS

We thank the organizers of the symposium for inviting us to participate. We are grateful to the following persons that allow us access to material under their care: K. Aaris, M. Carleton, A. Currant, J. Contreras, Y. Davies, D. Romero, E. Massoia, M. Piantanida, O. Scaglia, and K. Rosenlund. Robert Voss made valuable comments on an earlier version of this contribution. Finally, we would like to thank the following people: L. Borrero, S. Cirignoli, F. Cremonte, C. Galliari, P. Gonçalves, C. Manchini, F. Martin, I. Mora, J. Oliveira, R. Owen, J.L. Patton, D. Podesta and M.F. Smith. This work was supported by The American Society of Mammalogists (GD), The Rodhe Island Zoological Society (GD), The University of Michigan Museum of Zoology (GD), Rackham Graduate School though a Sokol Fellowship (GD), CONICET (UFJP), and Fundacion Antorchas (UFJP).

LITERATURE CITED

Avila-Pires, F. D.

   1972   A new subspecies of *Kunsia fronto* (Winge, 1888) from Brazil (Rodentia, Cricetidae). Revista Brasileira de Biologia 32:419-422.

Bianchi, N. O., O. A. Reig, O. J. Molina, and F. N. Dulout.

   1971   Cytogenetics of the South American akodont rodents (Cricetidae). I. A progress report of Argentinian and Venezuelan forms. Evolution 25:724-736.

Bianchini, J. C., and H. Delupi.

   1993   Mammalia. Fauna de agua dulce de la República Argentina (Zulma A. de Castellanos, dir.) 44:1-79.

Castro, D., and A. Gonzalez.

   2003   Una nueva especie de *Hoplopleura* (Phthiraptera, Anoplura) parásita de tres especies de *Bibimys* (Muridae, Sigmodontinae, Rodentia). Iheringia, Serie Zoología 93:183-188.

Contreras, J. R.

   1984   Nota sobre *Bibimys chacoensis* (Shamel, 1931) (Rodentia, Cricetidae, Scapteromyini). Historia Natural 4:280.

Dyzenchauz F. J., and A. I. Massarini.

   1999   First cytogenetic analysis of the genus *Bibimys* (Rodentia, Cricetidae). Zeitschrift für Saügetierkunde 64:59-62.

D'Elía, G.

   2000   Comments on recent advances in understanding sigmodontine phylogeny and evolution. Mastozoología Neotropical 7:47-54.

   2003   Phylogenetics of Sigmodontinae (Rodentia, Muroidea, Cricetidae), with special reference to the akodont group, and with additional comments on historical biogeography. Cladistics 19:307-323.

D'Elía G., E. M. González, and U. F. J. Pardiñas.

   2003   Phylogenetic analysis of sigmodontine rodents (Muroidea), with special reference to the akodont genus *Deltamys*. Mammalian Biology 68:351-364.

Engel, S. R., K. M. Hogan, J. F. Taylor, and S. K. Davis.
    1998   Molecular systematics and paleobiogeography of the South American
            sigmodontine rodents. Molecular Biology and Evolution 15:35-49.

Farris, J. S.
    1982   The logical basis of phylogenetic analyis. Pp. 7-36 in Advances in
            Cladistics: Proceedings of the Second Meeting of the Willi Hennig
            Society (N. Planick, and V. Funk., Eds.). Columbia University Press,
            New York.

Fitzinger, L.J.
    1867   Versuch einer naturlichen Anordnug der Nagethiere (Rodentia).
            Sitzungsberichte der Akademie der Wissenschaften in Wien. 16:57-168.

Geise, L., M. F. Smith, and J. L. Patton.
    2001   Diversification in the genus *Akodon* (Rodentia: Sigmodontinae) in
            southeastern South America: Mitochondrial DNA sequence analysis.
            Journal Mammalogy 82:92-101.

Gonçalves, P. R., J. A. de Oliveira, M. O. Corrêa, and L. M. Pessoa.
            Morphological and cytogenetic analyses of *Bibimys labiosus* (Winge,
            1887) (Rodentia, Sigmodontinae): implications for its affinities with the
            scapteromyine group. (This volume)

González, E. M., A. Langguth, and L. F. de Oliveira.
    1998   A new species of *Akodon* from Uruguay and Southern Brazil
            (Mammalia: Rodentia: Sigmodontinae). Comunicaciones Zoológicas del
            Museo Historia Natural Montevideo 191:1-8.

Gyldenstolpe, N.
    1932   A new *Scapteromys* from Chaco Austral, Argentine. Arkiv Zoology 24
            B:1-2.

Hershkovitz, P.
    1966   South American swamp and fossorial rats of the scapteromyine group
            (Cricetinae, Muridae) with comments on the glans penis in murid
            taxonomy. Zeitschrift für Saügetierkdunde 31:81-149.

Irwin. D. M., T. D. Kocher, and A. C. Wilson.
    1991   Evolution of the cytochrome *b* gene of mammals. Journal of Molecular
            Evolution 32:128-144.

Jernvall, J, and H. S. Jung.
    2000   Genotype, phenotype, and developmental biology of molar tooth characters. Yearkook of Physical Anthropology 43:171-190.

Jernvall. J., S. V. E. Keranen, and I. Thesleff.
    2000   Evolutionary modification of development in mammalian teeth: Quantifying gene expression patterns and topography. Proceedings of the National Academy of Sciences of the United States of America 97: 14444-14448.

Kluge, A. J., and J. S. Farris.
    1969   Quantitative Phyletics and the Evolution of Anurans. Systematic Zoology 18:1-32.

Massoia, E.
    1979   Descripcion de un género y especie nuevos: *Bibimys torresi* (Mammalia - Rodentia Cricetidae - Sigmodontinae - Scapteromyni). Physis C 38:1-7.

    1980a   El estado sistemático de cuatro especies de cricétidos sudamericanos y comentarios sobre otras especies congenéricas. Ameghiniana 17:280-287.

    1980b   Mammalia de Argentina -I- Los mamíferos silvestres de la provincia de Misiones. Iguazú 1:15-43.

    1988   Presas de *Tyto alba* en Campo Ramón, Departamento Oberá, Provincia de Misiones. Boletín Científico, Asociación para la Protección de la Naturaleza 7:4-16.

Massoia, E., J. C. Chébez y S. Heinonen Fortabat.
    1988   Depredación de mamíferos por *Tyto alba tuidara* en Teyú-cuaré, Departamento San Ignacio, Pcia. de Misiones. Boletín Científico, Asociación para la Protección de la Naturaleza 8: 7-13.

    1989a   Segundo análisis de egagrópilas de *Tyto alba tuindara* en el Departamento de Apóstoles, Província de Misiones. Boletín Científico, Asociación para la Protección de la Naturaleza 13:3-8.

    1989b   Análisis de regurgitados de *Tyto alba tuindara* de Los Helechos, Departamento Oberá, Provincia de Misiones. Boletín Científico, Asociación para la Protección de la Naturaleza 14:16-22.

Massoia, E., J. C. Chébez y S. Heinonen Fortabat.

1989c   Mamíferos y aves depredados por *Tyto alba* en el arroyo Yabebyrí, Departamento de Candelária, Provincia de Misiones. Boletín Científico, Asociación para la Protección de la Naturaleza 15:8-13.

1989d   Mamíferos y aves depredados por *Tyto alba tuidara* en Bonpland, Departamento Candelária, Provincia de Misiones. Boletín Científico, Asociación para la Protección de la Naturaleza 15:19-24.

McKenna, M. C., and S. K. Bell.

1997   Classification of Mammals above the Species Level. Columbia University Press, New York.

Miranda-Ribeiro, A.

1914   Historia natural. Zoología. Mamíferos. Commisão de Linhas Telegraphicas Estrategicas de Matto Grosso ao Amazonas 13 Annexo 5:1-49 + 3pp., 25pls.

Myers, P., B. L. Lundrigan, and P. K. Tucker.

1995   Molecular phylogenetics of oryzomyine rodents: the genus *Oligoryzomys*. Molecular Phylogenetics and Evolution 4:372-382.

Myers, P., J. L. Patton, and M. F. Smith.

1990   A review of the *boliviensis* group of *Akodon* (Muridae: Sigmodontinae), with emphasis on Peru and Bolivia. Miscellaneous Publications Museum of Zoology, University of Michigan 177:1-104.

Nachman, M. W., and P. Myers.

1989   Exceptional chromosomal mutations in a rodent population are not strongly underdominant. Proceedings of the National Academy of. Science USA 86:6666-6670.

Nixon, K. C.

1999   The Parsimony Ratchet, a new method for rapid parsimony analysis. Cladistics 15:407-414.

Nixon, K. C., and J. M. Carpenter.

1993   On outgroups. Cladistics 9:413-426.

Ortiz, P. E., U. F. J. Pardiñas, and S. J. Steppan.
2000 A new phyllotine (Rodentia: Muridae) from northwestern Argentina and relationships of the *Reithrodon* group. Journal of Mammalogy 81:37-51.

Paglia, A. P., P. De Marco, F. M. Costa, R. F. Pereira, and G. Lessa.
1995 Heterogeneidade estrutural e diversidade de pequenos mamíferos em um fragmento de mata secundária de Minas Gerais, Brasil. Revista Brasileira de Zoologia 12:67-79.

Pardiñas, U. F. J.
1995 Novedosos cricétidos (Mammalia, Rodentia) en el Holoceno de la Región Pampeana, Argentina. Ameghiniana 32:197-203.

1996 El registro fósil de *Bibimys* Massoia, 1979 (Rodentia) en la Argentina. Consideraciones sobre los Scapteromyini (Cricetidae, Sigmodontinae) y su distribuición durante el Plioceno-Holoceno en la region Pampeana. Mastozoología Neotropical 3:15-38.

1999a Fossil murids: taxonomy, paleoecology, and paleoenvironments. Quaternary of South America and Antarctic Peninsula 12:225-254.

1999b Los roedores muroideos del Pleistoceno tardío-Holoceno en la región pampeana (sector este) y Patagonia (República Argentina): aspectos taxonómicos, importancia bioestratigráfica y significación paleoambiental. Doctoral Dissertation, Universidad Nacional La Plata. ix + 283 pp.

Pardiñas, U. F. J., and E. P. Tonni.
2000 A giant vampire (Mammalia, Chiroptera) in the Late Holocene from the Argentinean pampas: paleoenvironmental significance. Palaeogeography, Palaeoclimatology, Palaeoecology 160:213-221.

Pardiñas, U. F. J., A. Cione, J. San Cristóbal, D. H. Verzi, and E. P. Tonni.
2004 A new last interglacial continental vertebrate assemblage in Central-Eastern Argentina. Current Research in the Pleistocene 21:1110112.

Peters, H., and R. Balling.
1999 Teeth - where and how to make them. Trends in Genetics 15:59-65.

Peters, W.

1860   Über einige merkwürdige Nagethiere (*Spalacomys indicus, Mus tomentosus* und *Mus squamipes*) des Königl. Zoologischen Museums. Abhandlungen der Königlichen Akademie der Wissenschften zu Berlin, Physikalische 139-158.

Reig, O. A.

1977   A proposed unified nomenclature for the enamelled components of the molar teeth of the Cricetidae (Rodentia). Journal of Zoology, London 181:227-241.

1980   A new fossil genus of South American cricetid rodents allied to *Wiedomys*, with ann assessment of the Sigmodontinae. Journal of Zoology, London 192:257-281.

Sbalqueiro, I. J., and A. P. Nascimento.

1996   Occurrence of *Akodon cursor* (Rodentia, Cricetidae) with 14, 15, and 16 chromosome cytotypes in the same geographic area in Southern Brazil. Brazilian Journal of Genetics 19:565-569.

Shamel, H. H.

1931   *Akodon chacoensis*, a new cricetine rodent from Argentina. Journal of Washington Academy of Sciences 21:427-429.

Sikes, D. S., and P. O. Lewis.

2001   Beta software, version 1. PAUPRat: PAUP* implementation of the parsimony ratchet. Distributed by the authors. Department of Ecology and Evolutionary Biology, University of Connecticut, Storrs, USA.

Smith, M. F., and J. L. Patton.

1991   Variation in mitochondrial cytochrome *b* sequence in natural populations of South American akodontine rodents (Muridae: Sigmodontinae). Molecular Biology and Evolution 8:85-103.

1993   The diversification of South American murid rodents: evidence from mitochondrial DNA sequence data for the akodontine tribe. Biological Journal of the Linnean Society 50:149-177.

1999   Phylogenetic relationships and the radiation of sigmodontine rodents in South America: Evidence from cytochrome b. Journal of Mammalian Evolution 6:89-128.

Steppan, S. J.
    1995   Revision of the Tribe Phyllotini (Rodentia: Sigmodontinae), with a phylogenetic hypothesis for the Sigmodontinae. Fieldiana: Zoology 80: 1-112.

Swofford D.
    2000   PAUP*: Phylogenetic Analysis Using Parsimony (*and other methods), 4.0. Sinauer Associates, Sunderland.

Thomas O.
    1884   On a collection of Muridae from central Peru. Proceedings of the Zoological Society of London 447-458.

Thompson, J. D., T. J. Gibson, F. Plewniak, F. Jeanmougin, and D. G. Higgins.
    1997   The Clustal X windows interface: flexible strategies for multiple sequence alignment aided by quality analysis tools. Nucleic Acids Research 24:4876-4882.

Winge, H.
    1887   Jordfundne og nulevende Gnavere (Rodentia) fra Lagoa Santa, Minas Geraes, Brasilien. E Museo Lundii 1:1-200.

Appendix 1. Specimens included in the phylogenetic analysis of cytochrome *b* gene sequences.

	Taxon	Catalog #[a]	Source[b]
	Ingroup		
1	*Abrothrix andinus*		AF108671
2	*Abrothrix longipilis*		U03530
3	*Abrothrix olivaceus*		AF027305
4	*Abrothrix xanthorhinus*		AF297902
5	*Abrothrix* sp.		AF297894
6	*Akodon aerosus*	MVZ 172870	*
7	*Akodon albiventer*	FMNH 129978	*
8	*Akodon azarae*	GD 264	This study
9	*Akodon boliviensis*		M35691
10	*Akodon cursor*	MAM 24	*
11	*Akodon iniscatus*	MVZ 182655	*
12	*Akodon juninensis*	MVZ 173038	*
13	*Akodon kofordi*	MVZ 171665	*
14	*Akodon lindberghi*	MN 48026	*
15	*Akodon lutescens*	MVZ 171612	*
16	*Akodon mimus*	MVZ 171752	*
17	*Akodon molinae*	AK 222	*
18	*Akodon mollis*	LSU 27007	*
19	*Akodon montensis*	MNHNP 2910	This study
20	*Akodon mystax*	MN 48041	*
21	*Akodon orophilus*	MVZ 173057	*
22	*Akodon serrensis*	MN 35927	*
23	*Akodon siberiae*	MSB 55209	*
24	*Akodon subfuscus*	MVZ 174109	*
25	*Akodon toba*		U03527
26	*Akodon torques*	MVZ 171720	*
27	*Andalgalomys pearsoni*		AF159285
28	*Andalgalomys roigi*		AF159286
29	*Andinomys edax*		AF159284
30	*Bibimys chacoensis*	CNP 756	This study
31	*Bibimys chacoensis*	GD 153	This study
32	*Bibimys labiosus*	MN 62062	This study
33	*Bibimys torresi*	MMP 3620	This study

Appendix 1. Continued.

	Taxon	Catalog #[a]	Source[b]
34	*Blarinomys breviceps*		AF108668
35	*Brucepattersonius iheringi*		AF108667
36	*Brucepattersonius soricinus*		MVZ183036
37	*Brucepattersonius* sp.	LG 108	*
38	*Calomys callosus*		AF159293
39	*Calomys lepidus*		AF159294
40	*Chelemys macronyx*		U03533
41	*Chilomys instans*		AF108679
42	*Chroeomys jelskii*		M35714
43	*Delomys sublineatus*		AF108687
44	*Eligmodontia morgani*		AF108691
45	*Eligmodontia puerulus*		AF159289
46	*Eligmodontia typus*		AF108692
47	*Geoxus valdivianus*		U03531
48	*Graomys domorum*		AF159291
49	*Graomys griseoflavus*		AF159290
50	*Irenomys tarsalis*		U03534
51	*Juscelinomys huanchacae*		AF133667
52	*Kunsia tomentosus*		AF108670
53	*Lenoxus apicalis*		U03541
54	*Loxodontomys micropus*		AF108690
55	*Microryzomys minutus*		AF108698
56	*Necromys amoenus*		M35711
57	*Necromys lasiurus*		U03528
58	*Necromys urichi*		U03549
59	*Oecomys bicolor*		AF108699
60	*Oecomys trinitatus*	LHE 579	*
61	*Oecomys superans*	MVZ 155004	*
62	*Oryzomys megacephalus*		AF108695
63	*Oxymycterus amazonicus*	LHE 603	*
64	*Oxymycterus delator*	UMMZ 133939	*
65	*Oxymycterus hiska*	MVZ 171518	*
66	*Oxymycterus nasutus*	MVZ 182701	*
67	*Oxymycterus paramensis*		U03536
68	*Oxymycterus* sp.	MVZ 183265	*

Appendix 1. Continued.

	Taxon	Catalog #[a]	Source[b]
69	*Phyllotis magister*		U86824
70	*Phyllotis xanthopygus*		AF108693
71	*Reithrodon auritus*		AF108694
72	*Rhipidomys macconnelli*		AF108681
73	*Salinomys delicatus*		AF159292
74	*Scapteromys aquaticus*	UMMZ 174991	This study
75	*Scapteromys tumidus*	MVZ 183269	This study
76	*Scolomys juruaense*		AF108696
77	*Sigmodon alleni*		AF155425
78	*Sigmodon arizonae*		AF155423
79	*Sigmodon hispidus*		AF108702
80	*Sigmodon mascotensis*		AF155424
81	*Sigmodon ochrognathus*		AF155422
82	*Tapecomys primus*		AF159287
83	*Thaptomys nigrita*		AF108666
84	*Thomasomys aureus*		U03540
85	*Thomasomys daphne*		AF108673
86	*Thomasomys gracilis*		AF108674
87	*Thomasomys ischyurus*		AF108675
89	*Thomasomys notatus*		AF108676
90	*Thomasomys oreas*		AF108677
91	*Thomasomys* sp.		AF108678
92	*Juliomys pictipes*		AF108688
93	*Juliomys* sp.		AF108689
	Outgroups		
94	*Alticola macrotis*		AF119273
95	*Arvicola terrestris*		AF159400
96	*Clethrionomys rutilus*		AB031581
97	*Dicrostonyx torquatus*		AF119275
98	*Ellobius fuscocapillus*		AF126430
99	*Eothenomys smithii*		AB037316
100	*Lemmus trimucronatus*		AF119276
101	*Microtus arvalis*		AF159403
102	*Myopus schisticolor*		AF119263

Appendix 1. Continued.

	Taxon	Catalog #[a]	Source[b]
103	*Ondatra zibethicus*		AF119277
104	*Phenacomys intermedius*		AF119260
105	*Synaptomys borealis*		AF119259
106	*Cricetulus griseus*		AB033693
107	*Mesocricetus auratus*		AF119265
108	*Phodopus campbelli*		AF119278
109	*Hodomys alleni*		AF186801
110	*Neotoma floridana*		AF186823
111	*Osgoodomys banderanus*		AF155383
112	*Peromyscus leucopus*		AF131926
113	*Reithrodontomys zacatecae*		AF176252
114	*Scotinomys teguina*		AF108705
115	*Tylomys* sp.	USNM 464887	This study

[a] Catalog numbers are given only for those taxa whose sequences were not retrieved from Genbank.

[b] For those sequences retrieved from Genbank, the corresponding accession numbers are given, otherwise sequence provenance is indicated. Asterisks indicate unpublished complete sequences kindly provided by James L. Patton and M. F. Smith (Museum of Vertebrate Zoology, Berkeley).

Appendix 2. Gazetteer of recording localities of the genus *Bibimys*. The first citation of each locality is given. "This paper" means the locality is first reported in this contribution. Specimens analyzed here are listed after the locality. Asterisks denote those specimens that were sequenced.

RECENT: ARGENTINA, *Province of Buenos Aires*: 1) Arroyo Talaveras, S 34º 04' W 59º 04' (Bianchini and Delupi, 1993). 2) Confluence Arroyo Las Piedras y Arroyo Las Cucarachas (Massoia, 1979). 3) Estación Experimental INTA Delta del Paraná, S 34º 09' W 58º 57' (Massoia, 1979): CEM 5067, skull and skin (holotype of *B. torresi*); CEM 1886, skull and skin (allotype of *B. torresi*); CEM 5015, skull and skin (paratype of *B. torresi*); CEM s/n, 12 incomplete skulls, 12 right mandibles, 10 left mandibles from owl pellets; MACN 20337; skull, MMP-Ma 3620* and MMP-Ma 3705, skull and skin. 4) Ingeniero Rómulo Otamendi, S 34º 13' W 58º 54' (Pardiñas 1996). 5) San Fernando, S 34º 26' W 58º 33' (González 1997). *Province of Entre Ríos*: 6) Isla Ibicuy, S 33º 44' W 59º 13' (Massoia 1980a). *Province of Chaco*: 7) 7 km North of General Vedia, S 26º 53' W 58º 36' (this paper): CNP 757, 12 incomplete skulls from owl pellets. 8) Cancha Larga, S 27° 04' W 58° 43' (this paper): CNP 756*, skull and skeleton. 9) General Vedia, S 26º 56' W 58º 40' (this paper): CNP 758, 39 incomplete skulls, two right maxillas, one left maxilla, 11 right mandibles, 14 left mandibles from owl pellets. 10) Las Palmas, S 27º 04' S W 58º 41' (Shamel 1931): USNM 236239, skull and skin (holotype of *B. chacoensis*). 11) Paso Mono, Estancia San Carlos, S ~27º 05' W ~58º 37' (this paper): MMP-Ma 2502, CFA 0911; CFA 1580, skull and skin. *Province of Misiones*: 12) 4 km N Loreto, S 27º 19' W 55º 32' (this paper): CNP 759, four incomplete skulls from owl pellets. 13) 11 de Noviembre, S 27º 28' W 55º 19' (this paper): CNP 760, one right mandible from owl pellets. 14) Apóstoles, S 27º 55' W 55º 46' (Massoia et al. 1989a). 15) Arroyo Yabebyrí, S 27º 17' W 55º 31' (Massoia et al. 1989c): CEM without number, 10 incomplete skulls, 11 right mandibles from owl pellets. 16) Bonpland, S 27º 29' W 55º 29' (Massoia et al. 1989d). 17) Campo Ramón, S 27º 28' W 55º 00', (Massoia 1988). 18) Estación Experimental INTA Cuartel Río Victoria (Massoia 1980b). 19) Los Helechos, S 27º 33' W 55º 03' (Massoia et al. 1989b). 20) El Dorado, S 26º 24' W 54º 35' (this paper): CNP 761, one incomplete skull from owl pellets. 21) Teyú Cuaré, S 27º 11' W 55º 39' (Massoia et al. 1988). 22) Santa Inés, S 27º 34' W 55º 49' (this paper): CNP 762, four incomplete skulls, three right maxilla, three left maxilla, six right mandibles, five left mandibles from owl pellets. BRAZIL, *State of Minas Gerais*: 23) Lagoa Santa, S 19º 39' W 43º 54' (Winge 1887): MZD (Lund collection without number), incomplete skull and skin (lectotype of *Bibimys labiosus*). 24) Mata do Paraíso (Viçosa), S 20° 45' W 42° 53' (Paglia et al. 1995): MN 62062* and MN 62063*. PARAGUAY, 25) See Myers et al. (in prep.): GD 153*, specimen in fluid.

FOSSIL: ARGENTINA, *Province of Buenos Aires*: 26) Balneario Menta, S 38° 00' W 57° 34', Late Holocene (Pardiñas 1999b): MLP 95-V-1-2, two left maxilla, two right mandibles, five left mandibles. 27) Centinela del Mar, S 38° 27' W 58° 14', Late Holocene (Pardiñas 1995): MLP 91-IV-15-2, four right maxilla, four left maxilla, three right mandibles, three left mandibles; MLP 91-IV-15-2, three right maxilla, one right mandible, three left mandibles. 28) Constitución, S 37° 57' W 57° 32', Late Pleistocene (Pardiñas et al. 2004): MLP s/n, isolated m1. 29) La Norma archaeological site, S 34° 55' W 57° 46', Late Holocene (Pardiñas 1999b): MLP s/n, left mandible. 30) Tixi Cave archaeological site, S 37° 57' W 58° 02', Late Holocene (Pardiñas 1995): MLP 84-X-20-10, right mandible; MLP 84-X-20-11, right mandible; MLP 84-X-20-12, left maxilla. BRAZIL, *State of Minas Gerais*: 31) locality not specified, Late Pleistocene-Holocene (this paper): BMNH, Claussen collection without number, right mandible.

MISIDENTIFICATIONS:

1. One unnumbered specimen from an unespecified locality between Capitán Solari and Colonia (S ~26° 50' W ~59° 37', Province of Chaco, Argentina) referred to *Bibimys chacoensis* by Contreras (1984), was later identified as *Akodon azarae* (C. Galliari, *com. pers.*);
2. The specimen MLP 24-V-77-1 from Canal 9 (S ~36° 45' W ~56° 45', Province of Buenos Aires, Argentina) mentioned by Bianchini and Delupi (1993) as *Bibimys torresi* was later identified as *Akodon azarae* by Pardiñas (1996).

A Revision of the Genera of Arboreal Echimyidae (Rodentia: Echimyidae, Echimyinae), With Descriptions of Two New Genera

Louise H. Emmons

Most recent taxonomic works have recognized four or five genera of echimyines, *Echimys, Makalata, Diplomys, Phyllomys* and *Isothrix* (McKenna and Bell, 1997; Leite, 2003). Three of these genera include divergent arrays of species that do not cluster with each other in parsimony analyses based on morphology. From a phylogenetic hypothesis based on morphological characters, a revision of the arboreal Echimyid rodents (Echimyidae, Echimyinae) is proposed. The echimyine taxa are reorganized into eight genera, six of them based on previously defined groupings and names: *Echimys, Phyllomys, Isothrix, Diplomys, Makalata,* and *Callistomys*. Two new genera are named and diagnosed: *Pattonomys*, including three species formerly grouped under *Echimys semivillosus*; and *Santamartamys*, a monotypic genus based on *Diplomys rufodorsalis*. The eight genera are diagnosed and illustrated. Phylogenetic relationships between the genera are poorly resolved; the genus *Isothrix* is of uncertain affinity, but it is provisionally retained within the echimyines. Taxa formerly placed in the subfamily Eumysopinae do not appear to be paraphyletic but, as their monophyly cannot yet be rejected, the subfamily is retained pending a more thorough revision. These hypotheses based on morphology are consistent with recently constructed phylogenetic trees based on molecular characters (Leite and Patton, 2002).

SYSTEMATICS OF THE ECHIMYIDAE

The family Echimyidae Gray, 1825 is the most diverse of the South American hystricognath rodents, with 20 genera and 78 species recently recognized (Woods, 1993). The taxonomic history of the family has been chaotic; a number of generic names have been proposed, several have been abandoned, and the contents of numerous genera remain highly unstable. There have been no modern revisions of the family and several of the currently recognized genera do not appear to represent coherent taxonomic units from either a phenetic or phylogenetic perspective. Using morphology, I re-examine the supraspecific taxonomy of arboreal echimyids (subfamily Echimyinae) and other taxa recently included in the Echimyidae in an attempt to bring more coherence to the systematics of this group and to provide morphological diagnoses for apparent taxonomic units. I first briefly summarize previous systematic views of the family and then describe the morphological character sets that I found to be most informative. The described characters are used to construct a phylogenetic hypothesis, from which I propose a revised generic classification.

Brief Taxonomic History

The family Echimyidae is of ancient origin in South America, found among the oldest fossil rodents of the late Oligocene (Patterson and Wood, 1982; Vucetich and Verzi, 1991). Recent work suggests that there was a diverse fauna of largely grassland (pampa) echimyids in the Miocene (Vucetich et al., 1993), but there is little early fossil record of the arboreal forest taxa. The family has recently been classified into three or four subfamilies (Table 1). The living Eumysopinae or Heteropsomyinae have been divided into six genera of terrestrial species and two of arboreal ones, with the greatest generic diversity among grassland taxa but the greatest number of species in the rainforest genus *Proechimys*. The Dactylomyinae includes three genera of specialized arboreal folivores of lowland and montane forests, while the Echimyinae has included five to six genera of large-bodied arboreal rats of forested or wooded habitats. Three small families or subfamilies (two of them monotypic) – Capromyidae, Chaetomyidae, Myocastoridae – have also been included by some authors.

When first discovered, the rat-like South American caviomorphs were placed in European rodent genera (e.g. *Myoxus chrysurus* Zimmermann 1780), but shortly thereafter they were recognized as distinct (*Echimys* Cuvier 1809). Nomenclatural confusion soon followed, with a rapid proliferation of names as new discoveries were sent to Europe and quickly described. As for many groups of South American mammals, the early taxonomy of echimyids was confounded by the use of separate generic classifications by French, German, Danish and English taxonomists, who named taxa or redefined their contents without viewing type material in other countries. Tate (1935) sifted though 150 years of publications in a heroic review of the taxonomic history of all caviomorph rodents. He reconciled many nomenclatural discrepancies and proposed a revised generic classification, but did not provide an extensive review of characters. His interpretation of the Echimyidae, in which he segregated the species of *Echimys* largely on the degree of hairiness of the tail (Table 2), left many questions unresolved. Tate's classification was adopted by most systematists until Cabrera (1961) again reviewed the family (together with all other South American mammals) and synonymized many named forms with little explanation. Cabrera's classification has been widely used until now (Honaki et al., 1983; Woods, 1993). Emmons and Feer (1990, 1997) proposed some revisions of the genera in field guide format, without explanation, including segregation of *Nelomys* from *Echimys* (1990); subsequently, these authors separated

Table 1. Recent classifications of the extant genera of Echimyidae. (*) refer to taxa
included by Patterson and Wood (1982) but not by Patterson and Pascual (1968).

Patterson & Pascual 1968		Patton & Reig 1989	
*Patterson & Wood 1982	Woods 1982	Woods 1993	McKenna & Bell 1997
HETEROPSOMYINAE	ECHIMYINAE	EUMYSOPINAE	HETEROPSOMYINAE
Carterodon	*Carterodon*	*Carterodon*	*Carterodon*
Clyomys	*Clyomys*	*Clyomys*	*Clyomys*
	Euryzygomatomys	*Euryzygomatomys*	*Euryzygomatomys*
Thrichomys	*Thrichomys*	*Thrichomys*	*Thrichomys*
Proechimys	*Proechimys*	*Proechimys*	*Proechimys*
Hoplomys	*Hoplomys*	*Hoplomys*[1]	*Hoplomys*
		Incertae sedis	
Lonchothrix	*Lonchothrix*	*Lonchothrix*	*Lonchothrix*
Mesomys	*Mesomys*	*Mesomys*	*Mesomys*
ECHIMYINAE		ECHIMYINAE	ECHIMYINAE
Echimys	*Echimys*	*Echimys*	*Echimys*
	Makalata	*Makalata*	*Makalata*
Diplomys	*Diplomys*	*Diplomys*	*Diplomys*
Isothrix	*Isothrix*	*Isothrix*	*Isothrix*
DACTYLOMYINAE	DACTYLOMYINA	DACTYLOMYINAE	DACTYLOMYINAE
Dactylomys	*Dactylomys*	*Dactylomys*	*Dactylomys*
Kannabateomys	*Kannabateomys*	*Kannabateomys*	*Kannabateomys*
Olallamys	*Olallamys*	*Olallamys*	*Olallamys*
MYOCASTORINAE			MYOCASTORINAE
Myocastor			*Myocastor*
extinct genera			
CHAETOMYINAE	CHAETOMYINAE	CHAETOMYINAE	
Chaetomys	*Chaetomys*	*Chaetomys*	
*CAPROMYINAE			
*Subfamily Plagiodontinae			
*Subfamily Capromyinae			
*Subfamily Myocastorinae			

[1] *Hoplomys* included within *Proechimys* by Patton and Reig (1989)

additional species of *Makalata* from *Echimys* (1997). Patton and Emmons (1985) reviewed the genus *Isothrix* and, later, Emmons and Vucetich (1998) segregated a new genus, *Callistomys*, from *Isothrix*. Leite (2003), and Emmons et al. (2002) revised the Brazilian Atlantic tree rats, *Phyllomys* Lund, and showed the name *Nelomys* to have been incorrectly applied to this taxon. Emmons (1997) presented an outline of many of the characters and conclusions given here.

Several prior classifications have focused on the larger picture of placing caviomorph families in the context of the fossil record (Patterson and Wood, 1982; Woods, 1982), while a series of recent studies on molecular genetic relationships by J. L. Patton and his students have clarified the systematics of taxa within several genera (*Proechimys, Mesomys, Dactylomys, Phyllomys*): da Silva and Patton (1993), Patton (1994), Lara et al. (1996), Patton et al. (2000), and Leite (2003). Leite and Patton (2002) have presented a molecular phylogeny with the most taxonomically dense sampling of echimyid diversity. Carvalho (1999) recently analyzed phylogenetic relationships of living and fossil echimyids, with emphasis on the subfamily Eumysopinae; some of his results are similar to those presented below.

MATERIALS AND METHODS

Phylogenetic Assumptions and Taxon Sampling

The initial phylogenetic hypotheses were that (1) the Echimyinae, as defined by Patton and Reig (1989), form a monophyletic clade within the family Echimyidae (Tables 1, 2), and (2) within the subfamilial clade Echimyinae, species should group as six monophyletic clades representing the genera *Isothrix, Diplomys, Echimys*, and *Makalata* (Patton and Reig, 1989; Woods, 1993; McKenna and Bell 1997), as well as *Phyllomys* and *Callistomys pictus* (Emmons and Vucetich, 1998; Emmons et al., 2002; Leite, 2003).

For parsimony analysis I used one species as representative of each taxon for Eumysopinae and Dactylomyinae. For nominal species of *Diplomys*, only *D. labilis* and *D. rufodorsalis* are included, as adequate material of *D. caniceps* was unavailable. Likewise, two species of *Isothrix, I. pagurus* Wagner 1945, and *I. bistriata orinoci* (Thomas, 1899) were included as representatives of that taxon. Because some recent classifications have included *Myocastor coypus* among the Echimyidae (Table 1), this species was included in the analysis to explore its possible membership in Echimyinae or other subfamilial clades. The octodontid rodents, represented by *Octodon degus*, were chosen as the outgroup for the Echimyidae because they appear to be the most closely related clade at the family level (Woods, 1982) if myocastorids are included within the Echimyidae.

Table 2. Recent species classifications of the subfamily Echimyinae.

Tate 1935	Cabrera 1961	Woods 1993
	all Eumysopinae & *Mesomys*	
Isothrix bistriata	*Isothrix bistriata*	*Isothrix bistriata*
Isothrix pagurus	*Isothrix pagurus*	*Isothrix pagurus*
	Isothrix picta	
	Diplomys caniceps	*Diplomys caniceps*
	Diplomys labilis	*Diplomys labilis*
	Diplomys rufodorsalis	*Diplomys rufodorsalis*
"HAIRY-TAILED GROUP":		
Echimys chrysurus	*Echimys chrysurus*	*Echimys chrysurus*
	Echimys blainvillei (syn. *medius, thomasi*)	*Echimys blainvillei* (syn. *medius)*
Echimys blainvillei		
Echimys pictus		*Echimys pictus*
Echimys lamarum		*Echimys lamarum*
Echimys grandis	*Echimys grandis* (syn. *rhipidurus*)	*Echimys grandis*
Echimys braziliensis	*Echimys braziliensis*	*Echimys braziliensis*
Echimys dasythrix	*Echimys dasythrix* (syn. *lamarum*)	*Echimys dasythrix*
Echimys rhipidurus		*Echimys rhipidurus*
Echimys saturnus	*Echimys saturnus*	*Echimys saturnus*
"NAKED-TAILED GROUP"		
Echimys semivillosus	*Echimys semivillosus* (syn *punctatus, carrikeri, flavidus*)	*Echimys semivillosus* (syn *punctatus, carrikeri, flavidus*)
Echimys punctatus		
Echimys carrikeri		
Echimys flavidus		
Echimys didelphoides	*Echimys didelphoides* (syn. *occasius, guianae, longirostris*)	*Makalata didelphoides* (syn. *occasius, guianae, longirostris*)
Echimys guianae		
Echimys longirostris		
Echimys obscura		
Echimys macrura	*Echimys macrurus*	*Echimys macrurus*
Echimys unicolor?	*Echimys unicolor*	*Echimys unicolor*
Echimys nigrispina	*Echimys nigrispina*	*Echimys nigrispina*
Echimys thomasi		*Echimys thomasi*
Echimys medius		
Echimys occasius		

No echimyine taxon other than *Phyllomys* (Leite, 2003; Emmons et al., 2002) has received recent systematic or taxonomic revision, and most genera and species lack recent diagnoses. Resolution of the many nomenclatural problems and revision of all taxa of echimyines is beyond the scope of this paper, which is limited to supraspecific relationships of the arboreal taxa.

Phylogenetic Analysis

All parsimony analyses were implemented using PAUP™ 4.0b10 (D. L. Swofford, 1999) heuristic search option with all characters treated as unordered and equally weighted. The data set was subject to 100 random addition replicates with TBR branch swapping. Bootstrap values (Felsenstein, 1985) were calculated using 1000 bootstrap replicates with random searching. Patterns of character change were explored using MacClade 3.04 (Madison and Madison, 1992). Minimal numbers of characters were retained for analysis; for example, I include only 15 cheektooth features, although one could define a great many more in a loph-by-loph analysis. All relevant tooth characters and all parts of the skull involved in the chewing apparatus probably evolve in concert and the characters thus do not represent independent evidence of evolutionary change.

Specimens Examined

The specimens of Echimyinae preserved in the following museums were examined: National Museum of Natural History, USA (USNM); American Museum of Natural History (AMNH); Field Museum of Natural History (FMNH); Museum of Comparative Zoology, Harvard University (MCZ); Museum of Vertebrate Zoology, Berkeley (MVZ); Louisiana State University (LSU); Philadelphia Academy of Sciences (PAS); Natural History Museum, London (BMNH); Muséum National d'Histoire Naturelle, Paris (MNHN); Muséum d'Histoire Naturelle, Geneva (MHNG); Museum für Naturkunde der Humboldt-Universität zu Berlin (ZMB); Senckenburg Museum, Frankurt (SMF); Naturhistorisches Museum, Vienna (NMW); Zoologisk Museum, Copenhagen (ZMC); Museo de Historia Natural, Lima (MHN); Escuela Politéchnica Nacional, Quito; Museu de Zoologia, São Paulo (MZSP); Museu Nacional, Rio de Janeiro (MNRJ); Colección Boliviana de Fauna (CBF); Museo de Historia Natural Nöel Kempff Mercado, Santa Cruz. The type material and a representative sample of specimens examined are listed below; many others were also examined.

Callistomys pictus: Nelomys pictus Pictet, 1841 MHNG 299.53, holotype; MNRJ 11027, 31545-6.

Diplomys: *Loncheres caniceps* Günther, 1876 BMNH 76.8.8.9, holotype. *Isothrix darlingi* Goldman 1913 USNM 179577, holotype; *D. labilis* USNM 296336-38, 305746, 460170-1, 396413, 339066, 335739-42, 178183, 457932.

Santamartamys: Isothrix rufodorsalis J. A. Allen, 1899 AMNH 14606, holotype; AMNH 34392.

Echimys: *Echimys saturnus* Thomas, 1928 BMNH 34.9.10.182, holotype; MCZ 41569; AMNH 71903, 98261; *Loncheres cristatus* Desmarest, 1817 MNHN 403, probable holotype; *E. chrysurus* USNM 460070, 549594-5, 549839-40, FMNH 93267, AMNH 96761-8.

Isothrix: Isothrix bistriata Wagner, 1845 NMW B 914, holotype; *Isothrix b. orinoci* USNM 406370, 406373-5, 496839; *Isothrix pagurus* Wagner, 1845 NMW B 913, holotype; USNM 555639; AMNH 95642, 95644, 95646-8, 95651.

Makalata: *Echimys didelphoides* Desmarest, 1817 MNHN 404, probable holotype; USNM 460069, 549593, 549837-8, 581981-2; *Loncheres macrura* Wagner, 1842 NMW B921, holotype; USNM 406378, 406380-2, 406384, 406386,406389-90, 406392, 406394-401, 496476, 496478, 496479-80, AMNH 92891-7, 93582-90, 96769-72, MVZ 153636-7, 157977; *Loncheres guianae* Thomas, 1888 BMNH 88.10.1.1, holotype; *Echimys armatus handleyi* Goodwin, 1962 AMNH 184813 holotype; *Echimys longirostris* Anthony, 1921 AMNH 42886, holotype; *Echimys armatus castaneus* J. A. Allen, and Chapman, 1883, AMNH 6001/4728, holotype; *Loncheres grandis* Wagner, 1845 NMW 920, holotype; AMNH 92907-9, 92912-17, 92938-50, 93594-608, 93610-25, 94021-25, MCZ 32352-3; *Echimys rhipidurus* Thomas, 1928 BMNH 28.7.21.89 holotype; AMNH 73231, 73267-8, 73270, 73273-4, 73276-7, 74084-6, 73791, BMNH 32.8.4.21, 32.8.4.22, 32.8.4.22a, FMNH 87243-49, 87251, 122991-2.

Pattonomys: Nelomys semivillosus I. Geoffroy 1838 MNHN 408b bis, possible holotype; USNM 280204-8; *Loncheres punctatus* Thomas, 1899: BMNH 98.12.1.18, holotype; BMNH 95.12.1.20, 52.12.3.21, 93.12.1.19; USNM 374741-2, 406402; *P. carrikeri* USNM 442715, 45629, 45636, 45643; *Loncheres flavidus* Hollister, 1914 USNM 63218, holotype; *Echimys occasius* Thomas, 1921 BMNH 21.2.15.6, holotype; BMNH 34.9.10.202; MCZ 37964; FMNH 84259; AMNH 98262, 71897, 68177.

Phyllomys: Phyllomys brasiliensis Lund, 1839 lectotype and other type material from Lagoa Santa, ZMC (not numbered); *Nelomys blainvilii* Jourdan, 1837 MHNG 250/19, lectotype; MNHN nos. 310, 402; A7789, MZUSP 6146, 6147; USNM 304580, MNHN 14868, MNRJ 1345, 1350, 1512, 1516, 1517, 1521, 1523, 1528, 1548, 1762, 2238; *Phyllomys dasythrix* Hensel, 1872 ZMB 38800 lectotype, 38794, 38799, paralectotypes; *Phyllomys kerri* Moojen, 1950 MNRJ 6241, holotype; *Nelomys lamarum* Thomas, 1916

BMNH 3.9.5.96, holotype; BMNH, 3.9.5.92, 3.9.5.93, 3.9.5.94, 3.9.5.95, 3.9.5.97, 3.9.5.98, 3.9.5.99, 3.9.5.101, 3.9.5.102, 3.9.5.103, 3.9.5.104; FMNH 35356; *Loncheres medius* Thomas, 1909 BMNH 3.4.1.84, holotype; *Loncheres nigrispina* Wagner, 1842 NMW B 918, holotype; BMNH 33.10.9.18, 33.10.9.19; FMNH 93045, 94358, 94359; MZUSP 175, 1950, 1951, 1952, 1953; *Mesomys thomasi* Ihering, 1897, MZUSP 47 lectotype; MZUSP 45, 51, 223, 526, 527, 532 FMNH 41360 (ex MZUSP 1408); BMNH 2.8.25.2 (ex MZUSP 224), paralectotypes; *Loncheres unicolor* Wagner, 1842, SMF 4319, holotype.

RESULTS

Character Descriptions

There are few sets of discrete and stable characters with which to diagnose supra-specific levels of echimyine taxa. Many morphological characters were rejected during this study as being too variable among individuals. As my interest was in understanding the Echimyinae, I did not make a detailed study of the characters of Eumysopinae or other caviomorphs that might better illuminate those taxa, although some are included in the analysis. Most taxa of echimyids have indeterminate growth, so that measurements are best compared only between like-aged individuals (Patton and Rogers, 1983; Pessôa and dos Reis, 1991; Leite 2003). This limits the usefulness of morphometrics to taxa for which large series of specimens are available. Unfortunately, this is rarely the case for arboreal species. Skull nomenclature follows Woods and Howland (1979), tooth nomenclature (Figure 1) follows Emmons and Vucetich (1998). The states of the characters given below and the associated matrix by taxon (Appendix 1) were used to construct a phylogenetic hypothesis by parsimony analysis and to support the resulting generic classification.

External Morphology:
Character 1. *Pelage of lower back spiny (1); bristly (2); stiff (3), or soft (4).* Echimyidae owe their name to the stiff, spiny or bristly guard hairs characteristic of many taxa. With a few exceptions, spines are visually inconspicuous; they lie flat and are surrounded by slender hairs and are best detected by running a finger lightly backward against the tips of the rump pelage. Spiny pelage (1) has its most robust guard hairs petiolate, stiff, and strongly flattened, with the tip forming a sharp, abruptly narrowing point in strong-spined species. The strongest spines usually occur on the lower back; they bend most readily at the petiolate base, so that they rise vertically when stroked backward. In bristly pelage (2), the tip of the most robust hairs is prolonged into a long, flexible hairlike process and the hair is narrower. When rubbed backward, the hairlike process bends back and the hair is felt as very stiff or prickly but it does not impale the finger; pushed further, the hair

bends at its base and stands upright. Stiff pelage (3) is resistant but not prickly when rubbed backward; the hairs bend in the middle but do not stand upright. A few taxa of echimyids are soft-furred (4); their guard hairs feel soft and bend easily if rubbed backward. Guard hairs of these four types are illustrated by Leite (2001) and Emmons et al. (2002).

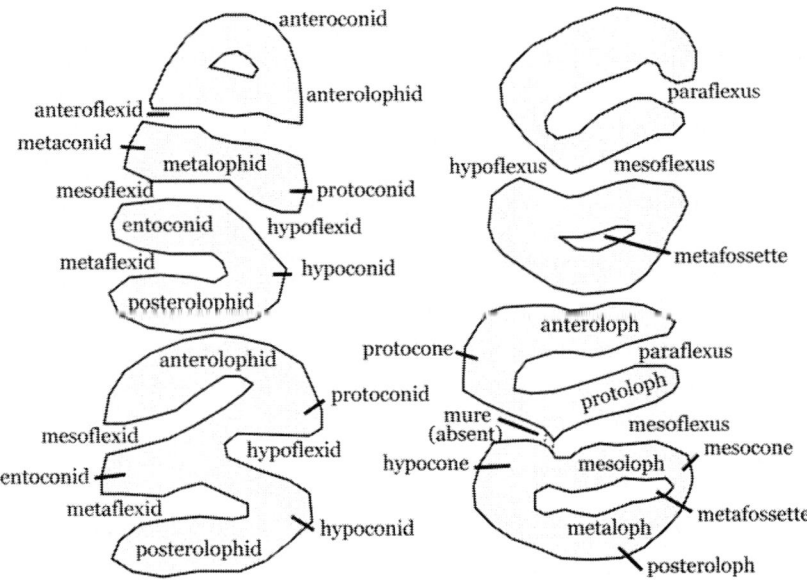

Figure 1. Terminology of the occlusal surface of the cheekteeth (after Emmons and Vucetich, 1998).

Character 2. *Guard hairs without dorsal sulcus (1) or with sulcus (2)*. Whether guard hairs are soft, stiff, or spiny, all but one of the taxa surveyed possess a dorsal, longitudinal groove that runs the length of the hair. The guard hairs are nearly circular in cross-section and it requires a microscope to detect the sulcus in the fine-haired *Callistomys pictus*. Of the taxa reviewed here, only *Myocastor coypus* lacks a sulcus; *O. degus* has a sulcus.

Character 3. *Crest of longer hair on crown and nape absent (1); or present (2)*. The crests of longer hairs on the crown and nape of some echimyine taxa are often composed of paler pelage bordered with contrasting darker stripes or are entirely black, contrasting with paler cheeks. Presumably, these crests can be erected in display, but there are no published descriptions of such behavior.

Character 4. *Guard hairs banded (1) or unbanded (2).* Dorsal guard hairs can be banded, usually with pale gray bases, dark subterminal bands, and pale tips, or unicolored.

Character 5. *Tail "naked" (1) slightly hairy (2) or well clothed with hair, usually with a terminal tuft (3).* Distal to the body fur the tail can range from "naked" or thinly clothed with inconspicuous short hairs that do not hide any scales; to slightly hairy, with conspicuous hairs throughout, but scales still partly visible; to "hairy" with the scales completely covered by dense long pelage. The tail tip may or may not have a pencil of long hair. Tail hairs of echimyines are monocolored and tend to curl away from the surface, which enhances visibility of the hairs. Tate (1935) strongly emphasized tail hairiness in his generic classification of the arboreal Echimyidae, however, the most polytypic taxa (*Makalata, Phyllomys*) have species with each state.

Character 6. *Tail tip colored as basal third of tail (1) sharply contrasting pale (2) or sharply contrasting black or darker (3).* Color of the tail tip, especially in hairy-tailed species with tufts (but not all such), can contrast sharply from the base: species in three taxa have white tail-tips, those in two taxa can have tails tipped with orange or black. When most specimens had unicolored tails but rare individuals had a small pale tip, the species was scored as (1).

Character 7. *Extension of body fur onto tail base less than two cm (1) or more than two cm (2).* Echimyids have either a significant (3-4 cm) extension of the dorsal pelage onto the tail base (species in most echimyine taxa) or the body fur ends abruptly where the tail joins the body (two echimyine species, many other echimyids).

Character 8. *Tiny tubercles cover naked plantar soles of feet (1) or tubercles only present between raised, smooth, well-developed pads (2) or tubercles absent (3).* Many caviomorphs have no distinct firm, smooth, raised plantar pads under the foot joints, but have the soles of the feet covered with tiny tubercles (illustrated for *Cuscomys* in Emmons 1999 [Figure 3]); this state (1) is found in all taxa of Abrocomidae and Capromyida as well as species of *Ctenomys, Octodon, Myocastor,* and others. It is this taxonomic distribution that suggests that it is the plesiomorphic condition. The Echimyidae possess a variable degree of retention of tiny tubercles on the soles, from entirely covering the soles (all dactylomyines, state 1) to strong development of smooth raised pads under weight-bearing joints, but tiny tubercles present around and between the pads (most echimyines, state 2); or smooth raised pads with no surrounding tubercles (*Diplomys* spp., state 3). The tubercles may provide friction for gripping smooth surfaces.

Character 9. *Hind foot without raised, smooth, firm, well-developed plantar pads (1), with five pads and the lateral metatarsal pad and first digital pads joined as a single pad (2) or six* *pads,with the lateral metatarsal pad and first digital pad not joined (3).* The number and development of the raised smooth pads under the foot joints varies among taxa (see character 8). Fluid preserved specimens were not available for all taxa, and the characters are difficult to see in skin preparations.

Character 10. *Mammae arranged in two pairs (1), three pairs (2), or four or more pairs (3).* With the possible exception of *S. rufodorsalis,* Echimyinae have two to five pairs of mammae found under the lateral body fur between elbow and hip in the dorsal pelage field, well above its line of demarcation from the ventral field. There are three principal patterns of mammae: (a) as many as 5 lateral pairs, which include three pairs about equally spaced along the sides in the dorsal pelage field, with another pair intercalated between each of the anterior and medial, and the medial and posterior pairs; (b) three lateral pairs of mammae that are about equally spaced along the sides in the dorsal pelage field (one anterior, one medial, and one posterior), and one inguinal pair on the ventral field; c) two functional lateral pairs which seem to correspond to the medial and posterior lateral pairs of (b), with an occasional inguinal pair. When pairs are reduced, the anteriormost lateral pair appears to the first one lost, followed by the inguinal pair. Because of their placement below dense dorsal pelage, unused lateral mammae are difficult to detect and mammae counts from skins of echimyids can be ambiguous.

Character 11. *Maxillary cheekteeth brachydont, with 4 roots (1), hypselodont with 3 roots (2), or hypsodont and unrooted (3).* Characters of the dental roots appear to distinguish higher (subfamilial) levels of Echimyidae. When the teeth are four-rooted and brachydont, (1), the cheekteeth are straight-sided; when the molars have three roots (2), the crown of the tooth is strongly bent laterally; when the teeth are hypsodont, with single, open roots, the crowns tend to be straight sided. Most echimyines have state 1, and eumysopines state 2.

Character 12. *Occlusal plane tip flattish relative to plane of palate (1), weak tip (2), or strong tip (3).* When the cheekteeth are 4 rooted (character 11) the occlusal plane may be parallel to the hard palate (state 1, only in dactylomyines) or tipped laterally with the labial maxillary gumline above the palatal line (state 2, Figure 2). When the teeth are three rooted and unilaterally hypsodont, the crown of the tooth may be straight sided and the occlusal plane parallel to the palate (state 1), or weakly (state 2), or strongly bent laterally (state 3, Figure 2) such that the maxillary occlusal surfaces face laterally. In this case the labial gumlines are about on a plane with the palatal gumline. When the maxillary occlusal plane is tipped laterally, the mandibular occlusal surfaces correspondingly tip medially for occlusion.

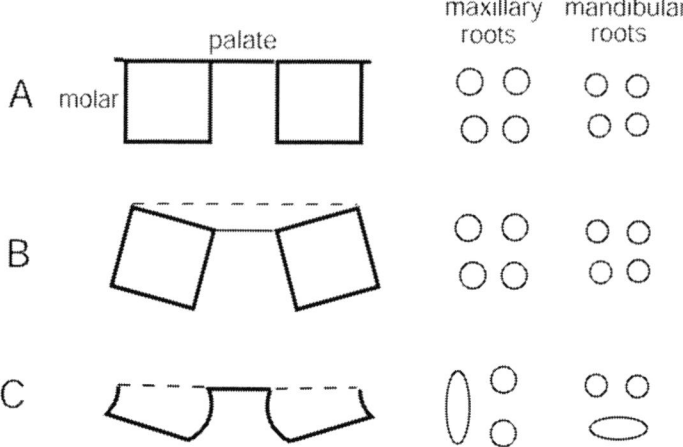

Figure 2. Schematic of maxillary cheek tooth angles in relation to palate, and associated molar root patterns.   (a) Occlusal plane parallel to palate (i.e., dactylomyines), (b) straight-sided teeth tipped, occlusal surface at an angle (most echimyines), (c) teeth bent, occlusal surface at angle (eumysopines, *Callistomys* spp.).

Character 13. *Upper incisor root originates within the maxillary root of zygoma, posterior to the root of the zygoma (1) or level with or outside zygoma (2).* The root of the upper incisors may extend into the infraorbital foramen to nearly contact the molar roots (1). The bulge at the base of the root forms beneath it the canal for the infraorbital nerve on the floor of the interior, medial surface of the inferior zygomatic root. Alternatively, the incisor root may be short, with its base level with or anterior to the vertical part of the maxillary root of the zygoma (2). In this case, the medial ventral wall of the infraorbital foramen is smooth. The condition of long incisor roots and a well-developed infraorbital canal, with a sheetlike dorsal extension of the maxillary over the canal to form a closed tube at maximal development, may be the plesiomorphic state, as it is present in octodonts.

Character 14. *Lower incisor root originates posterior to m3, high in coronoid process (1) below m3 (2) or anterior to m3 (3).* The lower incisor root can extend posteriorly high up the condylar process of the mandible, which is then extremely robust (state 1); or it can be much shorter, originating below or anterior to the third molar. The mandibular foramen lies posterodorsally to the incisor root, such that when the roots are long, it is high up near the condyloid process.

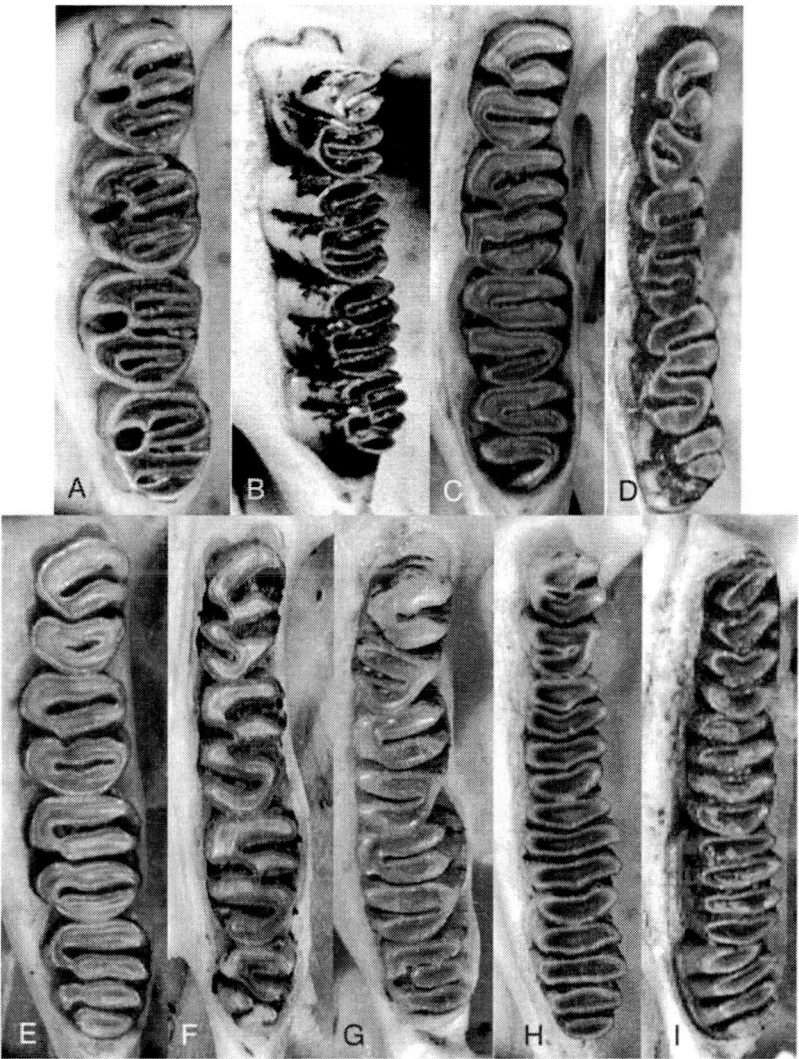

Figure 3. Occlusal aspect of left maxillary cheekteeth. (a) *Isothrix bistriata* USNM 460375, (b) *Callistomys pictus* BMNH 80.9.15.1, (c) *Makalata macrura* USNM 40638, (d) *Pattonomys occasius* FMNH 84259, (e) *Echimys chrysurus* USNM 549594, (f) *Pattonomys carrikeri* USNM 45642, (g) *Santamartamys rufodorsalis* AMNH 34392, (h) *Phyllomys nigrispina* MZSP 1951, (i) *Diplomys labilis* USNM 296337.

Character 15. *Occlusal surface of maxillary cheekteeth with one short lingual flexus/fossette and two labial flexi/fossettes (1,* Proechimys *pattern), one lingual flexus and three labial flexi (2,* Echimys *pattern), two lingual flexi, two labial flexi (3,* semivillosus *pattern), four separate and parallel laminae (4,* Phyllomys *pattern), none of these (5), 1 lingual flexus, 4 labial flexi/fossettes (6,* Mesomys *pattern), or polymorphic (2) and (3), (7).* All patterns refer to unworn or little-worn teeth; the pattern can change with wear. Teeth of state (2) are subcircular in shape with one re-entrant lingual flexus that penetrates less than a half-width of the tooth (Figure 3a). Teeth of state (3) are elongate, with two lingual flexi, the anterior one may completely divide tooth and connect to mid-tooth labial flexus, and two labial flexi (Figure 3d, f); teeth of state (4) are elongate, with three flexi that completely cross teeth and split them into 4 sub-parallel, laminar lophs (Figure 3h, i); state (5) refers only to the outgroup, *Octodon degus*, which has no enamel flexi or fossettes; while state 6 was found uniquely in *Mesomys* c.f. *hispidus*. Whether the posterior flexus opens lingually or labially is variable in *Makalata macrura*, such that an individual can have both patterns in the corresponding teeth on opposite sides of the mouth. In species with laminar teeth, the lophs often join with wear. The pattern in which this occurs, like the original pattern of lophs and flexi, depends upon the vertical depth of each part of each flexus. In unworn teeth, the eventual wear pattern can be predicted from the relative depth of the flexi as viewed from the side of the tooth.

Character 16. M3 *with four or more well-developed lophs (1) or with reduced posteroloph (2) or with three or fewer lophs (3) or polymorphic for (1) and (2) (4) or polymorphic for (2) and (3) (5).* The fourth or posteriormost loph of M3 can be well developed and extend the full width of the tooth (1, Figure 3e, i); or be reduced in size to half or less of the width of the anterior lophs of M3 (2, Figure 3c); or else completely absent (3, Figure 3d).

Character 17. M3 *mesoloph similar in size to protoloph (1 Figure 3c) or much shorter than protoloph (2 Figure 3d).*

Character 18. *Lower premolar anteroloph not triangular (1) or triangular with a flexid opening lingually or apparently so (2) or triangular with flexid open labially (3, Figure 4f) or triangular with flexid open posteriorly (4) or an enclosed triangle with a central fossetid (5, Figure 4c) or loph triangular or oblong, with no fossetid (6).* The anterior lophs of the unworn lower premolar have some of the most trenchant characters for distinguishing echimyid taxa (Figure 4). In most echimyines the anterior loph forms a roughly triangular, pointed or rounded structure, which varies in the orientation of its single flexid (2, 3, 4) or whether this is enclosed as a central fossetid (5) or is completely absent (6).

Character 19. *Lower premolar without separate (metalophid) bar in middle (1, Figure 4b, F) or bar is present (2, Figure 4c, E) or other (3).* I provisionally follow Vucetich and Verzi (1992) in considering the pentalophodont molariform tooth as plesiomorphic among echimyids, but I am uncertain of the homologies of the anterior lophs of p4. I consider the anteroconid and anterolophid (Figure 1) as two lophids, but where the metalophid bar is absent (compare Figure 4b to 4c), it is unclear whether it is the metalophid or the anterolophid that is lacking, or whether they have merged into a single loph. *Myocastor coypus* has premolars of a different form (state 3) not classifiable for this character.

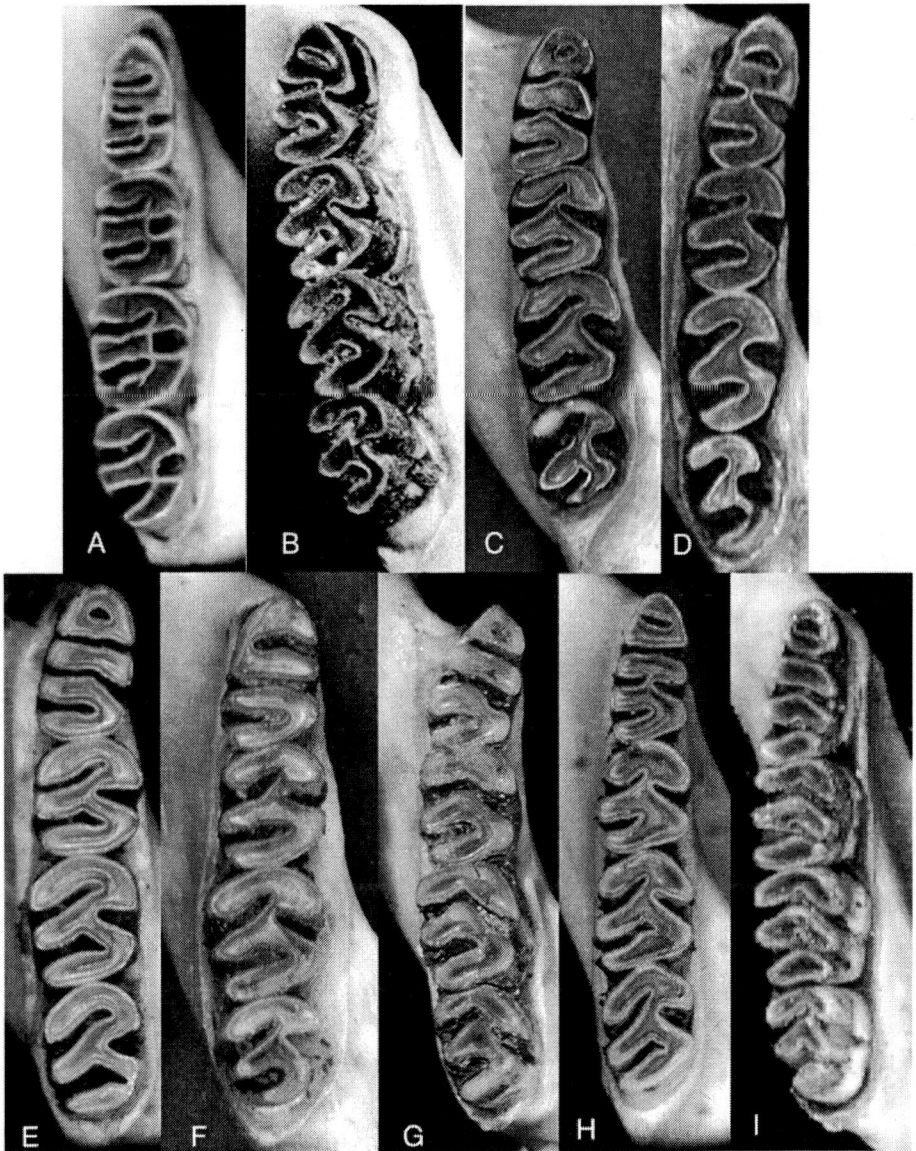

Figure 4. Occlusal aspect of the right mandibular cheek teeth. Specimens as in Figure 3.

Character 20. Anterior edge of crowns of m2-3 almost a straight line at right angle to tooth axis (1) or curved with radius like part of a circle encompassing tooth, or caret shaped (2, Figure 4, all) or a straight diagonal line relative to tooth axis (3).

Character 21. *Lower molar labial and lingual flexids slanted forward, medial end anterior to labial end, m2-3 metaflexid well separated from hypoflexid (1, Figure 4b-e) or labial and lingual flexids about 90° to tooth axis, metaflexid meets or almost meets hypoflexid in mid-tooth (2, Figure 4a).*

Character 22. *Lower molars with no separate laminar lophs (1, Figure 4a-f) or with one separate anteroloph (2, Figure 4g) or with three separate lophs (3, Figure 4i) or with one separate posteroloph (4).* State (1) is the condition of most caviomorphs and echimyids, and likely plesiomorphic. Only *Phyllomys* spp. share state (2), only *Diplomys* spp. have state (3), and only *Dactylomys* spp. state (4).

Character 23. *Hypoflexid slants posteriorly (1, Figure 4a) or slants weakly anteriorly (2, Figure 4d, e) or slants strongly anteriorly (3, Figure 4b, h).*

Character 24. *Mandibular toothrows strongly convergent anteriorly (1) slightly convergent anteriorly (2) or parallel or divergent (3).*

Character 25. *Relative length of upper toothrow/basilar length of Hensel very short, 20-22% (1), short 23-24.4% (2), intermediate 24.4-25.7% (3, Figure 11), or long, 26.9-31% (4, Figure 14).*

Character 26. *Lower incisors strongly curved (1, Figure 13) or straightish (2, Figure 11).*

Character 27. *Squamosotympanic fenestra a large open slit (1, Figure 5b) or a tube enclosed in bone (2, Figure 5a).* The squamosotympanic fenestra is an elongate open slit along the squamosal suture in some echimyids. This is likely to be the plesiomorphic state, as it is thus in capromyids, octodontids, abrocomids, and cuniculids. In most echimyini, the fenestra is closed posteriorly to form a tube that opens anteriorly. Functionally, this fenestra in caviomorphs is likely the same as the post-glenoid foramen of murids.

Character 28. *Masticatory foramen and foramen ovale acessorius (foa) are separated by a bony strut that is absent, or no masticatory foramen (1); a narrow strut (2, Figure 6e); a medium width strut (3, Figure 6b); a wide strut (4, Figure 6f); or are polymorphic 2 and 3 (5).* The strut of bone that separates the foa and masticatory foramina is narrow or very narrow, or the masticatory foramen is lacking, in most Eumysopinae, which likely possess the plesiomorphic state. It is narrow to absent in *Dactylomys* spp., and varies from narrow to wide in other echimyines.

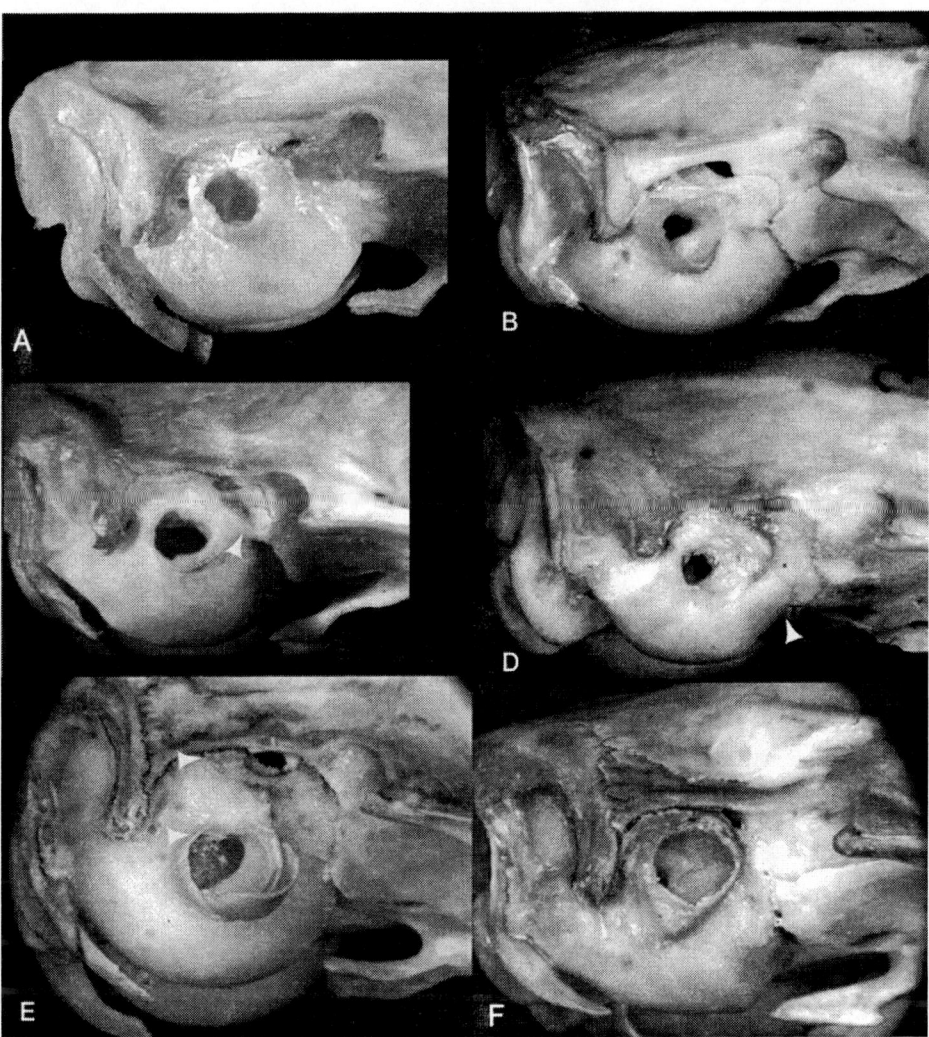

Figure 5. Auditory bullar region of Echimyids. (a) *Echimys chrysurus* USNM 549594 (reversed), (b) *Isothrix bistriata* USNM 406375, (c) *Makalata didelphoides* MDC572, (d) *Santamartamys rufodorsalis* AMNH 34392 (reversed), (e) *Phyllomys nigrispina* USNM 484508, (f) *Diplomys labilis* USNM 296337. Arrows denote crest on the ventral border of the squamosotympanic foramen, with depression below, in (c) and space above meatus in (e).

Character 29. *A small (interparietal) foramen between foa and masticatory foramen is present (1, Figure 6a) or absent (2, Figure 6c) or 3 polymorphic for (1) and (2).* This foramen was considered present in a specimen if it was found on both or only one side of a specimen.

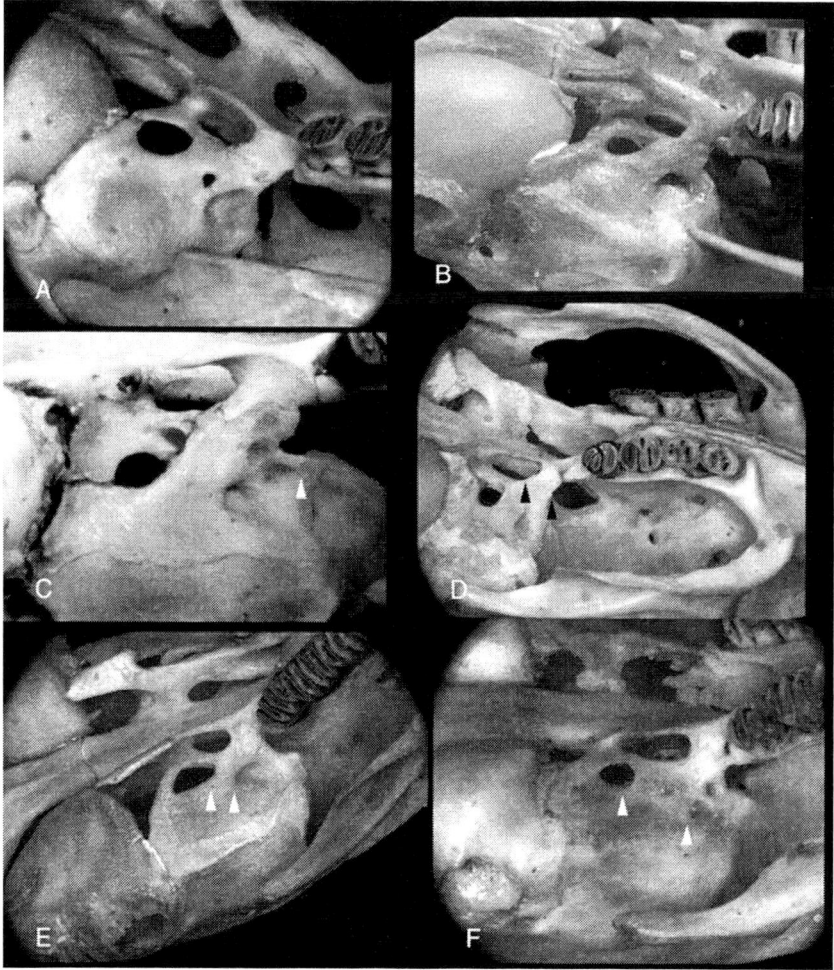

Figure 6. Alisphenoid region of (a) *Isothrix bistriata* USNM 406375, (b) *Echimys chrysurus* USNM 549594, (c) *Makalata grandis*, (d) *Pattonomys occasius* MVZ 37964, (e) *Diplomys labilis* USNM 296337, (f) *Santamartamys rufodorsalis* AMNH 34392. Arrows in (d-f) bracket the distance from the foramen ovale acessorius to the masticatory foramen; arrow in (c) points to the buccinator foramen.

Character 30. *Slant of lateral tube of auditory meatus straight out or slightly forward (1, Figure 5c) or slanted strongly forward or downward (2, Figure 5f) or slanted upward and backward (3).* Because the ear pinnae of echimyines are tiny and likely immobile, the direction of slant of the outer ear canal may be associated with the directionality of hearing; the functional significance of the different character states among echimyids is unclear, as all species are arboreal. Specimens often have bony rings external to the tubes (Figure 5e, on lower right of meatus), but these are usually lost in skinning or skull preparation and their presence/absence could not be scored. Nevertheless, bony rings seem more common in some taxa than in others (*Phyllomys* spp., *semivillosus* group).

Character 31. *Premaxillary and maxillary portions of septum within incisive foramen separate, maxillary portion dipping in dorsally (1, Figure 7a) or fused and maxillary portion dipping in dorsally (2) or broadly fused, in the same plane as rim of foramen (3, Figure 7b).* Characters of the incisive foramen have proved extremely useful for distinguishing *Proechimys* species (Patton, 1988). Among the Echimyinae, most have configuration (1), which is likely plesiomorphic, as it is widely distributed among caviomorphs. *Makalata* skulls can often be distinguished from other genera by this character alone. In very old animals, some fusion of the elements of state (1) can occur.

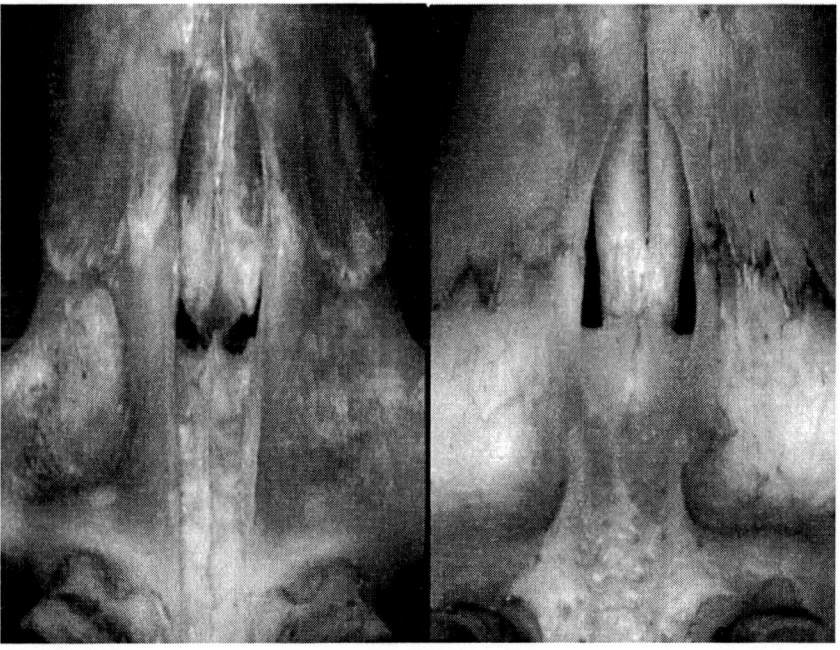

Figure 7. Incisive foramen of (a) *Pattonomys carrikeri* USNM 45629 and (b) *Makalata macrura* USNM 406380.

Character 32. *Anterior jugal hugely expanded in depth, more than 1/2 width of infraorbital foramen (1) or wide, but less than 1/2 width of infraorbital foramen (2) or narrow (3).* The most extreme expansion of the jugals is found in the semi-fossorial taxa *Euryzygomatomys spinosus, Carterodon sulcidens* and *Clyomys laticeps,* and in the semiaquatic *Myocastor coypus,* while the next widest condition is possessed by terrestrial taxa such as *Thrichomys apereoides* and *Proechimys longicaudatus,* and one arboreal species, *Callistomys pictus.* Most arboreal species have relatively slender jugals (Figure 12), suggesting an ecological association with jugal development.

Character 33. *Jugal fossa, anterior point diffuse and broad anteriorly (1) or comes to a sharp point (2, Figure 12).* State (1) is shared only by three terrestrial taxa and *C. pictus,* again suggesting an ecological association.

Character 34. *Inferior jugal process (jp) inconspicuous and forward of superior jp (1, Figure 17) or elongate and about level with or posterior to sjp (2, Figure 12).*

Character 35. *Jugal fossa, width of angle from upper rim to lower border of inferior process 10° or less (1), about 20° (2), about 30° (3, Figure 12), or about 40° or more (4, Figure 14).*

Character 36. *Infraorbital canal well developed with sharp crest or closed beneath a bony shelf (1) or present only as a groove (2) or completely absent (3) or polymorphic for 2 and 3 (4).* This canal or groove on the medial floor of the infraorbital canal was described and illustrated by Patton (1988) as a character of *Proechimys* spp. See notes for character 13. Most arboreal taxa lack a canal.

Character 37. *Ventral lip of squamosotympanic fenestra smooth, without a beaded rim or a depression ventrad (1, Figure 5f) or raised as a beaded rim, with distinct depression below it (2, Figure 5c arrow).*

Character 38. *Large palatal vacuities present in alisphenoid-basisphenoid region such that parapterygoids are freestanding (1) or small, distinct round openings of unfused sutures persisting to adulthood (2, Figure 11) or with sutures either completely fused in adults, or with only hairline slits present (3, Figure 12).* This character must be evaluated in adult specimens; vacuities in juveniles may fuse with maturity.

Character 39. *Buccinator foramen with no medial wall and open space below pterygoid (1) or with a bony shelf on foramen floor beside alisphenoid (2) or with a shelf and medial wall or partial wall forming a closed foramen (3).* This structure can only be seen in well-cleaned skulls, as it is deep in the medial cranium. It is viewed by holding the skull at eye level and sighting posteriorly through the infraorbital foramen along the plane of the buccinator foramen.

Character 40. *Mandibular foramen near top of the condylar process of mandible (1) or on low or mid ramus anterior to a bladelike condyloid ridge (2, Figure 8a) or at the base of ramus near toothrow on posterior edge of condyloid ridge (3, Figure 8c) or at base of ramus near toothrow on anterior side of condyloid ridge (4, Figure 8d).* See note for character 13.

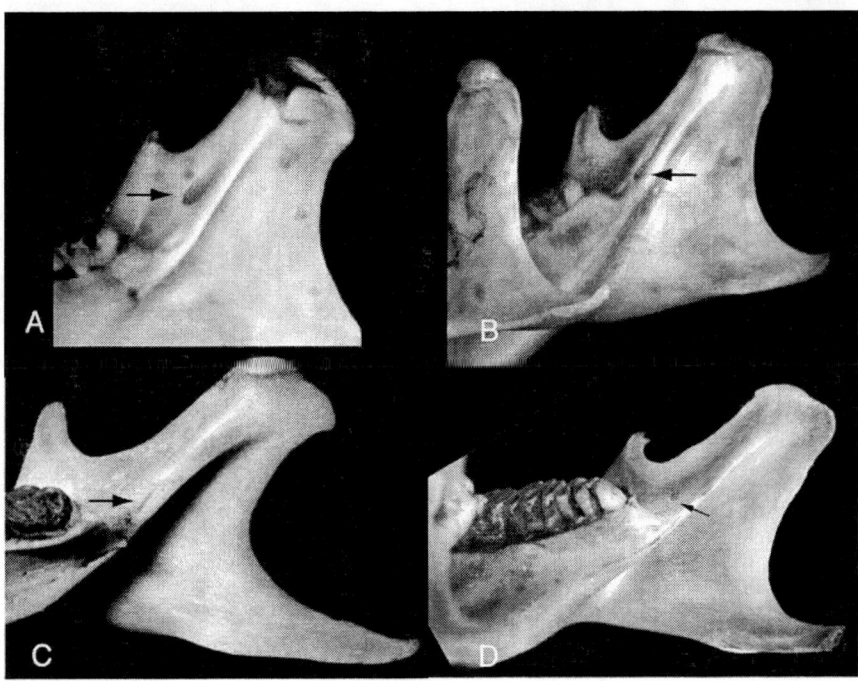

Figure 8. Position of mandibular foramen (arrows). (a) *Isothrix bistriata*, (b) *Echimys chrysurus*, (c) *Phyllomys medius*, (d) *Diplomys labilis*.

Character 41. *Angular process of mandible short posteriorly, about as long as condylar process (1, Figure 8b) or long, much longer than condylar process (2, Figure 8c).* The relative length of the angular process is gauged by setting the mandible on a flat surface and looking straight down on it from above, over the condylar process.

Character 42. *Dorsal rim of auditory meatus close to squamosal suture, space above meatus is narrower than width of auditory meatus (1, Figure 5a) or with wide flat space, almost as wide or wider than meatus, between it and the suture (2).* A wide space above the meatus is found only in *Phyllomys* species (Figure 5e, between arrows) and *Myocastor*.

Character 43. *Posterior maxillary notch of maxillary and palatine behind M3 enclosed as a foramen, with maxillary fused with suture to alisphenoid (1, Figure 6f) or open as notch, alisphenoid not fused to maxillary (2, Figure 6e) or polymorphic for 1 and 2 (3).*

Character 44. *Coronoid process of mandible higher than condylar process (1) or lower than condylar process (2).*

Character 45. *Fourth premolar deciduous (1) or not deciduous (2).* A diagnostic character of the family Echimyidae is the presence of a premolar retained throughout life that is not preceded by an erupted deciduous tooth. This tooth is likely the homologue of the deciduous premolar of other hystricognath rodents. I designate it as P4/p4, without judgment as to its homology. The bristle-spined porcupine, *Chaetomys subspinosus*, previously considered an echimyid, has a deciduous premolar that is replaced by a permanent tooth. A specimen in Senkenburg Museum, Frankfurt, (SMF 11045) has an unworn P4 beside well-worn molars, a state that does not occur in echimyids, where the cheekteeth always show decreasing wear from P4/p4 to M3/m3.

Character 46. *Squamosoparietal suture raised in a ridge extending across parietal (1) or smooth, with no raised ridge (2).*

Character 47. *Squamosal width at squamosotympanic foramen ≤ 1 mm (1) or 1 to ≤ 2 mm (2) or 2 to ≤ 3 mm (3).*

Phylogenetic Hypothesis

The above characters were used to construct a phylogenetic hypothesis. The bootstrap 50% majority rule consensus tree suggests the following relationships (Figure 9a): (1) a clade of taxa including all genera placed in Echimyinae by Patton and Reig (1989) and McKenna and Bell (1997) is weakly supported, but, unlike their classifications, it includes *Mesomys* sp., *Lonchothrix emiliae*, *Dactylomys boliviensis* and *Myocastor coypus*, (2) taxa formerly all synonymized under the genus *Echimys* (Tate 1935) group into four units, including *Makalata* species, *Echimys* species, *Phyllomys* species, and the *semivillosus* group of Caribbean coastal species, with which the Amazonian *occasius* associates, (3) *Isothrix* species, *Callistomys pictus*, *Diplomys labilis* and *rufodorsalis* do not cluster with others, (4) the eumysopines (sensu Patton and Reig, 1989) do not segregate as a monophyletic clade, but (5) the three semi-fossorial taxa cluster as a sister group to all the other Echimyidae.

Bootstrap analysis (Figure 9a) strongly supports clades of *Echimys* species and *Isothrix* species, as well as a clade including the three semifossorial taxa (*Carterodon sulcidens, Clyomys laticeps, Euryzygomatomys spinosus*). Clades of *Makalata* species and *Phyllomys* species are more weakly supported. McKenna and Bell (1997) placed

Myocastor among the Echimyidae, a view that seems supported by this character set. However, other characters of this taxon were not examined in detail and I consider the subfamilial affinity of *Myocastor* unresolved by this data set.

These results reveal some close similarities to patterns seen in molecular studies of a range of caviomorphs (Lara et al. 1996, Leite and Patton 2002, Figure 9b). Those analyses also grouped all taxa seen above the first node of Figure 9a and found little resolution at the base of the family, but indicated monophyly of the four included taxa of echimyines (*Echimys, Phyllomys,* and two taxa of *Makalata*). Like Figure 9a, the molecular analysis (Figure 9b) does not support monophyly of Eumysopinae sensu Patton and Reig (1989) and places *Euryzygomatomys* and *Clyomys* as a sister group to all other echimyids, which may be associated with *Myocastor*. My results also agree with the molecular hypothesis in showing separation of *Proechimys* from *Trinomys*. Leite (2003) recently analyzed complete cyt b gene sequences of *Phyllomys* from populations throughout the entire geographic range, with *Echimys chrysurus* and two *Makalata* species as outgroups. His molecular data set shows that the 33 individuals from populations of nine *Phyllomys* species form a strongly supported monophyletic clade with respect to the outgroups, but there are several divergent clades within *Phyllomys*. Of the four *Phyllomys* taxa analyzed here, Leite (2003) shows *P. lamarum* grouped in a clade with *P. blainvilii,* with *P. nigrispinus* in a more distant sister group to these, while *P. dasythrix* is widely divergent in a sister group to the clade including those three. This structure is partially mirrored in Figure 9a.

I find the molecular evidence compelling and I here conclude that on both morphological and molecular grounds there is as of yet no clear basis for assigning several taxa of Echimyidae into subfamilies with other genera, including *Myocastor, Dactylomys* spp., *Isothrix* spp. and *Mesomys/Lonchothrix* spp. Furthermore, as also concluded by Lara et al. (1996) and Leite and Patton (2002), a subfamilial clade Eumysopinae is not supported. With the caveats noted below, I provisionally define Echimyinae as the taxa grouped above *Proechimys* in Figure 9a, and grouped likewise above the arrow in Figure 9b. Nonetheless, I will follow recent authors (Table 1) and maintain the dactylomyines and *Myocastor* in their own subfamilies, pending more comprehensive and convincing systematic studies of those taxa. In this report I revise all the arboreal echimyine genera with the exception of *Mesomys* and *Lonchothrix* species.

A

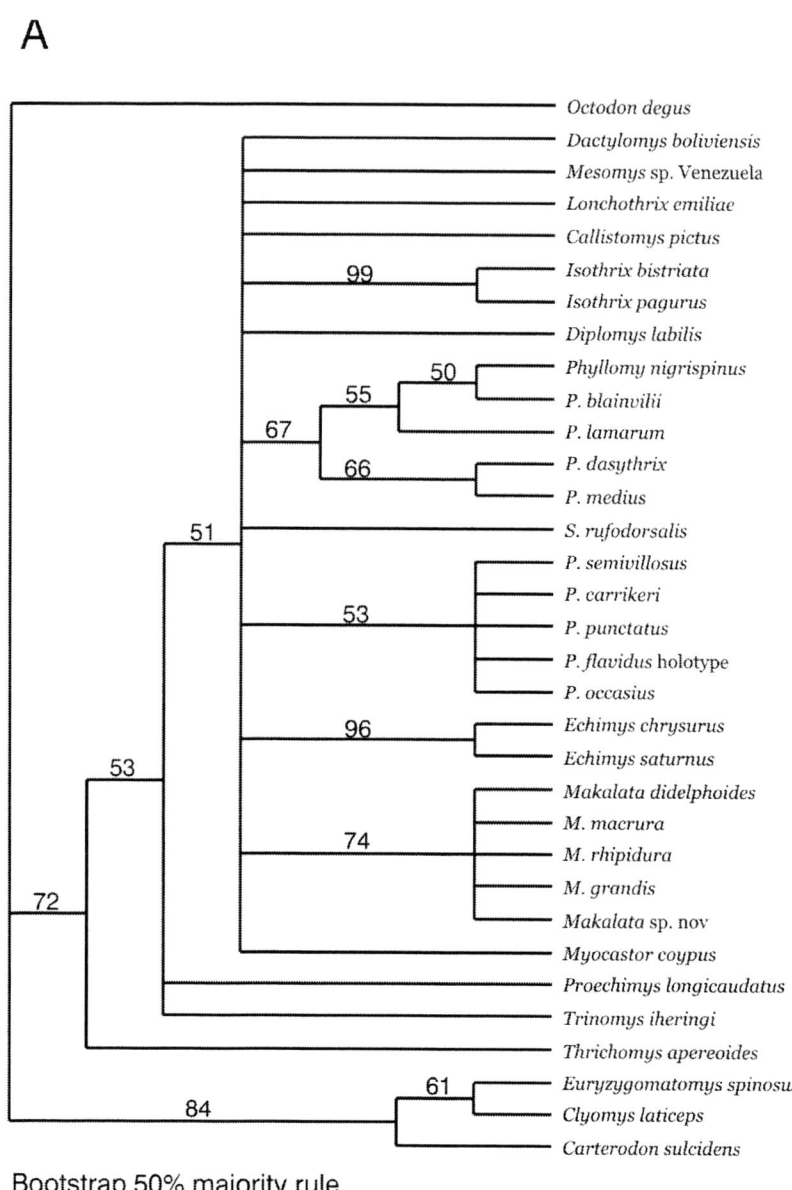

Bootstrap 50% majority rule

Figure 9a. Tree generated by 50% majority-rule bootstrap analysis of morphological characters. Bootstrap values from heuristic search in boldface italics; values ≥ 50% shown above branches.

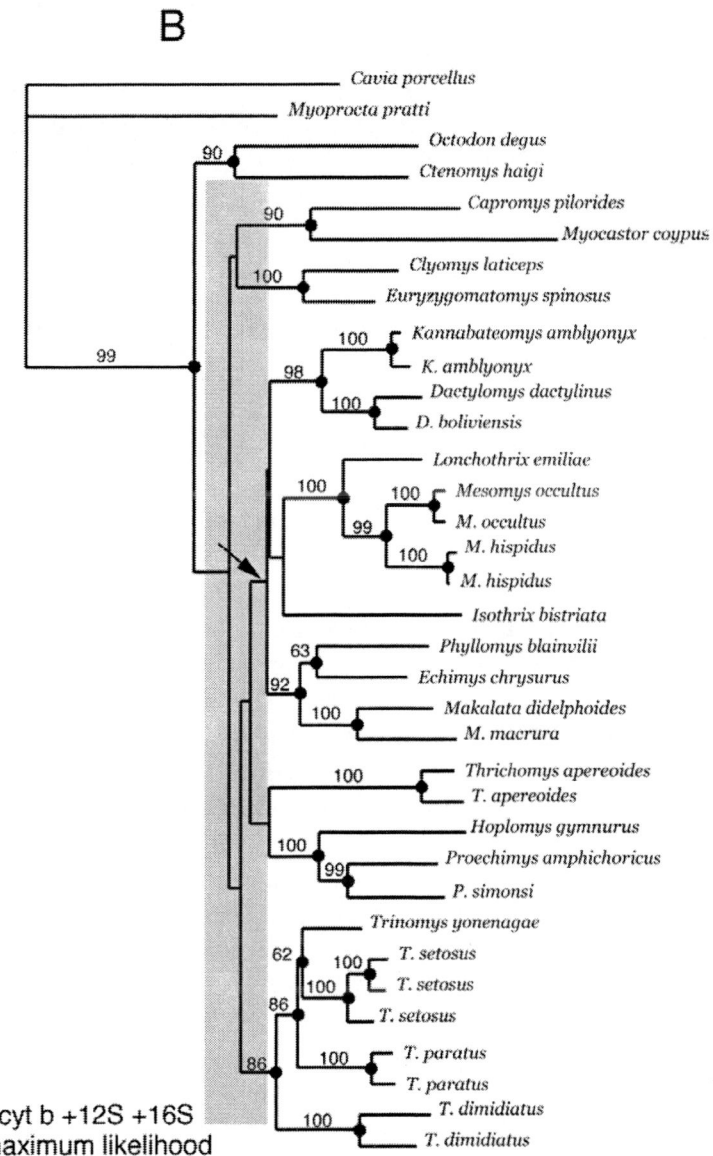

Figure 9b. Highest likelihood tree based on molecular characters of combined cyt b, 12S, and 16S genes. Percent bootstrap values shown above branches, gray area of tree has no resolution (from Leite and Patton, 2002).

The Family Echimyidae Gray, 1825

Before defining the genera of the subfamily Echimyinae, I briefly outline the most significant characters, primarily of dentition, that separate the Echimyinae from the other echimyid taxa provisionally grouped under Dactylomyinae and the likely paraphyletic Eumysopinae sensu Patton and Reig (1989). I retain the subfamily Eumysopinae pending a stronger case for rejecting it and substituting another, but I propose a tribal grouping within it, based on morphological characters and the molecular results of Leite and Patton (2002).

Subfamily Eumysopinae Rusconi, 1935

Diagnosis: Cheekteeth hypselodont (unilaterally hypsodont), with uppers on the lingual side and lowers on the labial side. Occlusal plane tipped at a sharp angle away from the palate, produced by bending of the tooth so that labial gum lines are close to the same plane as the palate (Figure 2c). Upper molars with three roots; the lingual, hypsodont side with a single, longitudinal, narrow root running the length of the tooth and the labial side with two small roots. Lower molars with three roots; one long narrow transverse root across the posterior side and two small anterior roots. Hypoflexids of m1-3 slant posteriorly (internal end more posterior to external, labial, end). Root of lower incisor reaches to M3 or posterior to it. Root of upper incisor originates within the orbit, posterior to the inferior zygomatic root. Upper cheekteeth subcircular in occlusal section, with one lingual flexus (hypoflexus) which penetrates less than half the width of tooth; no reentrant flexi from labial side, cheekteeth not split into separate lophs by deep transverse flexi. Toothrows short, less than 25% of basilar length of Hensel.

Remarks: The above characters are shared by many eumysopine taxa and may be plesiomorphic, as noted by Patton and Reig (1989). I list them here for heuristic comparison with characters of the echimyines described below. For brevity, below I use the term Eumysopinae to designate the group of taxa placed there by Patton and Reig (1989), without implying that I consider them to be a monophyletic group.

Tribe Euryzygomatomini

Included genera:
    Euryzygomatomys Goeldi, 1901
    Clyomys Thomas, 1916
    Carterodon Waterhouse, 1848

Diagnosis: Fossorially adapted rodents. Tail much shorter than head and body (about 50% or less of HB length), claws of forefoot greatly elongated. Root of lower incisor extends high above line of cheekteeth into ramus of mandible. Mandibular foramen high near top of ramus; coronoid process of mandible higher than condyloid process. Infraorbital canal strongly developed. Jugals often broadly expanded dorsoventrally, with the maxillary ascending portion of zygomatic arch usually broad. Basisphenoid above palatal notch completely surrounded by large vacuities.

Remarks: This cluster of monotypic genera likely deserves subfamilial rank, but detailed studies are needed of the molecular genetics and the morphology of both fossil and living forms.

Incertae Sedis

Genus *Proechimys* Allen, 1899
      synonym *Hoplomys,* Allen, 1908
Genus *Thrichcomys* Trouessart, 1880
Genus *Trinomys* Thomas, 1921
Genus *Mesomys* Wagner, 1845
Genus *Lonchothrix* Thomas, 1920

Subfamily Dactylomyinae Tate, 1935

Included genera:
    Dactylomys I. Geoffroy, 1838
    Kannabateomys Jentink, 1891
    Olallamys Emmons, 1988

Diagnosis: Arboreally adapted rodents. Pelage without spines or bristles. Tails much longer than head and body length. Feet with no raised, smooth palmar or plantar pads, palmar and plantar skin evenly and densely covered with tiny tubercles, toes elongate and slender. Cheekteeth brachydont, with four roots. Occlusal plane approximately parallel (in the same plane as) to palate. Cheekteeth extremely enlarged, broader than long, forming broad flat plates. Cheekteeth split by deep flexi into four lophs in two V-shaped pairs that open labially and come to a gradual point lingually. Lower molars split by transverse flexids into three lophs, an anterior, single, lamellar loph and a posterior, V-shaped pair of joined lophs. Alisphenoid with reduced or no bony bridge from foramen ovale to basisphenoid posterior to base of parapterygoid processes. Paraoccipital processes projecting laterally. Where diet is known, all taxa feed chiefly on bamboo.

Subfamily Echimyinae Gray, 1825

Included genera:
 Echimys F. Cuvier, 1809
 Phyllomys Lund, 1839
 Makalata Husson, 1978
 Diplomys Thomas, 1916
 Callistomys Emmons and Vucetich 1998
 Two new genera

Diagnosis: Arboreally adapted rodents with pelage spiny, bristly, or soft; tail as long as or longer than head and body length. Feet with raised, smooth plantar pads developed under joints. Molars brachydont, with four roots (except in *Callistomys*), but crowns often high. Posterior tip of lower incisor root terminates anterior to m3. Occlusal plane tipped away from plane of palate, achieved by tipping of straight-sided teeth, so that labial gumline is higher than palatal (Figure 2b, except in *Callistomys*). Cheekteeth with deep reentrant flexi/ids that may split teeth into separate, parallel laminae. One to three deep flexi present on labial side of maxillary cheekteeth. Maxillary zygomatic process and jugal usually slender. Toothrows long, over 25% of basilar length of Hensel.

Incertae Sedis

Isothrix Wagner, 1845

Genus *Isothrix* Wagner, 1845

Type species: *Isothrix bistriata* Wagner, 1845

Included species:

Isothrix bistriata Wagner, 1845
Isothrix pagurus Wagner, 1845
Loncheres [bistriata] orinoci Thomas, 1899
Isothrix [bistriata] negrensis Thomas, 1920
Isothrix sinnamariensis Vié et al. 1996

Diagnosis: Large arboreal rats; fur soft; tail covered with long, unbanded hair that curls outward like a bottle-brush but may not completely hide scales. Mammae numerous, three to five lateral pairs and usually one inguinal pair. Phallus of *I. bistriata* short and blunt, with deep longitudinal grooves on sides, lappet above urethra not salient from crater (Figure 10e). Maxillary toothrows short, ≤26.8% of basilar length of Hensel, with small, short, subcircular teeth (Figures 3a, 11). Maxillary cheekteeth not strongly tipped laterally as in other echimyines, but occlusal surface nearly on a plane parallel to the palate. All hypoflexi/ids of both upper and lower premolars and molars are oval to subcircular, with a mure; the adjacent lophs curve around the hypoflexi to nearly close at the rim of the tooth (Figures 3a, 4a). Mandibular cheekteeth with hypoflexids slanting slightly backward, with the internal end of the flexid slightly posterior to the external end (Figure 4a). Lower premolar pentalophodont, hypoflexid short, with a mure, and other flexids close to parallel and all opening lingually when unworn. Tooth quickly wears to a pointed oval with three parallel lingual fossettes and a round labial fossette derived from the hypoflexid. Cranium broad, with short, broad rostrum; zygomatic arches strongly bowed outward anterior to squamosal (Figure 11). Auditory tympanic bullae moderately inflated, auditory meatus medium-sized, near squamosal, short auditory tube strongly slanted forward. Squamosotympanic fenestra a long, open slit that reaches posteriorly to base of mastoid process (Figure 5b). Post-palatal notch deep, to middle of M2; incisive foramina large and wide. In pterygoid region, there is no shelf of bone behind base of parapterygoid (hamular) processes on the wall of the buccinator foramen. Below the presphenoid-basisphenoid suture a round vacuity connected to a slitlike vacuity along the side of the prespheniod is shaped like a written musical note. Base of mandible with strongly developed masseteric and pterygoid crests; mandibular foramen anterior to the condyloid ridge and not in a fossa (Figure 8a).

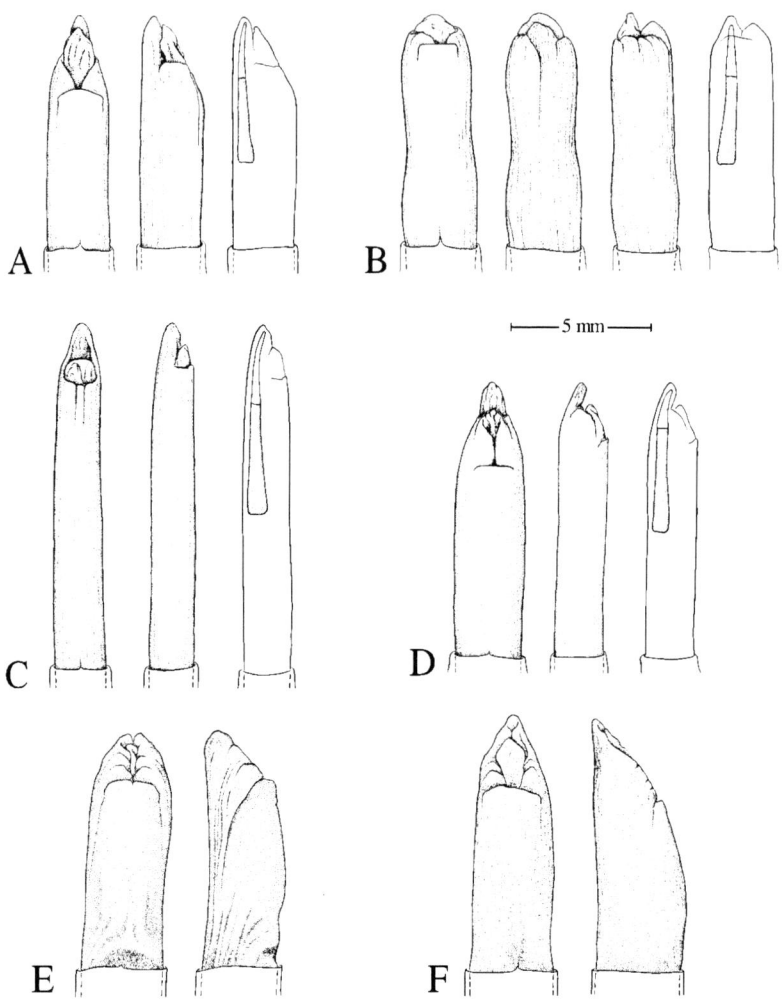

Figure 10. Preserved phalli, with prepuce folded back. Ventral (left) and lateral views. The unshaded (right) figures show the position of the rodlike bacula, with proximal bone shaded darkly and the distal cartilage shaded lightly. (a) *Makalata didelphoides* LHE 632, (b) *Echimys chrysurus* LHE 632 – ventral, dorsal, and lateral views, (c) *Pattonomys carrikeri* USNM 496508, (d) *Diplomys labilis* USNM 460179, (e) *Isothrix bistriata* MVZ 191300, (f) phallus of *Phyllomys medius* EDR 8 (from a photo by Yuri Leite, specimen courtesy of L. C. Machado Ribeiro, Museu de Zoologia de Universidade Católica do Paraná).

Remarks: Members of this genus, like *Callistomys* spp., have some characters shared with eumysopines but not with other echimyines, including small round cheekteeth with only the hypoflexi/ids open (other flexi/ids appearing as closed enamel folds) and an open squamosotympanic fenestra. Cyt b sequence data (Lara et al., 1996) fails to group *Isothrix* spp. with four echimyine taxa; its phylogenetic position is therefore equivocal. Molecular data has begun to resolve some species-level biogeographical questions within *Isothrix* (Patton and Emmons, 1985; Patton et al., 2000). I segregate *I. negrensis* and *I. orinoci* from *I. bistriata* because of the cranial distinctiveness of the former and the molecular distinctiveness of the latter, following the results of Patton and Emmons (1985) and Patton et al. (2000). However, as noted by those authors, additional work is needed and the geographic and morphological limits of the *bistriata* group of species need to be clarified.

Genus *Callistomys* Emmons and Vucetich, 1998

Type species: *Nelomys pictus* Pictet, 1843

Included species:
  Callistomys pictus (Pictet, 1843)
  Callistomys sp. (Emmons and Vucetich, 1998)
  syn. *Lasiuromys villosus* Winge, 1888

Diagnosis: Large, soft-furred, arboreally adapted rats with striking black and white pelage pattern; tail completely covered with hair. Guard hairs fine, underfur dense and wavy. Two pairs of lateral mammae. Cheekteeth apparently three-rooted, high crowned, P4 and M1 unilaterally hypsodont (lingual side of crown higher than labial side, crown curved outward); P4-M3 tetralophodont, with three labial flexi and one lingual flexus; hypoflexi and mesoflexi deep, P4 completely divided by the joined hypoflexus-metaflexus into two, U-shaped lophs, with no mure (Figure 3b). Hypoflexids of p4-m3 set at a strong oblique angle. Lower premolars tetralophodont, anteroconid and protoconid united, enclosing the anteroexternal flexid as a slitlike fossetid; anterior half of the tooth approximates a triangle; the hypoflexids and metaflexids do not join (p4 is not divided by a continuous flexid, (Figure 4b). Skull with jugals expanded dorsoventrally, lateral jugal fossa wide and diffuse anteriorly, not coming to a sharp point; anterior edge of fossa above P4. Superior zygomatic root of maxillary expanded posteriorly. Tympanic auditory bulla inflated. Angular process of mandible strongly projecting ventrally with respect to the inferior projection of the symphysis, such that an angle drawn between the ventral posterior tip of the angular process and the occlusal plane of the toothrow, with the apex at the anterior edge of the occlusal surface of p4, is greater than 30° (description from Emmons and Vucetich, 1998, crania are illustrated therein).

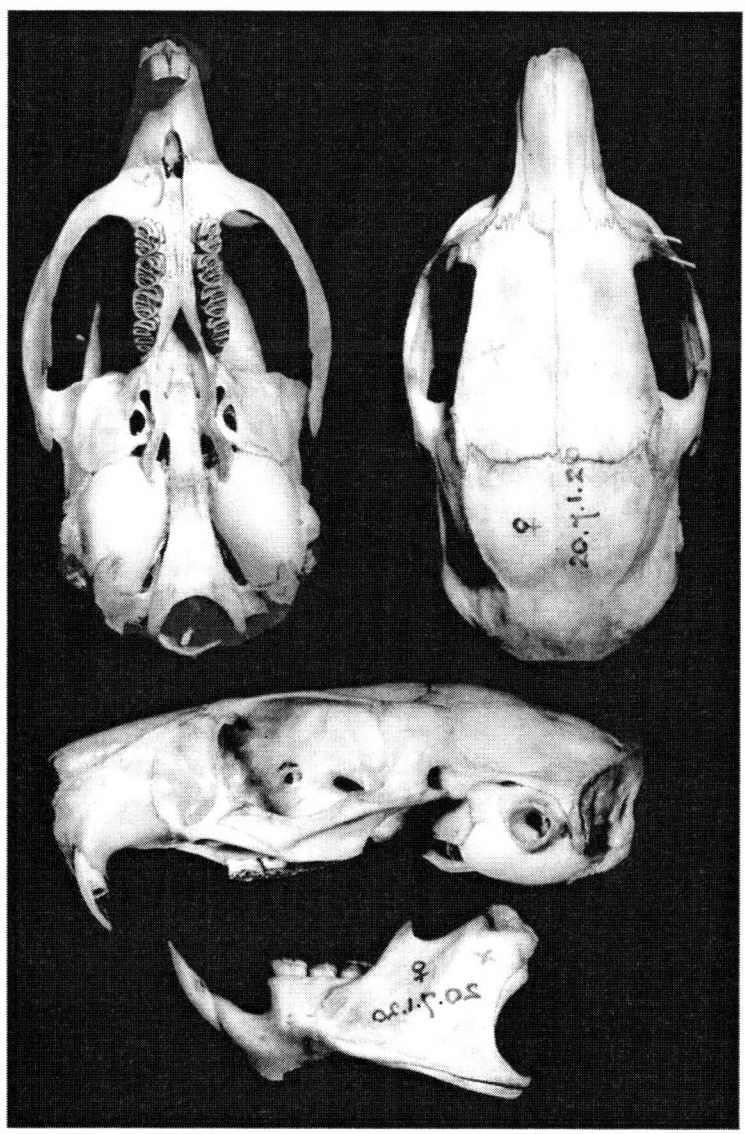

Figure 11. Cranium and mandible of *Isothrix negrensis* holotype BMNH 20.7.1.20.

Remarks: The characters of *C. pictus* were discussed by Emmons and Vucetich (1998). The following two major features distinguish this taxon from other Echimyinae and raise questions about whether it belongs in the subfamily: (1) possession of the "eumysopine" cheektooth root pattern and hypselodonty; and (2) greatly dorso-ventrally expanded jugals and anterior zygomatic arch. These features are in common with the Eumysopinae (Euryzygomatomini) but are not shared by any other living echimyine.

Genus *Makalata* Husson, 1978

Type species: *Echimys didelphoides* Desmarest, 1817

Included species: This genus includes two species groups:
didelphoides group
    Echimys didelphoides Desmarest, 1817
    Loncheres guianae Thomas, 1888
    Loncheres macrura Wagner, 1842
    Loncheres castaneus Allen and Chapman, 1893
    Echimys longirostris Anthony, 1921
    Echimys handleyi Goodwin, 1962
    Loncheres obscura Wagner, 1840 (?)
grandis group
    Loncheres grandis Wagner, 1845
    Echimys rhipidurus Thomas, 1928
    Makalata sp. nov.

Diagnosis: Arboreally adapted rats with short legs and long backs; pelage including spines (*didelphoides* group) or stiff bristles (*grandis* group); tail about as long as head and body length, sparsely (*didelphoides* group) to completely covered (*grandis* group) with stiff hair. Two pairs of functional lateral mammae and occasionally a third, apparently obsolete, pair. Phallus of *M. didelphoides* slender, heavily ridged on sides, with long pointed dorsal bacular papilla; urethral lappet anterior to it extruded, large, and pointed; ventral lip of crater forms a deep V-shaped notch proximally abutting a straight, horizontal fold (Figure 10a). Maxillary toothrows parallel or slightly divergent at either end (Figure 12), cheekteeth squarish to rectangular, longer than wide, with crowns of medium height, usually two lingually opening flexi (hypoflexus, metaflexus) and two labially opening flexi (paraflexus and mesoflexus), so that M1-2 have two, U-shaped lophs opening in opposite directions (Figure 12); but several species have one lingual and three labial flexi in some or all molars (Figure 3c). Para, meso, and metaflexi extending two-thirds or more across P4-M1-3, but hypoflexi short, crossing about one quarter of the width of the tooth. Lower molars with hypoflexids at an oblique angle, such that ectolophids at anterior tip of flexids are narrower than lophid posterior to tip (Figure 4c). Lower premolar pentalophodont, with anteroconid pointed; when unworn usually with flexid not enclosed by lophids (no central fossette), but opening posteriorly, such that anterior loph is an inverted V. Metaflexids of m1-2 bent posteriorly at internal tip, and mesoflexids bent forward; hypoflexid often curved forward. Lower incisors not strongly curved. Auditory tympanic bullae moderately inflated; auditory meatus medium sized to small; auditory tube short, outwardly or slightly forwardly directed; meatus close to squamosal suture. Squamosotympanic fenestra slitlike but short, with ventral lip raised in a ridge with a depression below (Figure 5c). Incisive foramen with maxillary septum broadly fused to premaxillary portion (Figure 7b). Mandibular foramen in a fossa with base on condyloid ridge; condyloid process of mandible short and dorsoventrally wide; masseteric crest strongly developed.

Remarks: The holotype of *Loncheres obscura* Wagner, 1840, collected by Spix in Brazil, has apparently been lost (Dr. Richard Kraft, pers. com.). It appears from the original description and figures to be a *Makalata* (Wagner 1840), but the name currently cannot be assigned to a specific population. Many of the *didelphoides* group species are colored warm brown with reddish rostrum forward of eyes; *grandis* group species are more blackish or dark-lined.

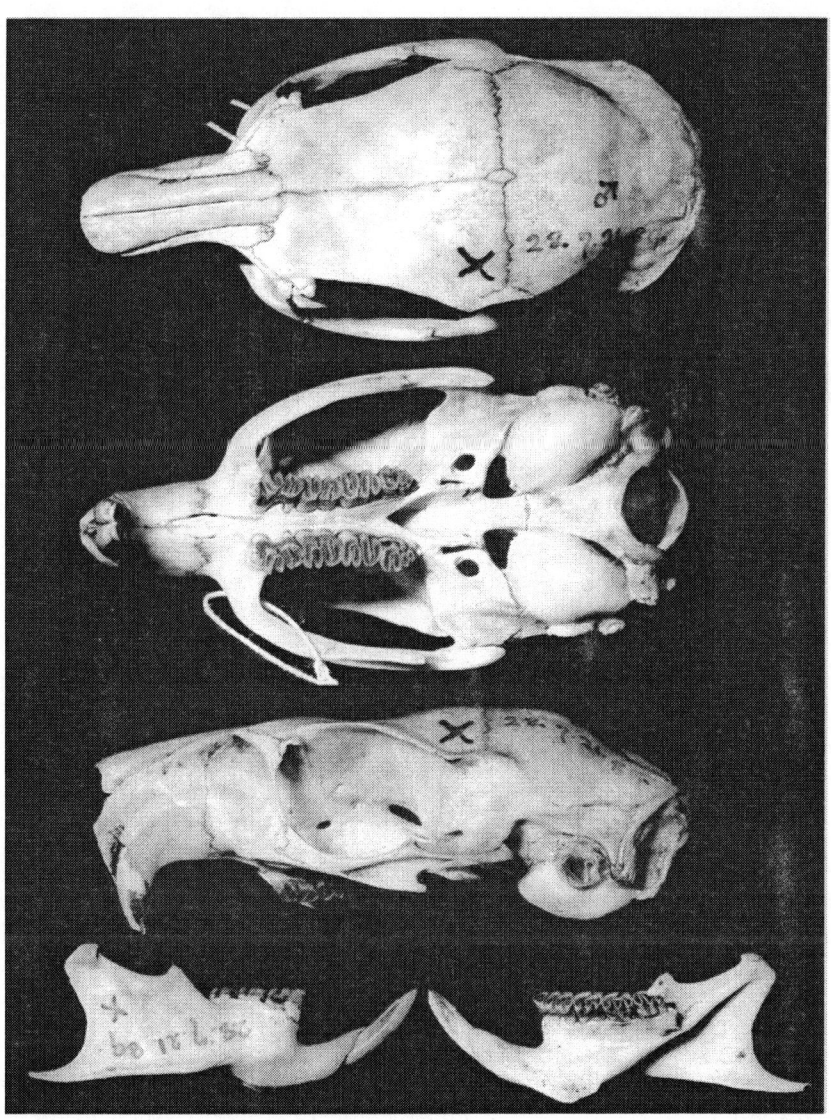

Figure 12. Cranium and mandible of *Makalata rhipidura*, holotype BMNH 20.7.1.20, mandible photo reversed.

Genus *Pattonomys*, gen. nov.

Type species: *Nelomys semivillosus* I. Geoffroy, 1838

Included species:
  Nelomys semivillosus I. Geoffroy, 1838
  Loncheres punctatus Thomas, 1899
  Echimys carrikeri J. A. Allen, 1911
  Loncheres flavidus Hollister, 1914
  Echimys occasius Thomas, 1921

Etymology: The genus is named to honor James L. Patton, whose many key contributions to echimyid systematics have brought light to a dark place. The name *Pattonomys* is in recognition of his outstanding generosity in sharing his knowledge with students and colleagues.

Diagnosis: Medium to large, arboreally adapted rats; pelage color generally gray on head and sides, often with yellowish cast; back tinged brown; tail uniformly reddish brown, lightly clothed throughout length with fine hairs. Pelage including strong spines, many on rump tipped whitish, imparting a speckled appearance. Two pairs of lateral mammae. Phallus long and slender, with long, pointed bacular papilla, a small urethral lappet, and with border of ventral crater wall a straight line, without a V- or U-shaped ventral fold (Figure 10c). Maxillary cheekteeth squarish to rectangular, longer than wide, posterior lophs rounded. Teeth have an uneven occlusal appearance due to unequal loph sizes (Figures 3f, 4f). P4-M3 always with two labial and two lingual flexi. Para- and mesoflexi short, most reach only to mid-tooth; hypoflexi long, likewise reaching midway across tooth, and opening at a wide angle. P4 and M1 always with a mure in the center of the tooth; M2-3 with or without a mure. Chief diagnostic features of the upper teeth include widely open flexi, some nearly as wide as the lophs, and anteroposteriorly expanded protocone and posteroloph (probably metaloph and posteroloph combined). These expanded lophs are accentuated with wear. The four lophs of P4-M2 are markedly unequal in length and width: the anteroloph is broad and squared lingually, tapering labially; and the posteroloph is pointed labially and strongly curved along the posterior margin of the tooth. The paraflexus and metaflexus of P4-M2 slant in markedly opposite directions, with the metaflexi slanting posteriorly from the labial to medial edges of the tooth (Figure 3f). Lower premolar tetralophodont, with two labial and two lingual flexids; tooth usually divided by one central flexid into two V-shaped lophs (Figure 4f), but it may have a central mure. Flexids of m1-2 form nearly straight-sided Vs. Lower incisors not strongly curved. Cranium short and broad, with expanded, winglike supraorbital shelf that curves upward from frontals; supraorbital region broad, especially developed in *P. carrikeri*, much narrower in *P.*

semivillosus (Figure 13). Tympanic auditory bulla moderately to considerably inflated, auditory meatus high, near squamosal, directed outward; auditory tubes short, their role taken by the formation of two overlapping bony rings, especially well developed in old individuals. Mastoid process short, usually to middle of auditory meatus, not extending ventrally beyond lower edge of meatus. Angular process of mandible slender; condyloid process narrow; mandibular foramen usually in a fossa on condyloid ridge; masseteric crest strongly developed anteriorly.

Remarks: The teeth of members of this genus are so distinctive that any single tooth is diagnostic except m1-3, and often these are too. The named forms of the genus were all synonymized under the name *semivillosus* by Cabrera (1961) and subsequent authors. However, all but *P. flavidus* are cranially distinctive and readily diagnosable. I consider them all to be valid species, but they merit further study. The form *P. flavidus* is cranially similar to coastal specimens of *P. carrikeri*, and may be synonymous, but as it also has distinctive features, such as an extremely deep mandible, I provisionally recognize it pending additional specimens and molecular genetic analysis. It is unlike the geographically closer *P. punctatus*.

Echimys occasius Thomas, 1921 was formerly placed in *Makalata* (Emmons and Feer, 1990), which it resembles in pelage color, smaller size, and geographic distribution in Amazonia. It does not possess several of the more diagnostic characters *Makalata*. In parsimony analyses, it groups with *Pattonomys*, where it is tentatively placed here. It may be a relictual form close to the common ancestor of *Makalata* spp. and other *Pattonomys*. Molecular studies are needed to better clarify its relationship. Thomas did not give the etymology of his name *occasius*. I surmise that from its purported type locality west of the Andes (likely erroneous), he derived the name from the Latin *occasus,* meaning western (to set as the sun), with the comparative superlative ending *ius*, or westernmost. The suffix therefore does not change with the gender of the generic epithet.

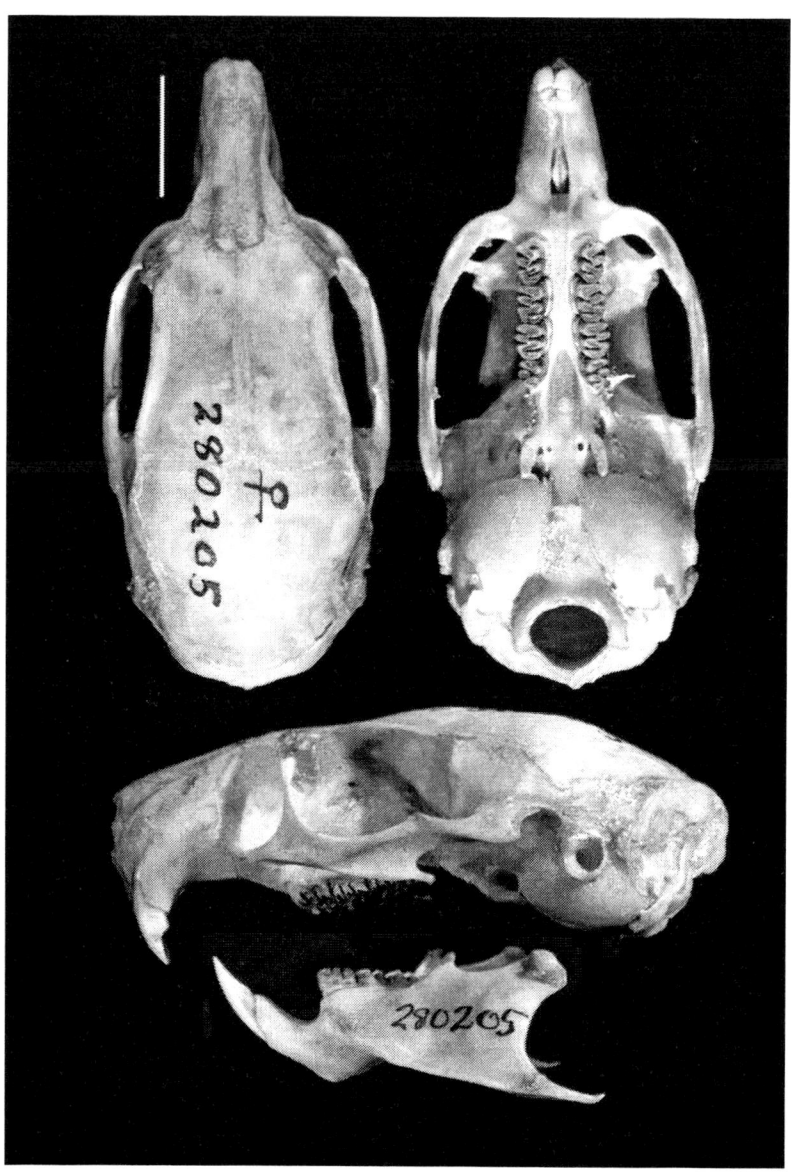

Figure 13. Cranium and mandible of *Pattonomys semivillosus* USNM 280205, scale 0.5 cm.

Genus *Echimys* F. Cuvier, 1809

Type species: Myoxus chrysurus Zimmermann, 1780

Included species:
  Echimys chrysurus (Zimmermann, 1780)
  Echimys saturnus Thomas, 1928

Diagnosis: Large arboreal rats with pelage including stout spines; tail longer than head and body length, completely, densely covered with hair that forms a pencil at tip. Four pairs of mammae, three lateral and one inguinal. Phallus of *E. chrysurus* stout, strongly ridged on sides, dorsal tip with short, triangular dorsal bacular papilla; ventral lip of urethral crater forms a short, U-shaped fold above a straight line lip (Figure 10b). Maxillary toothrows parallel, not converging anteriorly, cheekteeth rectangular, longer than wide, with tall crowns; only one lingual flexus, (the hypoflexus) that usually joins with mesoflexus to completely divide tooth into two, U-shaped lophs (no mure); three labial flexi (Figure 3e). The four lophs of P4-M2 are subequal in length and width and nearly parallel. Paraflexus and metaflexus long, about two-thirds or more of the width of the tooth, and parallel; hypoflexus or its trace, short, about one quarter to one third of the width of the tooth. Lower molars with hypoflexids reaching less than half way across tooth, opening at a wide angle, and scarcely, if at all angled forwards, such that lophids adjoining the tip of hypoflexid are about equal in width. Metaflexids of m1 and more prominently, m2, with angled internal tip with two points (perhaps outlining a mesoconid) and bending slightly posteriorly; mesoflexid with internal tip bending slightly anteriorly (Figure 4e). Lower premolar pentalophodont, completely divided by two flexids into the following three parts: (1) a symmetrical triangular anterior loph, rounded at its anterior tip and enclosing a subcircular fossetid, (2) a single, laminar, central metalophid bar, (3) posterolophid and entoconids of posterior third of tooth united labially into a single, V-shaped loph that opens lingually. Lower incisors not strongly curved. Auditory tympanic bulla small to medium-sized, with a small meatus on a short auditory tube directed straight outward at nearly a right angle to the skull axis; meatus near (less than half its diameter) squamosal suture (Figures 5a, 14, 15). Mastoid process short. Squamosotympanic fenestra small and subcircular, with no raised ventral lip, entirely anterior to auditory meatus. Incisive foramen with premaxillary part of septum free posteriorly. Jugal deep below post-orbital process, jugal fossa very broad, forming an angle of at least 40.° Mandibular foramen anterior to condyloid ridge, not in a fossa with base in ridge (Figure 8b); masseteric crest weakly developed.

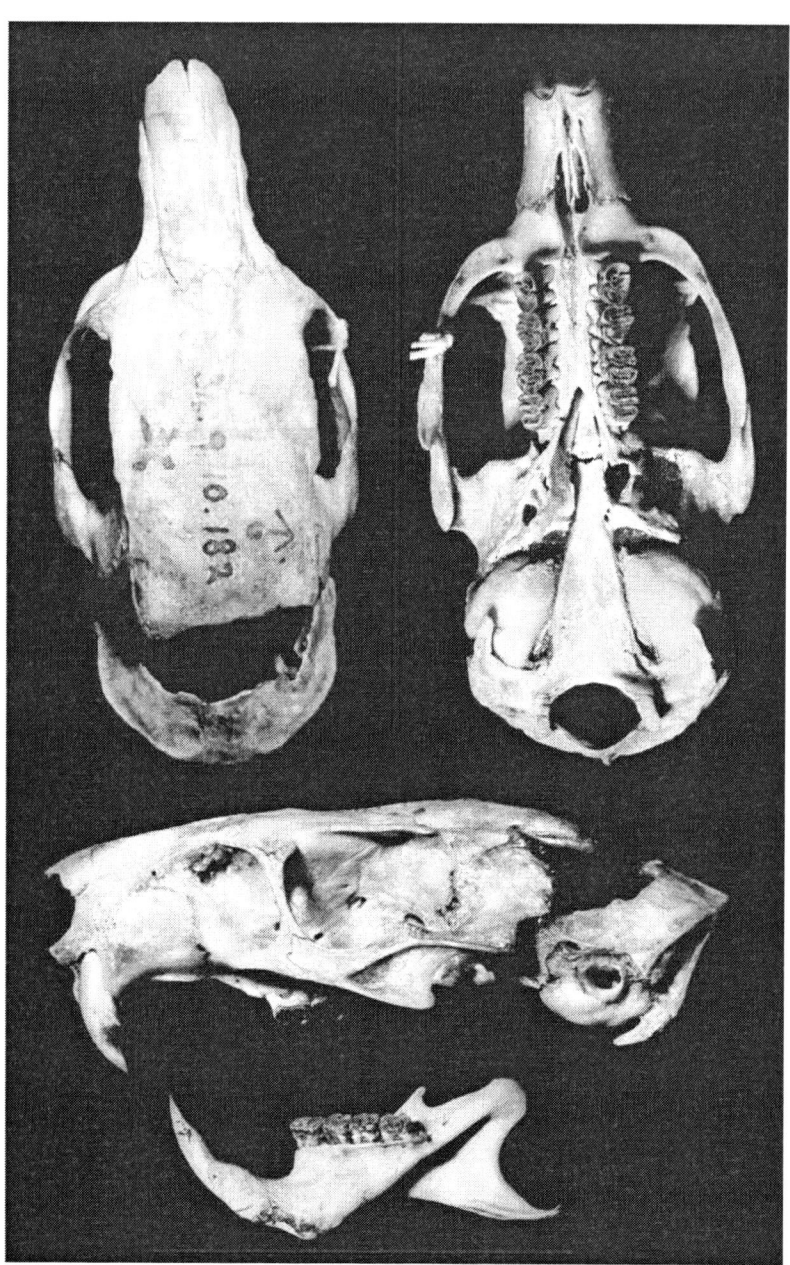

Figure 14. Cranium and mandible of *Echimys saturnus,* holotype BMNH 34.9.10.182.

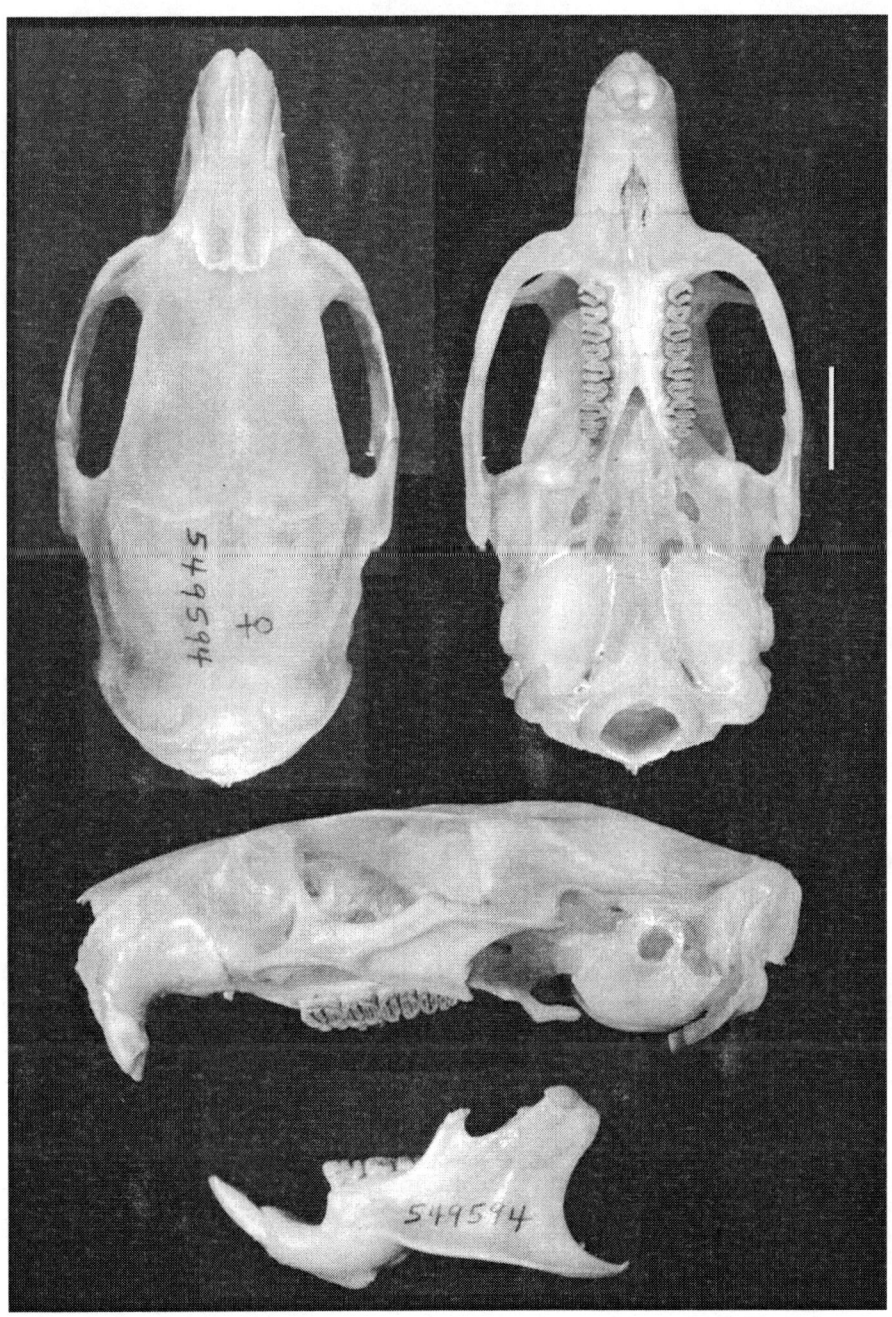

Figure 15. Cranium and mandible of *Echimys chrysurus* USNM 549594.

Genus *Phyllomys* Lund, 1839

Type species: *Phyllomys brasiliensis* Lund, 1840

Included species:
   Nelomys blainvilii Jourdan, 1837
   Phyllomys dasythrix Hensel, 1872
   Phyllomys kerri Moojen, 1950
   Nelomys lamarum Thomas, 1916
   Phyllomys lundi Leite, 2003
   Phyllomys mantiqueirensis Leite, 2003
   Loncheres medius Thomas, 1909
   Loncheres nigrispina Wagner, 1842
   Mesomys thomasi Ihering, 1897
   Loncheres unicolor Wagner, 1842
   Phyllomys pattoni Emmons, Leite & Costa, 2002

Diagnosis: Medium-sized to large arboreal rats with pelage including spines or bristles, or soft furred; tail sparsely to densely haired, with or without pencil or tuft at tip. Four pairs of mammae, three lateral and one inguinal. Phallus of *P. medius* robust, pointed, with triangular tip, large urethral lappet extrudes in an inverted pear-shape below pointed bacular papilla; ventral fold short and forming a U-shape with sides bordering lappet (Figure 10f). Maxillary teeth rectangular; longer than wide, unworn teeth completely split by three flexi into four nearly parallel laminae or lophs of quite uniform width; posterolophs crescent-shaped (Figure 3h). The two posterior lophs unite labially with wear; the anterior lophs unite lingually. Lower premolar pentalophodont, split by two flexids into three parts; unworn anterior loph including lingually opening flexid, this tending to close with wear to form a triangle with a central subcircular fossette (Figure 4h). Lower molars with narrow lophids; metaflexid long, often dividing tooth, hypoflexids at strong oblique angle. Lower incisors usually strongly curved. Auditory tympanic bullae with auditory meatus low, directed slightly forward, with a space as wide as meatus between meatus and squamosal (Figures 5e, 16). Additional rings of bone external to meatus often present. Mandibular foramen in fossa on base of condyloid ridge (Figure 8c); condyloid process narrow, angular process reflected ventrad with respect to occlusal plane of toothrow (Figure 16); masseteric crest weakly to moderately developed anteriorly.

Remarks: The taxonomy and status of all named forms of *Phyllomys* were recently reviewed by Emmons et al. (2002) and Leite (2003). Diagnoses and descriptions of all species listed above, with discussion of nomenclatural issues, are found therein. The systematics and phylogeography are reviewed in Leite (2001)

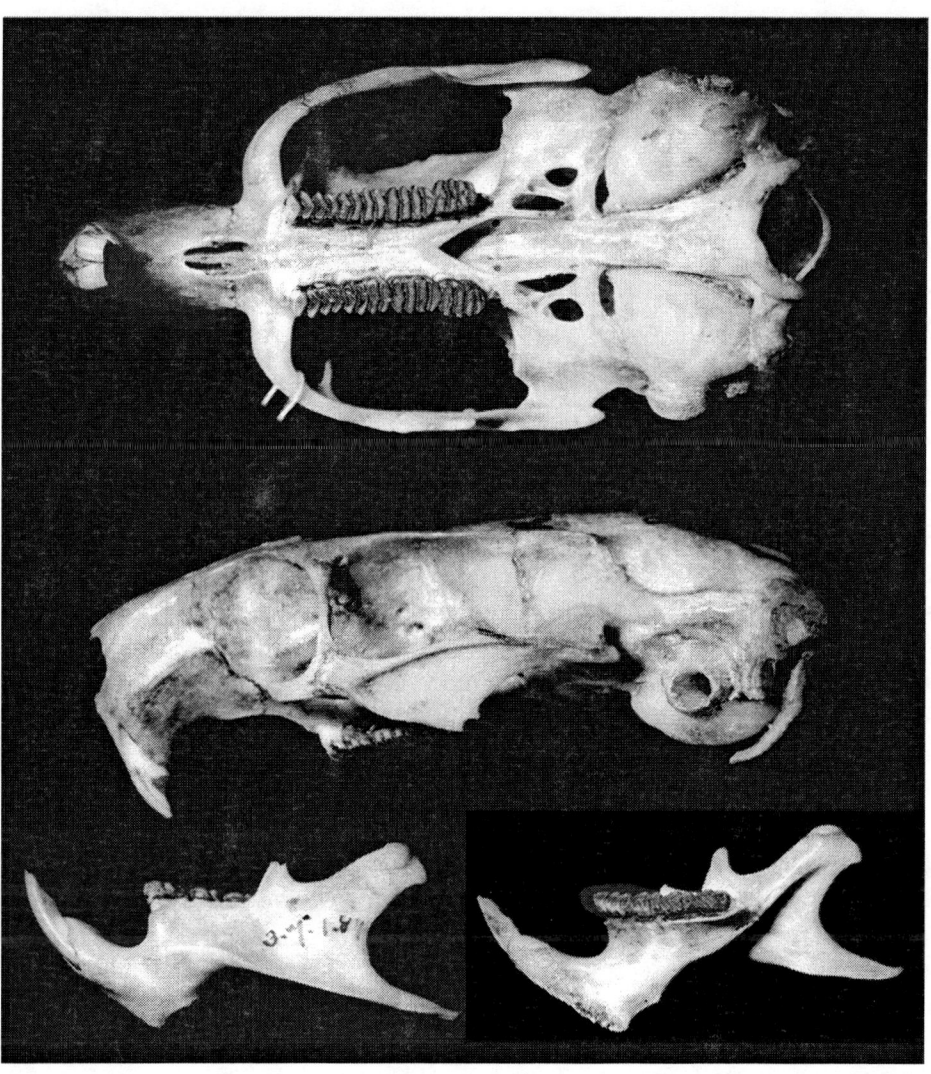

Figure 16. Cranium and mandible of *Phyllomys medius*, holotype MBNH 3.7.1.84.

Genus *Diplomys* Thomas, 1916

Type species: *Loncheres caniceps* Günther, 1877

Included species:
 Loncheres caniceps Günther, 1877
 Loncheres labilis (Bangs, 1901)
 synonym: *Isothrix darlingi* (Goldman, 1913)

Diagnosis: Large arboreal rats without spines, but guard hairs flattened and slightly stiff, with deep dorsal sulcus; guard hairs 1.8-2.0 cm long on dorsum; underhairs few, fine and inconspicuous. Tail with dorsal body fur extension 1.5-5 cm onto base; distal part fully but moderately clothed with stiff, bristlelike, brown hairs, scales visible beneath. Two pairs of lateral mammae. Feet without tiny tubercles between pads. Phallus without deep lateral ridges, dorsal tip with a long, slender, bacular papilla; inconspicuous urethral lappet, ventral surface with a long, longitudinal slit distal to a short transverse fold (Figure 10d). Maxillary cheekteeth large, toothrow relatively long, teeth as in *Phyllomys* spp., divided by three flexi into four laminae; laminae parallel in M1-3; P4 with labial ends of the two anterior lophs bent forward (Figure 3i). Flexi of even depth, such that with wear, three straight fossettes are centered on molars. In mandibular cheekteeth, all flexids may traverse all lophs in unworn teeth (Figure 4i); but with wear the metaflexid divides m1-3 into a free posterolophid; middle loph (entoconid + hypoconid) usually connected to anterolophid at near its midpoint by a short mure, forming a highly distinctive pattern of two, crescent-shaped lophids connected by a central stem, like an Inuit knife. Hypoflexids at an oblique angle (Figure 4i). Lower premolar split into three or four parts by two or three flexids; anterior loph a rounded triangle enclosing a flattened oval transverse fossette, metalophid a separate lamina, and entoconid and posterolophid either joined labially, or divided into separate parts by metaflexid (Figure 4i). Lower incisors strongly curved (Figure 17). Cranium long and narrow, dorsal surface curved throughout length from nasals to occiput when viewed laterally (Figure 17); auditory tympanic bullae not inflated; auditory meatus close to squamosal; auditory tubes well developed, directed strongly forward, so that inner ear bones not visible from side (Figure 5f). Alisphenoid with distance between foramen ovale and masticatory foramen usually very short (Figure 6e, between arrows); ventral roof of this canal often nearly obsolete. Hamular process of pterygoid with prominent, anteriorly directed spur (Figure 6e). Large oval sphenopalatine vacuity present below presphenoid-basisphenoid suture (Figure 17). Squamosotympanic fenestra tubelike, opening forward, without depression on squamosal below (Figure 6e). Jugal narrow below postorbital process, with ventral process inconspicuous and well forward of dorsal process (Figure 17). Angle of sigmoid notch between angular and condyloid processes of mandible shallow;

mandibular foramen on side of ramus, far anterior to condyloid ridge, not in a fossa (Figure 8d). Masseteric crest large.

Remarks: There is much variability in which of m1-3 the "Inuit knife" wear pattern appears.

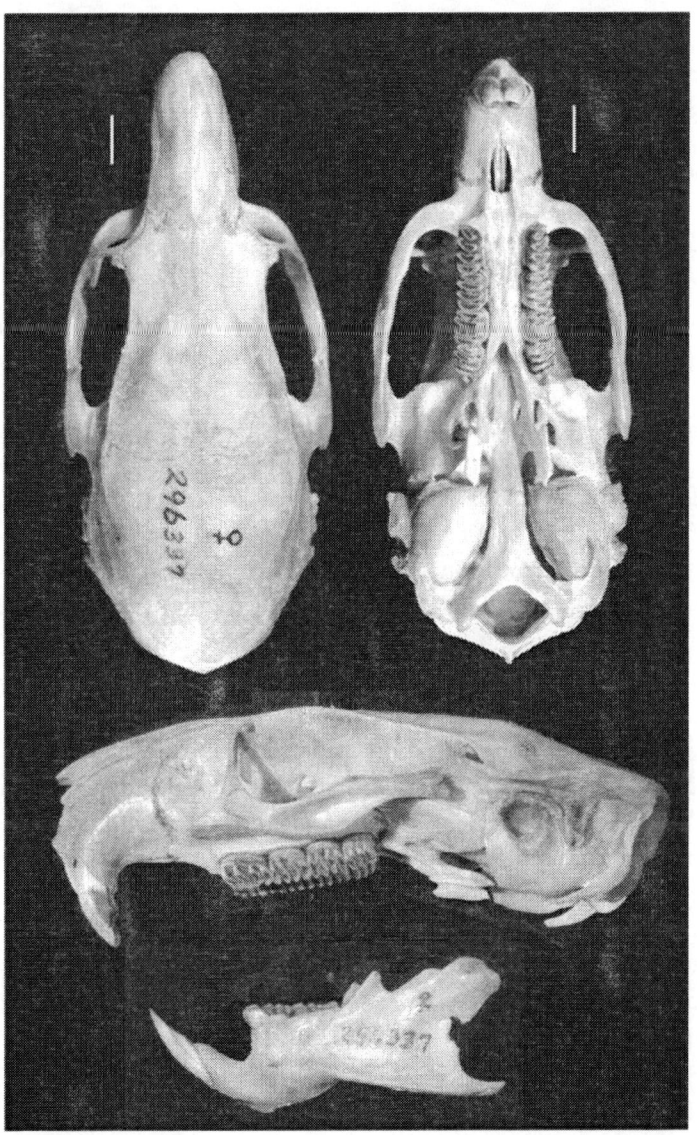

Figure 17. Cranium and mandible of *Diplomys labilis* USNM 296337.

Santamartamys gen. nov.

Type species: *Isothrix rufodorsalis* Allen, 1899
  synonym.– *Diplomys rufodorsalis* (Tate, 1935)

Included species: Only one species is recognized, it is known from only two specimens.

Etymology: Named for the type locality and origin of both known specimens, the Sierra Nevada de Santa Marta, Colombia.

Diagnosis: A medium-sized, bright rust-red arboreal rat with long, lax overhairs, 2.0-3.8 cm in length on dorsum; pelage not stiff or bristly, underfur dense, wavy, gray. Overhairs so slender that they are difficult to distinguish from other pelage. Dense woolly pelage covers legs to ankles and wrists. Crest of long hair on crown between ears. Longest genal and mystacial vibrissae 5 cm, vibrissae present on wrist. Two pairs of lateral mammae on abdominal edge of lateral pelage, in ventral pelage field. Tail robust, covered with extension of dorsal pelage for 2.5 cm, well clothed distally with fine hairs, basal half brown, distal half pure white. Feet without tiny tubercles between pads, pollux with a nail. Maxillary cheekteeth rectangular, longer than wide, teeth and toothrow relatively long, bowed inward (Figure 18); P4-M3 split by a deep flexus into two parts, anterior lophs with flexi opening labially and posterior lophs with flexi opening lingually; protocones broad, conferring a wishbone shape to joined anteroloph/protoloph pairs, especially on P4 (Figure 3g). Lower molars split by meso- and hypoflexid into two parts, a curved, laminar anterolophid, and a somewhat wishbone-shaped entoconid/posterlophid (Figure 4g). Lower premolar split by two flexids into a small closed triangular anterior loph apparently lacking a fossette; a laminar metalophid, and a wishbone-shaped posterior loph that opens lingually. Hypoflexids strongly oblique; protoconids large and squarish. Lower incisors strongly curved (Figure 18). Cranium conspicuously curved in dorsal profile, rostrum short and broad. Auditory tympanic bullae small and flattened; merging with alisphenoid at a shallow angle (Figures 5d, 6f, 18). Auditory meatus small, placed high near the squamosal; auditory tube short, strongly directed anteriorly; mastoid process extremely short (Figure 5d). Bony bridge between foramen ovale and masticatory foramen exceptionally long (Figure 6f, between arrows); there seems to be a small oval vacuity below the presphenoid-basisphenoid suture (Figure 18). Condyloid process of mandible deep; angle of sigmoid notch between angular and condyloid processes of mandible shallow (Figure 18); mandibular foramen in a fossa beside the condyloid ridge. Masseteric crest of lower edge of mandible poorly developed and shallow, pterygoid shelf small.

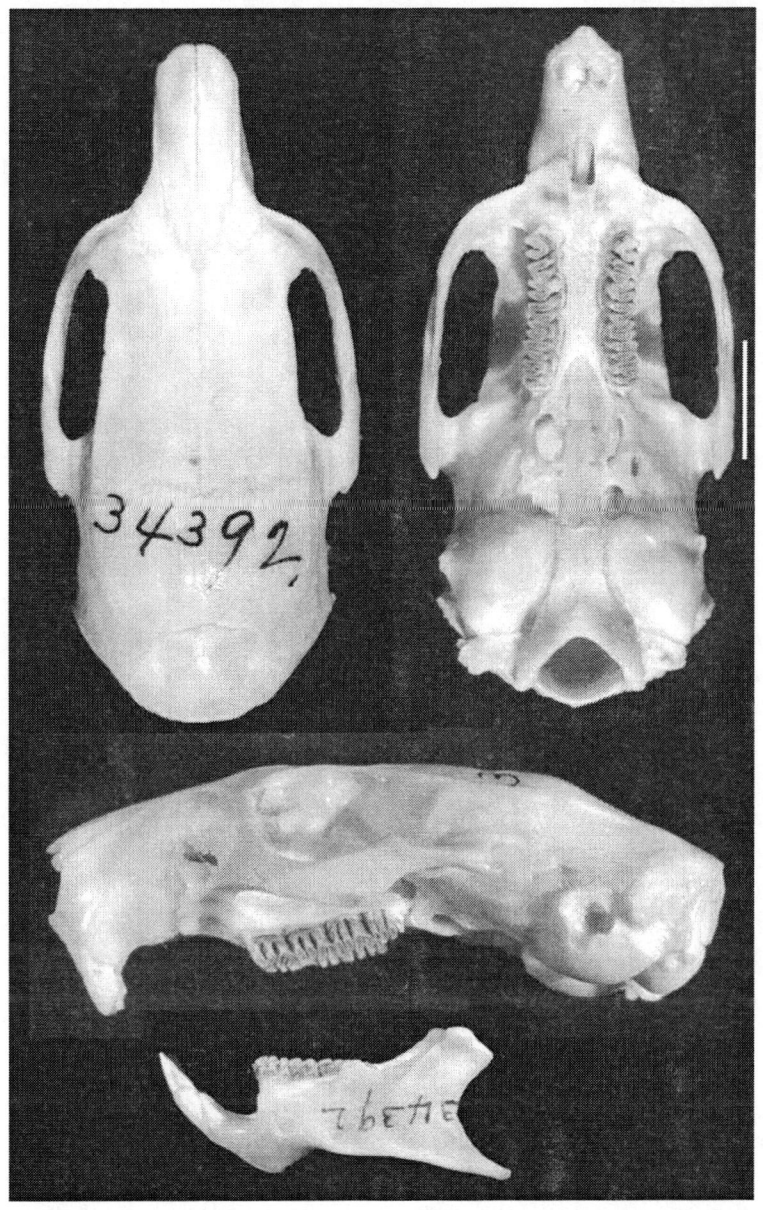

Figure 18. Cranium and mandible of *Santamartamys rufodorsalis* AMNH 34392.

Remarks: *Santamartamys rufodorsalis* seems closely allied to members of the genus *Diplomys*, where it has recently been placed. It shares with *Diplomys* a number of probable apomorphies, as well as a close geographic affinity. I segregate it as a genus primarily because of its distinctive cheektooth occlusal morphology, which is different from all other living Echimyidae. This taxon also has unique pelage, mammae placement, and bullar and alisphenoid configurations. The descriptions above were largely based on AMNH 34392, a young female with little-worn dentition. The teeth of the holotype, AMNH 14606, are extremely worn. Because only two specimens are known, the diagnosis is expanded with descriptive material to aid those unable to view the specimens.

Comparisons Between Genera

Pelage: All species in the genera *Isothrix*, *Santamartamys*, *Callistomys*, and all Dactylomyines are soft furred, while the guard hairs of *Diplomys* spp. and the eumysopine *Thrichomys* spp. are flattened and somewhat stiff, but nearly soft. Most members of other Echimyid genera are spiny or have stiff, bristly dorsal guard hairs, which is likely the plesiomorphic condition for the family. However, it is noteworthy that species in all three of the largest echimyid genera, *Proechimys*, *Phyllomys*, and *Makalata*, range from heavily spined to stiff- or soft-furred (*Phyllomys* only), thus spininess is not evolutionarily stable.

Mammae: *Echimys* and *Phyllomys* species have three lateral pairs of mammae that are about equally spaced along the sides in the dorsal pelage field (one anterior, one medial, and one posterior), and one inguinal pair on the ventral field. Species of *Callistomys*, *Makalata*, *Pattonomys*, *Diplomys*, and *Santamartamys rufodorsalis*, have only two functional lateral pairs that seem to correspond to the medial and posterior lateral pairs of *Echimys* spp. Some individuals of *Makalata* spp. have an inguinal pair of mammae that appears to be non-functional. *Isothrix pagurus* can have as many as 5 lateral pairs, including the three standard pairs, with another pair intercalated between the anterior and medial and the medial and posterior pairs. *Isothrix bistriata* can have three lateral pairs: medial, posterior, and one between these, and an inguinal pair. The patterns typical of *Isothrix* spp. are thus not found among other genera. The lateral mammae of *S. rufodorsalis* are placed in the ventral pelage field on the edge of the dorsal field, a unique condition among the Echimyinae for which the pattern mammae placement is known. This observation needs confirmation as it is based on a single specimen but, because the species has (with *C. pictus*) the longest, densest, and finest fur of the genus, I speculate that the mammae may have "migrated" to the abdominal field because the unusually dense pelage interferes with nursing.

Phallus: Too few species and individuals – often only one individual per genus - were examined for confident intergeneric comparison but, based on the material seen, the phallus of *Phyllomys medius* is very similar to that of *Makalata didelphoides* (Figure 10). The phallus of *Pattonomys carrikeri* is uniquely long, slender, and of simple structure at the tip, while that of *Diplomys labilis* is remarkable for its smooth, unridged exterior and elongated, digitlike tip. The phallus of *Echimys chrysurus* almost lacks the bacular papilla seen in other genera so it is shorter and blunt rather than pointed at the tip. The phallus of *Isothrix bistriata* appears to have a much broader tip above the ventral fold than do those of other taxa and the tip is grooved, rather than forming a solid papilla. I did not examine the specimens of *I. bistriata* and *P. medius*, which were illustrated from photos supplied by Yuri Leite. Leite (2003) illustrated the phallus of several *Phyllomys* species.

Squamosal and auditory region: This cranial region has some of the most useful characters for diagnosing the genera of Echimyinae. The low placement of the auditory meatus, with an expanse of bone wider than the meatus between it and the squamosal suture (Figure 5), distinguishes *Phyllomys* species from members of all other genera, although there is some species-level variation in width of the supra-meatal area within *Phyllomys*. *Santamartamys rufodorsalis* is distinct from members of all other genera in possessing a highly flattened and ossified junction between the small, flattened, auditory tympanic bulla and the alisphenoid (Figure 18). The squamosotympanic fenestra is an elongate open slit along the squamosal suture in *Isothrix* spp. alone among Echimyinae, a condition that they share with the echimyids *Euryzygomatomys*, *Carterodon*, *Clyomys*, and some *Proechimys* and *Myocastor*. In the other echimyines, the suture is closed and the foramen opens from a bony tube (Figure 5). In *Makalata* species the tubelike development is most accentuated, with a slight to well developed bony crest or ventral lip to the foramen that is more or less continuous with the rim of the auditory meatus. There is a distinct depression on the squamosal ventrad to this crest (Figure 5c). This crest and depression are universally developed only among *Makalata* species, although one individual of *P. semivillosus* had a similar crest. In *Makalata* species, the mastoid process is long (usually reaching ventrad of the lower rim of the auditory meatus) and is free of the cranium at its tip, whereas in other echimyines it is short and adpressed to the cranium, enclosed medially in bone.

Alisphenoid and parapterygoid region: The strut of bone which separates the foramen ovale acessorius and masticatory foramen is narrow to absent in *Dactylomys* spp., narrow in *Diplomys* spp., *Echimys chrysurus* and *Callistomys pictus*, and exceptionally wide in *Santamartamys rufodorsalis* (Figure 6). In the other Echimyinae it is wide. The buccinator foramen is enclosed by a medial wall, partly enclosing it below the pterygoid process, in *Santamartamys*, *Echimys*, *Pattonomys*, and some *Makalata* and *Phyllomys* species. It has a dorsal shelf partly enclosing it in some species of

Phyllomys and *Makalata*.   In *Isothrix*, *Diplomys*, and *Dactylomys* species and all eumysopines, the foramen is open and unenclosed medially.

Incisive foramen: In *Makalata* species, the premaxillary portion of the septum in the incisive foramen is broad and broadly fused to the maxillary; in all other genera, including all Eumysopinae other than some *Proechimys* and *Trinomys*, the premaxillary portion is slender and dips dorsally, leaving a gap between it and the maxillary (Figure 7).

Occlusal pattern of the cheekteeth: The molar roots and occlusal patterns of lophs and flexi of the cheekteeth provide important characters for distinguishing the echimyid taxa defined here, despite some intraspecific variation.

The maxillary cheekteeth of *Isothrix* spp. are small and nearly circular in occlusal outline (Figure 3a), with a short, subcircular to oval hypoflexus less than half the width of the tooth. The para-, meso- and metaflexi are short (about 2/3 the width of the tooth) and parallel sided. In all other echimyines, the molariform teeth are large and roughly square (some *Makalata* spp., *Echimys* spp.) or rectangular (all others). In unworn teeth of *Diplomys* spp. and *Phyllomys* spp., three flexi completely cross each molar, separating the lophs into four approximately parallel laminae (Figure 3h, i). In *C. pictus*, *Echimys* spp., *Makalata* spp. and *S. rufodorsalis*, all four cheekteeth are each split by the joined mesoflexus and hypoflexus into two halves, each of which is a pair of joined lophs (Figure 3). The paired lophs of some *Makalata*, *C. pictus*, and all *Echimys* spp. approach a laminar design (nearly parallel, straight lophs and flexi). In contrast, *Pattonomys* species have a unique occlusal pattern, characterized by the presence of a mure about 2/5 to 1/2 of the distance across the tooth between hypoflexus and mesoflexus.   The hypoflexus and mesoflexus are short and wide, and the hypocone, protocone and protoloph are broad antero-posteriorly.

In mandibular cheekteeth, all taxa but *C. pictus* and *Pattonomys* species have a distinctly pentalophodont p4. In *Isothrix* spp., the anterolophid is visible as a small projection in unworn teeth, with wear it merges with the anteroconid and becomes indistinguishable (Figure 4a).   In the other taxa, the metalophid of p4 is a bar entirely separated from the adjacent lophids by transverse flexids that completely cross the tooth (Figure 4). *C. pictus* differs from *Makalata* species in lacking a metalophid bar (Figure 4b). Lower premolars of *Pattonomys* spp. differ from those of all other taxa in the combination of lacking both a metalophid bar and a mure medial to the hypoflexid (Figure 4f). The lower cheekteeth of *Isothrix* spp. differ from those of all other echimyine genera in having sub-circular hypoflexids, and meso- and metaflexids nearly at right angles to the axis of the tooth row and nearly closed at the lingual margins by curvature of the tips of the lophids (Figure 4a). The hypoflexids of *Isothrix* spp. are slightly angled in the opposite direction from those of all other taxa (Emmons and Vucetich, 1998), where the medial end is anterior to the labial opening of the flexid. *Diplomys* spp. are unique among echimyines in always having all lower cheekteeth completely split into three laminae by three flexids that completely traverse the tooth, (Figure 4i). *Santamartamys rufodorsalis* has the mesoflexid and hypoflexids of m1-3 united to traverse the tooth, creating one free anterior loph, but in contrast to *Diplomys* spp., the metaflexids do not traverse the teeth. A similar condition is sometimes observed in *Phyllomys* species. In all the other genera the lophids of m1-3 are W-shaped and differ chiefly in the angle and breadth of the hypoflexids, which are steeper and narrower in *Callistomys* and *Phyllomys* species.

Incisors and mandible: In Echimyinae, Dactylomyinae, and Eumysopinae other than Euryzygomatomini, the lower incisor roots originate below the molar toothrow, and the mandibular foramen is low on the ramus (Figure 8). The mandibular foramina of *Dactylomys* spp., *Diplomys* spp., *Isothrix* spp., *Echimys* spp., and *S. rufodorsalis* are above and anterior to the condyloid ridge, while the foramen of *Makalata* spp. is either directly on the spine of the condyloid ridge, or above it; at the bottom of a small fossa on the ridge, while those of *Phyllomys* spp. are close behind the molar toothrow and on the condyloid ridge. In *Echimys* spp. and *Makalata* spp., the condyloid ridge retains a sharp keel almost to the articulation on the posterior tip of the condyloid process, whereas among the other genera the ridge tends to flatten out and merge with the process farther below the condyle (Figure 8). The angular process of the mandible is short, reaching posteriorly about the same length as the condylar process in Eumysopini, *Isothrix* spp., *Diplomys* spp., and *C. pictus*. It is longer than the condylar process in Euryzygomatomini, Dactylomyinae and all the other echimyines.

DISCUSSION

The Echimyinae, as I propose here (Table 3), include seven or eight genera and about 38 species of arboreal rats. Several species await description and there are likely more than 50 extant species, but the number of species now grouped under most echimyid genera, including *Isothrix*, *Dactylomys*, *Mesomys*, *Makalata*, and *Pattonomys*, will be unclear until each has been revised. In both morphological (here) and molecular (Lara et al., 1996, Leite and Patton, 2002) analyses, the relationships between genera are poorly resolved, remaining consistent with the conclusion drawn by Lara et al. (1996) and reconfirmed by Leite and Patton (2002) that the genera in the family Echimyidae:

> "... may well represent a star-phylogeny, with an origin in the late Miocene resulting in a set of polytomous relationships reflecting that cladogenic history rather than to inadequate data. ... True polytomous relationships, as opposed to demonstrably dichotomous ones, are an expected outcome of

rapid and near-simultaneous divergence of multiple lineages. As a consequence, resultant taxa are likely to be composites of shared-primitive and uniquely derived characters, and relationships based on any character set will be difficult to establish" (p. 411).

This appears to be the case for several morphological characters, with the species in two of the eight echimyine genera, *Callistomys* and *Isothrix*, possessing suites of composite characters that make their affinities ambiguous; the same is true for *Mesomys*, *Dactylomys*, *Thrichomys*, and others. Although the phylogenetic relationships between the echimyine genera I define here may prove difficult or impossible to resolve if they evolved as a star-phylogeny, the monophyly of the generic groupings I propose will be testable with molecular techniques when tissues become available, as has recently been the case for *Phyllomys* (Leite, 2003).

Living Echimyinae can be placed in the following sequence of development of cheekteeth with increasingly parallel and separate laminae: *Isothrix*, *Pattonomys*, *Makalata*, *Santamartamys*, *Callistomys*, *Echimys*, *Phyllomys*, and *Diplomys*. Of major systematic interest is whether the laminar molars of *Phyllomys* species are synapormorphic with those of *Diplomys* species. I conjecture that they are not, and are homoplasies, chiefly because the closest living relative to *Diplomys* is clearly *S. rufodorsalis*, which does not have a laminar occlusal pattern; instead, *S. rufodorsalis* has a unique occlusal pattern, albeit with similarities to that of *Diplomys* species. It seems unlikely to have evolved by reversal from the laminate pattern of *Diplomys* spp., although it could have done so. The near laminarity of the teeth of *Echimys* species may reflect a closer phylogenetic relationship of *Echimys* to *Phyllomys* than of *Echimys* to either *Makalata* or *Pattonomys*, as is suggested by the molecular data sets of Leite and Patton (2002; Figure 9b). Too little is known of the diet of any species to evaluate whether, as might be predicted, increasing cheektooth laminarity is associated with increasing folivory or a diet including coarse vegetation such as stems and petioles.

The echimyine genera are segregated geographically, with *Phyllomys* and *Callistomys* species restricted to the Atlantic forests of eastern Brazil, *Echimys*, *Makalata*, and *Isothrix* species found only in Amazonia (including the Guianas), *Pattonomys* and *Santamartamys* species restricted to the northern coastal forests of Colombia and Venezuela (with the exception of the West-Amazonian *P. occasius*), and *Diplomys* isolated alone in Central America and coastal Colombia and Ecuador. A number of divergent taxa, often in genera that are monotypic or with few species, are endemic to small regions around the periphery of the continent, including *C. pictus*, *S. rufodorsalis*, *D. caniceps and D. labilis*, and, to a lesser extent, *P. occasius*. The Dactylomyine species of *Dactylomys peruanus* and both *Olallamys* species are likewise narrow endemics of small peripheral Andean regions. This centrifugal distribution of highly differentiated and now relictual forms suggests a long history of radiation in the Neotropical forests, with likely centers of speciation in the central

Table 3. A classification of the Echimyinae. Genera are arranged in order of increasing laminarity of the maxillary cheekteeth, without implication that this represents a phylogenetic series.

INCERTAE SEDIS

Genus *Mesomys*

    M. hispidus

    M. stimulax

    M. occultus

    Several other species

Genus *Lonchothrix*

    L. emiliae

Genus *Isothrix:*

    I. bistrata

    I. orinoci

    I. negrensis

    I. pagurus

    I. sinamariensis

ECHIMYINAE

Genus *Pattonomys*

    P. semivillosus

    P. punctatus

    P. carrikeri

    P. flavidus

    P. occasius

Genus *Makalata:*

    Didelphoides group:

    M. didelphoides

    M. guianae

    M. castanea

    M. handleyi

    M. longirostris

    M. macrura

    M. obscura?

Rhipidura group:

    M. grandis

    M. rhipidura

    M. sp. nov.

Genus *Santamartamys:*

    S. rufodorsalis

Genus *Callistomys:*

    C. pictus

Genus *Echimys:*

    E. chrysurus

    E. saturnus

Genus *Phyllomys:*

    P. brasiliensis

    P. blainvilii

    P. lamarum

    P. lundi

    P. mantequeirensis

    P. dasythrix

    P. unicolor

    P. medius

    P. thomasi

    P. nigrispinus

    P. kerri

    P. pattoni

Genus *Diplomys:*

    D. caniceps

    D. labilis

Amazonian and Atlantic coastal regions. These peripheral taxa are those currently at most risk of extinction. Several taxa are known from fewer than 10 specimens: *S. rufodorsalis* has not been reported since the two specimens were collected in the late 1800's, its locality is now severely degraded and the species could even be extinct. *Phyllomys unicolor* is known only from the holotype collected in 1824 and is likewise possibly extinct. *P. brasiliensis* is represented by only three specimens, one of which is recent (Emmons et al., 2002). *C. pictus* is rare in a small region of intensely exploited habitat. *E. saturnus* occupies a small geographic area of unknown extent at the Andean base of Ecuador and N. Peru; this area currently has considerable forest, but there are few specimens and less information. As I conjectured earlier for an arboreal murid on the Sunda Shelf (Emmons, 1993), specialized arboreal morphology and habits, such as that of folivory, might allow the persistence of primitive or relictual taxa in the presence of invasions or radiations of more advanced taxa such as murid rodents. This might partially explain the high diversity of rainforest arboreal echimyid genera (twelve) relative to terrestrial forms (two genera, each with many species).

The three Amazonian genera of echimyines overlap widely in geographic range, suggesting differentiated ecological roles. Of these, *Isothrix bistriata* and all *Makalata* species seem restricted to waterside and floodplain forests (Emmons and Feer, 1990; Patton et al., 2000), while *Echimys chrysurus* has a wider habitat range in both floodplain and terra firme interfluvial forests. *Dactylomys* species in Amazonia are restricted to bamboo-dominated lowland and/or floodplain habitats or to montane forests, where they feed on bamboos (Emmons, 1981; pers. obs. of *D. bolivianus, D. dactylinus* and *D. peruanus*). Consistent with their large, tall-crowned, laminar or near-laminar teeth and large to enormous hindgut fermentation compartments (Dactylomyines, Emmons 1981; pers. observation of specimens of all Amazonian genera), nearly all large-bodied arboreal Echimyidae may be strongly folivorus. Nonetheless, I have watched *E. chrysurus* feed on fruit and *M. didelphoides* feed on leaves and fruit. Floodplain forests, on younger alluvial soils than forests on adjacent uplands, may provide plant resources with higher protein, mineral, or other nutrient content or fewer toxic secondary compounds than those of terra firme. Howler monkeys (*Alouatta seniculus*), the most folivorus of Amazonian primates, are likewise chiefly restricted to floodplain forests in Amazonia (Peres, 1997; personal observation), but they use terra firme in the Guiana region (north of the Rio Amazon and east of the Rio Negro), where, interestingly, *Isothrix pagurus* (unlike *I. bistriata* in Amazonia) is also found in terra firme forests. The cause of this habitat change in Guianan forests is not understood, but I speculate that from deposition of marine aerosols, terra firme Guianan forests are more mineral-rich then are forests of Central Amazonia, where many herbivores, including *Alouatta* spp., frequent both natural and artificial sodium sources and mineral licks (Emmons, unpublished data).

In the only two echimyine genera known to include sympatric congeners - *Phyllomys* (Yuri Leite, pers. com.) and *Makalata* - sympatry seems restricted to members of different species groups, such as *Makalata grandis* with *M. macrura*. Too little is known of the ecology of echimyines to lend much insight into either their biogeography or evolution; both merit further research.

ACKNOWLEDGEMENTS

This work was done over the course of many years, with the help and support of a great many colleagues. For their hospitality and for facilitating access to specimens in their care, I thank Paulina Jenkins (BMNH), Hans Baagoe and Mogens Anderson (ZMC), Michel Tranier and Jacques Cusin (MHNP), Manuel Ruedi (MHNG), Luis Flamarion and Leandro Salles (MNRJ), Guy Musser and Robert Voss (ANMH), Bruce Patterson (FMNH), Victor Pacheco (MHNP), James Patton; (MVZ), Dieter Kock (ZSM), Maria Rutzmoser (MCZ), and Barbara Hertzig (NMW). My travel to the BMNH was generously supported by the American Museum of Natural History, facilitated by Guy Musser. I am grateful to Alfred Gardner, Sharon Jansa, James Patton and particularly Robert Voss for meticulous critical reviews that improved the manuscript. To Sharon Jansa I owe a special debt for her help with the parsimony analyses. For stimulating discussions, moral support, and help in resolving problems or providing material throughout the years that this work was in progress, I thank Alfred Gardner, Yuri Leite, Guíomar Vucetich, Robert Voss, and especially James Patton, whose generosity in sharing ideas, information and parts of his own data were instrumental in completing this work. Figure 10 was drawn by Karolyn Darrow.

LITERATURE CITED

Cabrera, A.
   1961   Catálogo de los mamíferos de America del Sur. Revista del Museo Argentino de Ciencias Naturales "Bernardino Rivadavia" e Instituto Nacional de Investigación de las Ciencias Naturales Zoologia 4:309-732.

Cuvier, F.
   1809   Extrait des premiers Mémoires de M. F. Cuvier, sur les dents des mammifères considérées comme charactères génerics. Nouveau Bulletin des Sciences, par la Société Philomatique. 1:393-395.

da Silva, M. N. F., and J. L. Patton

    1993   Amazonian phylogeography: mtDNA sequence variation in arboreal echimyid rodents (Caviomorpha). Molecular Phylogenetics and Evolution 2:243-255.

Emmons, L. H.

    1981   Morphological, ecological, and behavioral adaptations for arboreal browsing in *Dactylomys dactylinus* (Rodentia, Echimyidae). Journal of Mammalogy 62:183-189.

    1993   A new genus and species of rat from Borneo (Rodentia: Muridae). Proceedings of the Biological Society of Washington 106:752-761.

    1997   A revision of the genera of arboreal echimyid rodents (Echimyidae, Echimyinae). Seventh International Theriological Congress, Acapulco, Mexico. Paper read and Abstracts, p. 96.

    1999   A new genus and species of abrocomid rodent from Peru (Rodentia: Abrocomidae). American Museum Novitates 3279:1-14.

Emmons, L. H., and F. Feer

    1990   Neotropical rainforest mammals: A field guide. University of Chicago Press, Chicago, IL.

    1997   Neotropical rainforest mammals: A field guide. 2nd ed. University of Chicago Press, Chicago, IL.

    1999   Mamíferos de los bosques húmedos de América tropical: Una guía de campo. Editorial F.A.N, Santa Cruz, Bolivia.

Emmons, L. H., Y. L. R. Leite, D. Kock, L. P. Costa

    2002   A review of the named forms of *Phyllomys* (Rodentia: Echimyidae) with the description of a new species from coastal Brazil. American Museum Novitates 3380:1-40.

Emmons, L. H., and M. G. Vucetich

    1998   The identity of Winge's *Lasiuromys villosus* and the description of a new genus of echimyid rodent (Rodentia: Echimyidae). American Museum Novitates 3223:1-12.

Felsenstein, J.
   1985   Confidence limits on phylogenies: An approach using the bootstrap. Evolution 39:783-791.

Honacki, J. H., K. E. Kinman, and J. W. Koeppl
   1982   Mammal species of the World: A taxonomic and geographic reference. The Association of Systematics Collections, Lawrence, KS.

Jourdan, C.
   1837   Mémoire sur quelques mamifères nouveaux. Comptes Rendues Hebdomadaires des Séances de l'Académie des Sciences (Paris) 5 (juillet-décembre 1837): 521-524.

Lara, M. C., J. L. Patton, and M. N. F. da Silva
   1996   The simultaneous diversification of South American Echimyid rodents (Hystricognathi) based on complete cytochrome b sequences. Molecular Phylogenetics and Evolution 5:403-413.

Leite, Y. L. R.
   2003   Evolution and systematics of the Atlantic tree rats, genus *Phyllomys* (Rodentia, Echimyidae), with description of two new species. University of California Publications in Zoology 132:1-118.

Leite, Y. L. R., and J. L. Patton
   2002   Evolution of South American spiny rats (Rodentia, Echimyidae): the star-phylogeny hypothesis revisited. Molecular Phylogenetics and Evolution 25:455-464.

Lund, P. W.
   1839   Coup d'oeil sur les espèces éteints de Mamifères du Brésil; extrait de quelques mémoires présentés à l'Academie royale des Sciences de Copenhague. Annales de Sciences Naturelles, ser. 2, 11:214-234.

   1840   Nouvelles recherches sur la faune fossile du Brésil. Annales de Sciences Naturelles, ser. 2, 13:310-19.

   1840a   Blik paa Brasiliens Dryerverden för Sidste Jordonvaeltning. Tredie Afhandling: Fortsaettelse af Pattendryne. Det Kongelige Danske Videnskabernes Selskabs Naturvidenskabelige og Mathematiske Afhandlinger 8:217-272.

Maddison, W. P., and D. R. Maddison
    1992   MacClade: Analysis of phylogeny and character evolution. Version 3.0. Sinauer Associates, Sunderland, MA.

McKenna, M. C., and S. K. Bell
    1997   Classification of mammals above the species level. Columbia University Press, New York.

Patterson, B., and R. Pascual
    1968   New Echimyid rodents from the Oligocene of Patagonia, and a synopsis of the family. Breviora 301:1-14.

Patterson, B., and A. E. Wood
    1982   Rodents from the Deseadan Oligocene of Bolivia and the relationships of the Caviomorpha. Bulletin of the Museum of Comparative Zoology 149:371-543.

Patton, J. L.
    1988   Species groups of spiny rats genus *Proechimys* (Rodentia: Echimyidae). Fieldiana Zoology 39:305-345.

Patton, J. L., and L. H. Emmons
    1985   A review of the genus *Isothrix* (Rodentia, Echimyidae). American Museum Novitates 2817:1-14.

Patton, J. L., M. N. F. da Silva, and J. R. Malcolm
    1994   Gene genealogy and differentiation among arboreal spiny rats (Rodentia: Echimyidae) of the Amazon basin: a test of the riverine barrier hypothesis. Evolution 48:1314-1323.

    2000   Mammals of the Rio Juruá and the evolutionary and ecological diversification of Amazonia. Bulletin of the American Museum of Natural History 244:1-306.

Patton, J. L., and O. A. Reig
    1989   Genetic differentiation among echimyid rodents, with emphasis on spiny rats, genus *Proechimys*. Pp. 75-96 in Advances in Neotropical Mammalogy (K. H. Redford and J. F. Eisenberg, eds.). Sandhill Crane Press, Gainesville, FL.

Patton, J. L., and M. A. Rogers
  1983   Systematic implications of non-geographic variation in the spiny rat genus *Proechimys* (Echimyidae). Zeitschrift für Saügetierkunde 48:363-370.

Peres, C.
  1997   Primate community structure at twenty western Amazonian flooded and unflooded sites. Journal of Tropical Ecology 13:381-405.

Pessôa, L. M., and S. F. dos Reis
  1991   Cranial intraspecific differentiation in *Proechimys iheringi* Thomas (Rodentia: Echimyidae). Zeitschrift für Saügetierkunde 56:34-40.

Santos Carvalho, G. A. dos
  1999   Relações filogenéticas entre formas recentes e fóseis de Echimyidae (Rodentia: Hystricognathi) e aspectos da evolução da morfologia dentária. MS thesis, Univ. Fed. Rio de Janeiro, Rio de Janeiro.

Swofford, D. L.
  1998.   *PAUP**. Phylogenetic Analysis Using Parsimony (*and Other Methods). Sinauer Associates, Sunderland, MA.

Tate, G. H. H.
  1935.   The taxonomy of the genera of Neotropical hystricoid rodents. Bulletin of the American Museum of Natural History 68:295-447.

Thomas, O.
  1921   New *Sigmodon, Oryzomys,* and *Echimys* from Ecuador. Annals and Magazine of Natural History 7:448-450.

Voss, R. S., and R. Angermann
  1997   Revisionary notes on Neotropical porcupines (Rodentia: Erethizontidae). 1. Type material described by Olfers (1818) and Kuhl (1820) in the Berlin Zoological Museum. American Museum Novitates 3214:1-44.

Vucetich, M. G., M. M. Mazzoni, and U. F. J. Pardiñas
  1993   Los rodeores de la formación Collón Cura (Mioceno Medio), y la ignimbrita Pilcaniyeu, Cañadon del Tordillo, Neuquen. Ameghiniana 30:361-381.

Vucetich, M. G., and D. H. Verzi
    1991   Un nuevo Echimyidae (Rodentia, Hystricognathi) de la edad
            Colhuehuapense de Patagonia y consideraciones sobre la sistmatica de
            la familia. Ameghiniana 28:67-74.

Wilson, D. E., and Reeder D. M. (eds.)
    1993   Mammal species of the world: A taxonomic and geographic reference.
            2nd ed. Smithsonian Institution Press, Washington, D.C.

Woods, C. A.
    1979   Adaptive radiation of capromyid rodents: Anatomy of the masticatory
            apparatus. Journal of Mammalogy 60:95-116.

    1982   The history and classification of the South American Hystricognath
            rodents: Reflections on the far away and long ago.   Pymatuning
            Laboratory of Ecology Special Publication 6:377-392.

    1991   Suborder Hystricognathi. Pp. 771-806 in Mammal species of the world:
            A taxonomic and geographic reference (D. E. Wilson and D. M. Reeder,
            eds.). 2nd ed. Smithsonian Institution Press, Washington DC.

Woods, C. A., and E. B. Howland
    1979   Adaptive radiation of capromyid rodents: anatomy of the masticatory
            apparatus. Journal of Mammalogy 60:95-116.

Appendix 1. Character matrix for morphological analyses of relationships among echimyine rodents. Each character is identified by the number assigned to it in the text (e.g., character 1 = pelage of lower back). Character states correspond to those identified for each trait in the text

Taxon	Character															
	1	2	3	4	5	6	7	8	9	10	11	12	13	14	15	16
Octodon degus	4	2	1	1	3	3	1	2	3	2	3	1	1	1	5	3
Dactylomys boliviensis	4	2	2	1	1	1	1	1	1	1	1	1	2	3	2	3
Euryzygomatomys spin.	2	2	1	1	2	1	1	3	3	2	2	2	1	1	1	3
Carterodon sulcidens	2	2	1	1	2	1	1	3	?	2	2	1	1	1	1	1
Clyomys laticeps	2	2	1	1	2	1	2	3	2	1	2	1	1	1	1	3
Trichomys apereoides	3	2	1	1	3	1	2	3	3	2	2	2	1	2	1	3
Proechimys longicaud	2	2	1	1	2	1	2	3	3	2	2	2	1	2	2	1
Trinomys iheringi	1	2	1	1	2	1	2	2	3	?	2	2	1	2	2	3
Mesomys sp Venez.	1	2	1	1	3	1	2	3	3	2	2	2	1	2	6	3
Lonchothrix emiliae	1	2	1	1	3	1	2	?	?	?	2	3	1	2	2	1
Callistomys pictus	4	2	2	2	3	2	2	2	?	1	2	3	2	3	2	1
Isothrix bistriata	4	2	2	1	3	3	2	2	3	3	1	2	1	3	2	1
Isothrix pagurus	4	2	2	1	3	1	2	2	2	3	1	2	1	3	2	1
Diplomys labilis	3	2	1	2	3	1	1	3	2	1	1	3	2	3	4	1
Phyllomys nigrispinus	2	2	1	1	2	1	1	2	3	3	1	3	2	3	4	1
P. blainvilli	2	2	1	1	2	1	1	2	?	3	1	3	2	3	4	5
P. lamarum	2	2	1	1	1	1	1	2	?	?	1	3	2	3	4	2
P. dasythrix	3	2	1	1	1	1	1	2	?	?	1	3	2	3	4	1
P. medius	3	2	1	1	1	1	1	2	2	?	1	3	2	3	4	1
S. rufodorsalis	3	2	2	2	3	2	1	3	2	1	1	3	2	3	3	1
P. semivillosus	1	2	1	1	1	1	1	2	3	2	1	3	2	3	3	2
P. carrikeri	1	2	1	1	1	1	1	2	3	2	1	3	2	3	3	2
P. punctatus	1	2	1	1	1	1	1	2	3	?	1	3	2	3	3	2
P. flavidus holotype	1	2	1	1	1	1	1	2	3	?	1	3	2	3	3	2
Echimys chrysurus	1	2	2	2	3	2	2	2	2	3	1	3	2	3	2	1
Echimys saturnus	1	2	2	2	3	2	2	2	2	?	1	3	2	3	2	1
Makalata didelphoides	1	2	1	1	3	1	1	3	2	2	1	3	1	3	3	1
M. macrurua	1	2	1	1	2	1	1	3	2	2	1	3	1	3	7	1
M. rhipidura	2	2	1	1	3	1	1	2	3	2	1	3	1	3	3	2
M. grandis	2	2	1	1	3	1	1	2	2	?	1	3	1	3	3	4
Makalata sp nov	2	2	1	1	2	1	1	2	3	2	1	3	1	3	3	2
P. occasius	1	2	1	1	1	1	2	2	3	2	1	2	2	3	3	3
Myocastor coypus	4	1	1	1	1	1	1	1	1	3	2	3	1	2	2	1

Appendix 1 continued.

| | | | | | | | | | Character | | | | | | | |
Taxon	17	18	19	20	21	22	23	24	25	26	27	28	29	30	31	32
Octodon degus	?	6	1	3	1	1	3	2	4	1	1	1	1	1	1	2
Dactylomys boliviensis	1	6	1	3	1	4	1	1	4	1	2	1	1	1	1	3
Euryzygomatomys spin.	?	2	1	1	2	1	1	1	2	1	1	1	1	1	1	1
Carterodon sulcidens	?	1	1	1	2	1	1	1	4	1	1	1	1	3	1	1
Clyomys laticeps	?	4	1	1	2	1	1	1	2	1	1	1	1	1	1	1
Trichomys apereoides	?	1	1	2	1	1	1	1	2	1	2	2	1	1	1	2
Proechimys longicaud.	?	2	1	1	1	1	1	1	2	1	2	2	1	1	1	2
Trinomys iheringi	?	2	1	1	1	1	1	2	3	1	2	1	1	1	2	2
Mesomys sp Venez.	?	5	1	1	1	1	1	3	1	1	2	2	1	1	1	3
Lonchothrix emiliae	?	5	1	1	1	1	1	3	1	2	2	5	1	1	1	3
Callistomys pictus	1	2	1	2	1	1	3	3	4	1	2	3	1	1	1	2
Isothrix bistriata	1	3	1	2	2	1	1	3	3	1	1	2	2	2	1	3
Isothrix pagurus	1	3	1	2	1	1	1	3	3	1	1	2	3	2	1	3
Diplomys labilis	1	5	2	2	1	3	3	3	4	1	2	2	1	2	1	3
Phyllomys nigrispinus	1	2	2	2	1	2	3	3	4	1	2	2	2	1	1	3
P. blainvilli	1	2	2	2	1	2	3	3	4	1	2	2	2	1	1	3
P. lamarum	1	2	2	2	1	2	3	3	4	1	2	2	2	1	1	3
P. dasythrix	1	2	2	2	1	2	3	3	4	1	2	2	1	1	1	3
P. medius	1	2	2	2	1	2	3	3	4	1	2	2	1	1	1	3
S. rufodorsalis	1	6	2	2	1	2	3	3	4	1	2	3	2	2	1	3
P. semivillosus	2	6	1	2	1	1	3	3	4	2	2	3	1	1	1	3
P. carrikeri	2	6	1	2	1	1	3	3	4	2	2	3	1	1	1	3
P. punctatus	2	6	1	2	1	1	3	3	4	2	2	2	1	1	1	3
P. flavidus holotype	2	6	1	2	1	1	3	3	?	2	2	2	2	1	1	3
Echimys chrysurus	1	5	2	2	1	1	2	2	4	2	2	3	3	1	1	3
Echimys saturnus	1	5	2	2	1	1	2	2	4	2	2	3	2	1	1	3
Makalata didelphoides	1	4	2	2	1	1	3	3	4	2	2	4	1	1	3	3
M. macrurua	1	2	2	2	1	1	3	3	4	2	2	4	2	1	3	3
M. rhipidura	1	4	2	2	1	1	3	3	4	2	2	4	1	1	3	3
M. grandis	1	4	2	2	1	1	3	3	4	2	2	4	1	1	3	3
Makalata sp nov	1	4	2	2	1	1	3	3	4	2	2	4	2	1	3	3
P. occasius	2	2	1	2	1	1	2	3	4	2	2	4	1	1	1	3
Myocastor coypus	1	5	3	1	1	1	1	1	4	1	1	2	1	3	2	1

Appendix 1 continued.

Taxon	33	34	35	36	37	38	39	40	41	42	43	44	45	46	47
Octodon degus	2	1	2	1	1	1	1	2	2	1	1	2	1	2	2
Dactylomys boliviensis	2	2	3	3	1	3	1	2	2	1	1	2	2	1	2
Euryzygomatomys spin.	2	2	3	1	1	1	1	1	2	1	1	1	2	2	2
Carterodon sulcidens	2	1	2	1	2	1	1	1	2	1	1	1	2	?	?
Clyomys laticeps	1	2	2	1	1	1	1	1	2	1	1	1	2	2	2
Trichomys apereoides	1	1	2	2	1	1	1	2	2	1	1	2	2	2	1
Proechimys longicaud.	2	1	2	2	1	3	1	2	1	1	2	2	2	1	1
Trinomys iheringi	2	2	4	2	1	3	1	2	1	1	2	2	2	2	1
Mesomys sp Venez.	2	2	2	3	1	3	1	2	1	1	1	2	2	2	3
Lonchothrix emiliae	2	2	3	3	1	3	1	2	1	1	1	2	2	2	3
Callistomys pictus	1	1	3	?	1	2	1	3	2	1	1	1	2	2	3
Isothrix bistriata	2	1	3	3	1	2	1	2	1	1	2	2	2	1	1
Isothrix pagurus	2	1	3	3	1	2	1	2	1	1	2	2	2	1	1
Diplomys labilis	2	1	2	3	1	2	1	2	2	1	2	2	2	1	2
Phyllomys nigrispinus	2	2	3	3	1	3	2	4	2	2	3	2	2	1	2
P. blainvilli	2	2	4	3	1	3	1	4	2	2	2	2	2	2	2
P. lamarum	2	2	3	3	1	3	1	4	2	2	2	2	2	1	2
P. dasythrix	2	2	4	3	1	3	3	4	2	2	2	2	2	1	2
P. medius	2	2	4	3	1	3	?	4	2	2	2	2	2	1	2
S. rufodorsalis	2	1	3	3	1	3	3	2	2	1	1	2	2	1	2
P. semivillosus	2	2	3	3	1	3	1	3	2	1	2	2	2	2	2
P. carrikeri	2	2	4	3	2	3	3	4	2	1	3	2	2	2	2
P. punctatus	2	2	4	3	1	3	3	3	2	1	1	2	2	2	2
P. flavidus holotype	2	2	4	3	1	3	1	4	2	1	1	2	2	2	2
Echimys chrysurus	2	2	4	3	1	3	3	3	2	1	2	2	2	2	3
Echimys saturnus	2	2	4	3	1	3	3	2	2	1	2	2	2	1	3
Makalata didelphoides	2	2	3	3	2	3	3	2	2	1	2	2	2	1	3
M. macrurua	2	2	3	4	2	3	3	3	2	1	2	2	2	1	3
M. rhipidura	2	2	3	4	2	3	3	3	2	1	2	2	2	1	2
M. grandis	2	2	4	3	2	3	3	3	2	1	2	2	2	1	3
Makalata sp nov	2	2	3	3	2	3	3	3	2	1	2	2	2	1	2
P. occasius	2	2	3	2	2	3	1	3	2	1	3	2	2	2	2
Myocastor coypus	1	2	1	3	1	2	1	4	2	1	1	2	2	3	3

Diversification of Small Mammals in the Atlantic Forest of Brazil: Testing the Alternatives

Márcia C. Lara, Lena Geise, and Christopher J. Schneider

Hypotheses of speciation in the tropics generate expected patterns of species distributions and phylogenetic relationships. Here, we review phylogeographic predictions from the main models of speciation in the tropics and compare them to observed patterns of mtDNA variation in three genera of small mammals from the Atlantic Forest of Brazil. The patterns of mtDNA variation among the genera are idiosyncratic, suggesting that large-scale external environmental factors have not affected all taxa similarly. However, we find some support for relative isolation among the three main mountain ranges in the Atlantic Forest domain consistent with the Montane Isolates Hypothesis. Also, there is some support for the Ecological Gradients Hypothesis in the distribution of sister species/clades in different habitats within mountain ranges. Speciation in the Atlantic Forest is likely to involve multiple mechanisms and additional, fine-scale studies of numerous groups are needed to differentiate among alternative hypotheses.

SPECIATION AND TROPICAL FOREST DIVERSITY

The Atlantic rain forest of coastal Brazil is recognized as one of the world's most threatened biodiversity hotspots and preserving what remains of the Atlantic Forest is a high priority among conservation organizations (McNeely et al., 1990). In general, current conservation efforts in rainforests worldwide seek to conserve the ecological conditions that accommodate high endemic species diversity. While this goal is laudable, it has recently been emphasized that it is also important to conserve the evolutionary processes that generate and maintain species and ecological diversity (Crandall et al., 2000; Balmford et al., 1999; Smith et al., 1993; Erwin, 1991a). The single largest impediment to effectively conserving the evolutionary processes that generate diversity in rainforests is our lack of knowledge concerning those processes. Here, we review the four major models of diversification that have been the mainstays of explanations for tropical diversity. We then review available data from vertebrates in the Atlantic Forest that are pertinent to discriminating among these hypotheses. None of the studies to date have been designed explicitly to test alternative hypotheses of diversification in the Atlantic Forest, but they provide critical insight into the patterns of genetic and species diversity that are necessary for designing detailed analyses to test alternative models of diversification.

Theories of Rainforest Speciation

The evolutionary mechanisms responsible for generating the high biodiversity of tropical rainforests have been a topic of debate for decades (Erwin, 1991b; Stebbins, 1974; Mayr, 1963). Generally, models of speciation for tropical faunas have emphasized the geographic context of speciation and focused on divergence in allopatry. The single exception is Endler's model of parapatric divergence in response to diversifying selection across ecological gradients (Endler, 1982a).

Allopatric models: There are three prominent models for allopatric diversification in tropical rainforests, namely the Pleistocene Refugia Hypothesis (PRH) (Mayr and O'Hara, 1986; Vanzolini and Williams, 1970; Haffer, 1969), the Riverine Barrier Hypothesis (RBH) (Wallace, 1852) and the Montane Isolate Hypothesis (MIH) (Moreau, 1966). All three models invoke vicariant mechanisms - either the contraction of rainforests to isolated refugia during Pleistocene glacial maxima (PRH and MIH) or the dissection of continuous habitat by the formation of large rivers such that populations on either side are isolated (RBH). In each scenario, the geographically isolated populations diverge due to genetic drift and/or natural selection because they are insulated from the homogenizing effects of gene flow.

The Pleistocene Refuge Hypothesis, which has been the most prominent model of allopatric speciation (Whitmore and Prance, 1987; Vanzolini and Williams, 1970; Haffer, 1969), stresses the role of large, isolated forest refugia in generating new taxa and high species diversity (Mayr and O'Hara, 1986; Haffer, 1969). The PRH depends on the premise that climatic change caused rainforests to contract, leaving refugia separated by dry forests or savanna (Prance, 1982). Since its inception, the PRH has been controversial (Cracraft, 1994; Endler, 1982a and 1982b; Mayr and O'Hara, 1986; Haffer, 1969). It reflects a long history of equating centers of biodiversity with centers of species origin (Willis, 1922; Diels, 1908; Darwin, 1858). Criticisms of this model include (1) uncertainty about whether Amazonian rainforests contracted or just changed in composition (Colinvaux and de Oliveira, 2001; Colinvaux et al., 2000; Colinvaux et al., 1996), (2) bias caused by differences in sampling effort (well studied areas tend to have more described species and are therefore identified as refugia and/or centers of endemism; Nelson et al., 1990), (3) circular reasoning in the location of contact zones defined by sister taxa relationships (Mayr and O'Hara, 1986; Endler 1982b), (4) a lack of discussion of alternative models that provide equally good explanations for biogeographic patterns (Endler, 1982a; Endler, 1982b), and (5) difficulty in identifying *a priori* spatially and temporally bounded refugia.

Although early debates focused on the effects of Pleistocene events, particularly those of the last glacial cycle or two (e.g., Diamond and Hamilton, 1980; Haffer, 1969), the PRH has more recently been extended to Tertiary events on the assumption that the climate oscillations throughout this period were also of sufficient amplitude and duration to promote rainforest fragmentation and speciation (Haffer, 1993). Generally, the importance of glacial periods to speciation is an active area of debate (Moritz et al., 2001; Arbogast and Slowinski, 1998; Avise and Walker, 1998; Klicka and Zink, 1998; Schneider et al., 1998; Klicka and Zink, 1997; Zink, 1997).

Important, but rarely discussed, are the ecological conditions within proposed refugia and the associated processes of divergence (e.g., Vanzolini and Williams, 1981). If ecological conditions among refugia differ substantially then populations may be expected to diverge in response to divergent natural selection, whereas if ecological conditions remain similar, populations may diverge only by drift.

The second allopatric model, the Montane Isolate Hypothesis (MIH), postulates that isolation in montane rainforest remnants during climatic dry periods led to divergence and speciation (with new species subsequently expanding their range to the lowlands). First introduced by Moreau (1966) based on distribution of closely related birds on isolated mountains in West Africa, the MIH proposes divergence of populations simply as a result of isolation on mountains and does not specify whether divergence results primarily from drift or divergent selection.

The third allopatric model, the Riverine Barrier Hypothesis (RBH), postulates that geographic isolation by river systems causes divergence and speciation (Wallace, 1852). The common observation that the boundaries of closely related species or subspecies often coincide with the major rivers of Amazonia has been taken as evidence for the RBH. For example, several primates (Hershkovitz, 1977), some passerine birds (Cracraft and Prum, 1988; Haffer, 1974; Haffer, 1969), and many Amazonian lizards (Avila-Pires, 1995) have distributions bounded by rivers. However, congruence between patterns of phenotypic diversity and distribution of rivers is inconsistent (Capparella, 1992). A problem shared with the PRH is that distributional data alone are consistent with multiple speciation hypotheses. Difficulties with the RBH arise from the fact that large rivers, because they inhibit dispersal, may be convenient meeting points for species that diverged elsewhere (Patton et al., 1994) and, thus, have little or nothing to do with the process of speciation. Furthermore, the strength of the barrier to gene flow is expected to diminish toward the upper reaches of a river (Peres et al., 1996; Ayres and Clutton-Brock, 1992) and drainage patterns are dynamic and change over evolutionary time.

Non-allopatric models: Non-allopatric models of divergence include divergence among contiguous populations (parapatric divergence; Mayr, 1963) as well as classical sympatric divergence. The most prominent parapatric model is the

Ecological Gradients Hypothesis (EGH). This model postulates that divergent selection across strong environmental gradients is sufficient to cause differentiation and speciation despite the presence of gene flow (Orr and Smith, 1998; Smith et al., 1997; Endler, 1977; Rice and Hostert, 1993). The EGH implies that transitional environments (altitudinal gradients or ecotones) are areas in which divergent selection and speciation occur. Consequently, the EGH predicts that habitat transitions, rather than rivers or refuge boundaries, are areas of contact between sister taxa. Recent emphasis on current (Tuomisto et al., 1995) and, possibly, historical (Colinvaux et al., 1996) heterogeneity of vegetation structure within Amazonia suggests that transitional environments are a common component of rainforest landscapes and may provide opportunity for the EGH to operate within rainforests. The observation that hybrid zones are commonly located in ecotones (Endler, 1982a and 1982b), and that sharp ecological gradients along the lower slopes of mountain ranges are often the areas of highest species diversity (Fjeldså, 1994; Duellman, 1978) are consistent with the EGH, but distributional data alone are open to multiple interpretations (e.g., Mayr and O'Hara, 1986; Endler, 1982b)

Testing Alternative Hypotheses

The hypotheses outlined above generate a series of predicted patterns at both the interspecific (phylogenetic) and intraspecific level (Figure 1). In general, allopatric models predict that sister species should be found on either side of current or historical barriers to dispersal whereas the ecological gradients model predicts that sister species should occupy adjacent but different habitats (e.g., altitudinal replacement by sister species). If we assume that the processes that generate species are the same as the processes that result in divergence of populations, then analyses of population divergence can be particularly enlightening. Simultaneous comparisons of populations occupying different habitats in geographically isolated regions provide strong evidence of the processes important in divergence. Sampling of populations across habitats within and among geographically isolated regions provides for a direct test of the relative importance of geographic isolation and ecological gradients in divergence (see Figure 1). If, for a given level of genetic divergence, reproductive isolation or morphological divergence is greater among habitats than within habitats, a role for divergence in response to ecological gradients is indicated. If reproductive or morphological divergence between populations isolated in similar habitats is as great as that among isolates that occupy different habitats, then little role for adaptive divergence across ecological gradients is indicated.

a) Population Distribution   b) Population Genetic Test   c) Phylogenetic Test

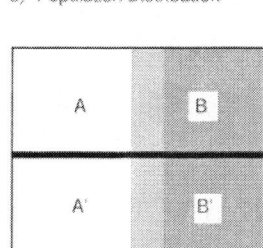

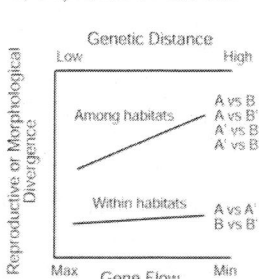

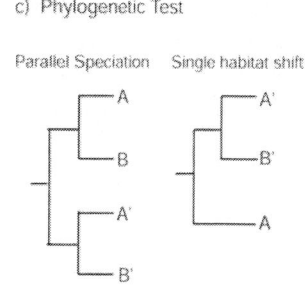

Figure 1. Tests for the effect of selection in divergence among populations (modified from Orr and Smith, 1998; Schluter and Nagel, 1995). (a) Four regions are defined. Shading indicates different habitats and the bold horizontal line indicates a partial or complete barrier to gene flow (e.g., geographic distance or a physical barrier). Populations A and A' occupy one habitat and populations B and B' occupy a different habitat. (b) By comparing morphological divergence in fitness-related traits or reproductive divergence among populations, relative roles of drift and selection in divergence can be evaluated. If selection is driving population divergence then, for a given level of genetic divergence, greater reproductive isolation (or morphological divergence) is expected among populations from different habitats than among populations occupying similar habitats. If the degree of reproductive divergence is similar within and among habitats, then factors acting independent of the environment (e.g., drift) are indicated. (c) Historical relationships among populations provide an additional test of the hypothesis that selection is important in speciation. Populations A and B (also A' and B') are sister groups that occupy different habitats while populations A and A' occupy similar habitats but are not sister groups. In parallel speciation (Schluter and Nagel, 1995) similar adaptive divergence of populations occurs independently two or more times. In the single habitat shift scenario, A is the ancestral habitat and a single shift to habitat B occurs. If reproductive divergence is greater between A and B (and A' and B'), than between A and A' (and B and B') then selection is implicated in divergence.

The Atlantic Forest

Rainforests in the Atlantic Forest region of southeastern Brazil are concentrated in four main mountain ranges and the coastal lowlands (Figure 2), covering coastal plains, coastal mountains and the inland plateau (Brasil, 1992). The rainforests in the region are classified into wet evergreen forest (*ombrófila densa*) and drier semi-deciduous rainforest. To the west, the semi-deciduous forest grades into the dry savanna-like vegetation of the Cerrado. The two rainforests in Brazil - the Atlantic Forest and the Amazon Forest - are separated by a wide area of Cerrado and semi-arid vegetation (*Caatinga*). The Atlantic Forest originally stretched 3,500km along the coast of Brazil, covering 1,360,000 km². It is estimated that only 8% of the original vegetation cover remains today (Conservation International do Brasil et al., 2000). In some states deforestation is even greater, with the state of Espirito Santo having only 1% of its original forest remaining.

The degree of endemism in the Atlantic Forest is high: 50% of tree species and 40% of mammal species are endemic to the Atlantic Forest (Câmara, 1991). In the Atlantic Forest, 18% of birds, 30% of reptiles, 26% of amphibians, and 38% of fish are endemic (Conservation International do Brasil et al., 2000). Of the 6 genera and 20 species and subspecies of primates in the Atlantic Forest, 2 genera and 18 species are endemic, with 14 species listed as endangered or vulnerable (Rylands et al., 1997; Câmara, 1991). It is estimated that at least 30% of all plant and animals species are endangered (Miranda and Mattos, 1992), with mammals alone accounting for 14% of the threatened species. Due to its high levels of deforestation and high diversity, the Atlantic Forest is considered one of the three top-priority areas for conservation (Myers et al, 2000).

Of the four major mountain ranges in the Atlantic Forest, the Serra do Espinhaço is the most northerly (Figure 2), running north-south with the Rio Jequitinhonha cutting through it at more or less its mid point. The highest elevation in Serra do Espinhaço is 2,062m above sea level, making it the lowest of the four major mountain chains in this area. It is also the most inland, encompassing the Atlantic Forest as well as the Cerrado domains. The Serra da Mantiqueira runs northeast-southwest, defining the path of two major rivers: the Rio Doce on its northwest slope, and the Rio Paraiba do Sul along its southern margin. It is the highest of the four ranges, with the highest point at 2,787m. The Serra do Mar (alt. 2,310m) is the most eastern mountain range, on or very close to the coast, where, at many places, it literally ends on the beach. The Serra dos Órgãos is the most localized of the mountain chains, and may be regarded as an extension of the Serra do Mar. The Serra dos Órgãos is restricted to the state of Rio de Janeiro, with a maximum altitude of 2,275m. Due to the orientations of Serra da Mantiqueira, Serra do Mar and Serra dos Órgãos, all of which parallel to the coast, precipitation is highest on the eastern slopes, with more than 1,500mm of rainfall a year (maximum 4000mm per year); this is where the evergreen broadleaf forest is found. The height of the mountains results in a rain shadow on the western side of the ranges, where the forest is drier, with a one to two month dry season, creating the semi-deciduous forest.

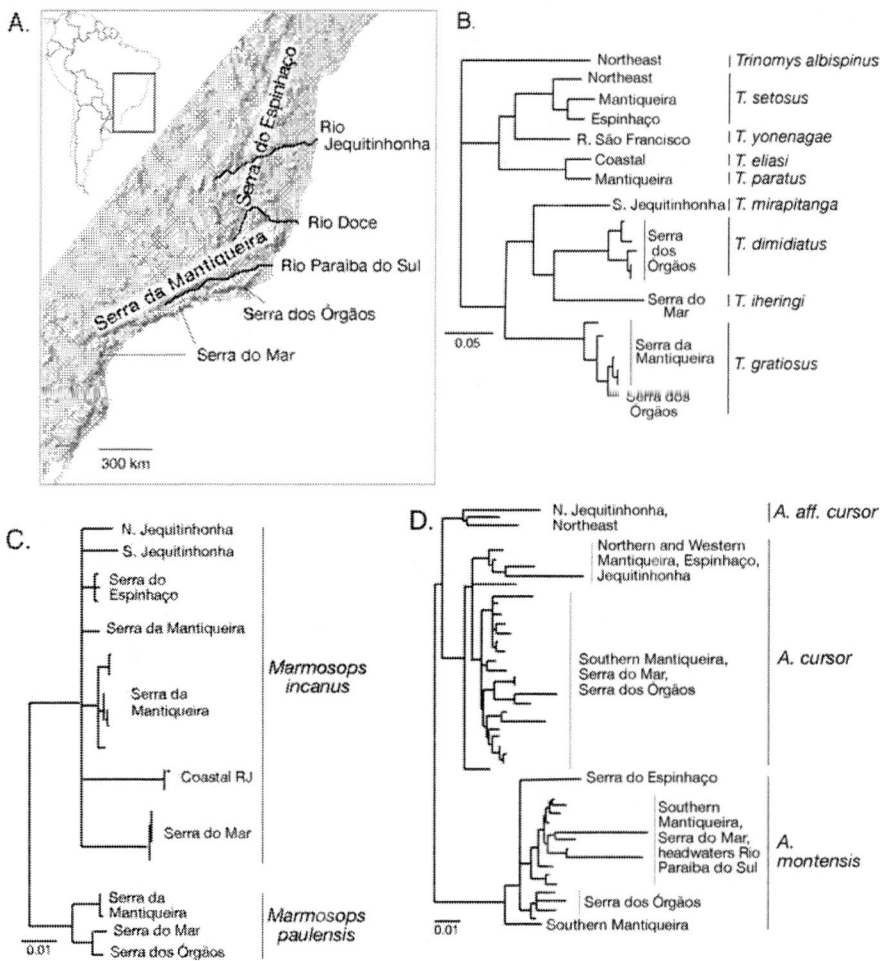

Figure 2. (a) Political and relief map showing the four major mountain ranges and major rivers of the Atlantic Forest region. (b) Phylogeny of *Trinomys* from Lara and Patton (2000). (c) Phylogeny of *Marmosops* from Mustrangi and Patton (1997). (d) Phylogeny of *Akodon* from Geise et al. (2001).

APPLICATION OF THE MODELS OF SPECIATION TO THE ATLANTIC FOREST

Pleistocene Refuge Hypothesis

Molecular systematic studies on diverse taxa from several rainforest regions indicate that the vast majority of speciation events among recognized species occurred prior to the Pleistocene, thus rejecting the Pleistocene Refuge Hypothesis (PRH) as a significant contributor to the current high diversity and endemism of vertebrates in tropical rainforests (Patton and Costa, 2003; Moritz et al., 2001; Lara and Patton, 2000; Moritz et al., 1997; Patton et al., 1997; Lara et al., 1996; Heyer and Maxson, 1983). Moreover, studies at the population level (Schneider and Moritz, 1999; Schneider et al., 1999) found that even long-term (pre-Pleistocene) geographic isolation had little effect on phenotypic divergence in Australian endemic rainforest lizards. This suggests that geographic isolation, even for long periods of time, may be insufficient to generate phenotypic divergence and speciation.

The PRH has played a prominent role in discussions of the biogeographic history of the Atlantic Forest (Brown, 1987; Jackson, 1978). Based on patterns of species diversity and endemism, several refuges have been proposed (Câmara, 1991). Data bearing on the importance of Pleistocene refugia in speciation in the Atlantic Forest are few, but one of the earliest molecular systematic studies (Heyer and Maxson, 1983) examined albumin immunological distances among species of frogs from the Atlantic Forest and found that the depth of divergence among species far exceeded expectations from Pleistocene speciation. More recently, analyses of mtDNA variation in several small mammals have demonstrated a similar pattern of ancient divergence (Patton and Costa, 2003; Mustrangi and Patton, 1997; Patton et al., 1997; Lara et al., 1996; but see Leite, 2002 for two cases of recent divergence in the Atlantic tree rat *Phyllomys*). While divergence among most recognized species antedates the Pleistocene, recent rainforest refugia may provide the opportunity for allopatric divergence among populations of widespread species and refugia may thus still play a role in generating biodiversity.

Riverine Barrier Hypothesis

Phylogenetic and phylogeographic analyses of several groups of South American vertebrates are equivocal with respect to the Riverine Barrier Hypothesis (RBH). Sister species (or sister populations) are not always found on either side of major rivers, as predicted by the RBH. In several cases, however, sister groups are found on either side of large rivers, which is consistent with the RBH (Patton et al., 2000; Ron, 2000; Patton et al., 1997; Avila-Pires, 1995; Cracraft and Prum, 1988; Vanzolini, 1988). The RBH has been explicitly tested in only one instance. An extensive study of eight rodent species along the Rio Juruá in the Brazilian Amazon region found that genetic divergence between conspecific populations across the river was much less than that between populations along the same side of river that were isolated historically by a substantial geographic barrier (Patton et al., 2000; Patton and da Silva, 1998; Patton et al., 1997; Patton et al., 1994). These results indicate that the Rio Juruá was not a significant barrier to gene flow in small mammals, as required by the RBH (but see Peres et al., 1996 for a riverine effect in primates).

The RBH has not been suggested as a significant contributor to diversification of the Atlantic Forest fauna, although large rivers such as the Rio Paraiba do Sul, Rio Doce, and Rio Jequitinhonha (Figure 2) may separate lowland portions of the Atlantic Forest. The Rio Jequitinhonha and Rio Doce have been identified as limits to regional faunas, but no studies to date have determined whether large rivers in the Atlantic Forest result in population divergence or speciation (cf. Patton et al., 1994). To a large degree, riverine barriers in the Atlantic Forest correspond to breaks between mountain ranges and, therefore, for all taxa except those confined to lowland rainforests, riverine barriers may be considered as part of the Montane Isolate Hypothesis.

Montane Isolate Hypothesis

The Montane Isolate Hypothesis (MIH) has gained some support, especially in Africa (Roy, 1997; Fjeldså, 1994). However, Schneider and Moritz (1999) found that long-term geographic isolation among montane rainforest isolates in Australia did not result in significant phenotypic evolution or speciation. The MIH has not been examined explicitly in the Atlantic Forest even though the mountainous topography of the region suggests that montane isolation is a plausible mechanism for geographic isolation. In general, the MIH has been lumped together with the PRH in discussions of Atlantic Forest biogeography (e.g., Jackson, 1978). Some sigmodontine rodents such as *Delomys collinus* (Bonvicino and Geise, 1995), *Brucepattersonius* spp., *Akodon mystax* and *A. serrensis* occur only at elevations above 2,000 m in the Serra do Mar (Bonvicino et al., 1997) and Serra da Mantiqueira (Geise, pers. obs.). For these taxa, montane isolation is almost certainly a reality, but it is not clear whether montane isolation has been the driving force in speciation. Again, recent analyses of mtDNA variation in small mammals are informative (Geise et al., 2001; Lara and Patton, 2000; Mustrangi and Patton, 1997). As discussed below, at least two small mammal genera (*Marmosops* and *Trinomys*), and possibly a third (*Akodon*), show patterns of mtDNA variation that are consistent with, but do not provide a sufficient test of, the montane isolate hypothesis.

Phylogeography of *Trinomys, Marmosops* and *Akodon*

We reanalyzed previously published mtDNA data from the Atlantic Forest spiny rat *Trinomys* (Lara and Patton, 2000), the slender mouse opossum *Marmosops* (Mustrangi and Patton, 1997), and the grass mouse *Akodon* (Geise et al., 2001) and found patterns consistent with montane isolation (Figure 2). In *Trinomys*, variation is primarily structured by habitat, with two main clades of species occupying wet evergreen forest and semi-deciduous forest, respectively. The species *T. gratiosus*, *T. iheringi*, *T. dimidiatus* and *T. mirapitanga* form a strongly supported clade and inhabit wet evergreen forest in the coastal lowlands, the Serra do Mar, Serra dos Órgãos, and Serra da Mantiqueira. In contrast, the species *T. setosus*, *T. yonenagae*, *T. paratus*, and *T. eliasi*, which also form a strongly supported clade, inhabit semi-deciduous forest in the Serra do Espinhaço and Serra da Mantiquera. The exception is *T. yonenagae*, which is also part of this clade but is found along the Rio São Francisco outside of the Atlantic forest domain and is not pertinent to our discussion.

Within the wet evergreen forest clade, there is a clear pattern of species occupying different mountain ranges (Figure 2). *T. dimidiatus* occupies the Serra dos Órgãos and is the sister species to *T. iheringi*, which occupies the Serra do Mar. These two species are part of a clade that is the sister group to *T. gratiosus*, which occupies the Serra da Mantiqueira and spills over into the Serra do Mar. These patterns of phylogenetic relationship among species are consistent with divergence of species in different mountain ranges while largely retaining an ancestral ecological affiliation with wet evergreen forest.

Patterns in the clade that inhabits semi-deciduous forest are less consistent with the MIH. Populations of *T. setosus* occupy both the Serra da Mantiqueira and Serra do Espinhaço as well as adjacent lowland forest. The sister species *T. eliasi* and *T. paratus* both occupy semi-deciduous forest but *T. eliasi* is restricted to lowlands (although it is known from only two localities), while *T. paratus* is more widespread and ranges up to at least 600m in the Serra da Mantiqueira.

Moreau's original hypothesis (Moreau, 1966) postulated that Pleistocene glacial periods resulted in isolation of rainforest taxa on mountains where rainforest persisted. The depth of divergence between *T. dimidiatus* and *T. inheringi* (11% difference in mitochondrial cytochrome-b; Lara and Patton, 2000) suggests pre-Pleistocene divergence and, hence, the species of *Trinomys* appear to be much older than the Pleistocene. Nonetheless, the presence of sister species on adjacent mountain ranges is consistent with one prediction from the MIH.

Two of the six species of small arboreal marsupials in the genus *Marmosops* are found in the Atlantic Forest. *Marmosops incanus* and *M. paulensis* are sister species (Mustrangi and Patton, 1997) that occupy both wet evergreen and semi-deciduous forest from the coast to over 1500 meters elevation. *M. incanus* occurs from the lowlands to ca 800 m in the Serra do Mar, Serra da Mantiqueira and Serra do Espinhaço. *M. paulensis* occurs primarily above 800 m in the southern Serra da Mantiqueira, Serra do Mar, and Serra dos Órgãos.

In *M. incanus*, Mustrangi and Patton (1997) identified several well-supported lineages defined by mtDNA sequences. These lineages correspond well with each of the major mountain ranges – one lineage is found in the Serra do Espinhaço, another in the Serra da Mantiqueira and a third in the Serra do Mar. Three additional lineages are found in the coastal lowlands of Rio de Janeiro extending into the Serra do Mar, the lowlands on either side of the Rio Jequitinhonha, and at Linhares (50 m elevation) in the state of Espirito Santo. The relationships among mitochondrial clades are not well resolved so we cannot determine if sister groups of mtDNA haplotypes are found in adjacent mountain ranges, as predicted by the MIH. The depth of divergence among mtDNA clades in *M. incanus* is less than among species of *Trinomys*, but is still so great (average of 6.4%) as to suggest pre-Pleistocene differentiation.

Sampling of *M. paulensis* by Mustrangi and Patton (1997) was limited but they identified two main lineages of mtDNA haplotypes - one found in the southern Serra da Mantiqueira and another in the Serra do Mar and Serra dos Órgãos. The depth of divergence among these lineages (5.9%) is similar to that seen among mtDNA lineages in *M. incanus* and the phylogeographic patterns of the two species are remarkably concordant, suggesting that similar environmental factors have structured mtDNA variation in both species (Mustrangi and Patton, 1997).

If speciation in montane isolates was responsible for the origin of *M. incanus* and *M. paulensis*, we would predict that each species would be distributed in adjacent montane areas. The data are ambiguous in this regard because both species occupy the Serra da Mantiqueira and Serra do Mar, although the more southerly distribution of *M. paulensis* and its affinity for cooler, higher regions in the northern part of its range argues for a more southern origin for this species. While the presence of distinct mtDNA lineages in both *M. incanus* and *M. paulensis* that are distributed among the major mountain ranges is consistent with the effects of montane isolation, studies designed specifically to test this and other hypotheses of diversification in this group are needed.

Akodon is a small sigmodontine rodent that is found in a wide range of habitats in the Atlantic Forest region, including lowland and upland forest (both wet evergreen and semi-deciduous forest), restinga, and both lowland and high altitude grasslands. Of the three species in the *cursor* complex, two species are common in the Atlantic Forest. *Akodon cursor* is primarily associated with wet evergreen forests across a wide altitudinal range, occurring from sea level to 800 m altitude in all of the main mountain ranges. *A. montensis* occupies a wide range of habitats in the four main mountain ranges above 300 m (Geise, 1995). Where *A. montensis* and *A. cursor* are sympatric in the state of Rio de Janeiro, *A. montensis* replaces *A. cursor* at altitudes above 800 m.

Akodon shows relatively shallow mtDNA divergence within (0 - 5%) and among species (c. 4 - 15%) suggesting that it is a relatively recent taxon in the region (Geise et al., 2001). In contrast to *Marmosops* and *Trinomys*, *Akodon* shows little structure in mtDNA variation among regions and among habitats. The shallow divergence and presence of shared haplotypes among localities suggests high levels of gene flow, or, more likely, recent colonization in the Atlantic Forest. Interestingly, *A. montensis* shows greater geographic structure among mountain ranges than does *A. cursor*, which is consistent with it being a more montane species in which gene flow among mountain ranges is restricted. Even though divergence is very shallow and there is sharing of haplotypes among mountain ranges in both *A. cursor* and *A. montensis*, the haplotypes show some structure among mountain ranges, but the ranges do not contain exclusive haplotype groups. In *A. cursor*, a large clade of haplotypes is found in the southern Serra da Mantiqueira, Serra do Mar, and Serra dos Órgãos. A second, smaller clade is found in the northern and western Serra da Mantiqueira, Serra do Espinhaço, and lowlands of the Rio Jequitinhonha. Where the Serra do Espinhaço and Serra do Mantiqueira join, representative haplotypes of both groups are found.

Although sampling is not as extensive for *A. montensis*, a pattern of haplotype variation similar to that in *A. cursor* is apparent. One monophyletic group of mtDNA haplotypes is found in the Serra dos Órgãos and another in the Serra do Mar, extending around the headwaters of the Rio Paraiba do Sul and into the southern Serra da Mantiqueira. The Serra do Espinhaço are represented by a single site in the southern part of this range that contains a distinct mtDNA haplotype, as does another site in the far southern Serra da Mantiqueira.

A third group of *Akodon* (*aff. cursor*) is the sister group to *A. cursor*. This third group is found north of the Rio Jequitinhonha into northeast Brazil. The MIH predicts that sister species should occupy adjacent mountain ranges. While there is some indication of montane isolation in the distribution of mtDNA haplotypes, the relationship among the three closely related species of *Akodon* presented here is not consistent with the MIH. It is likely that the distribution of mtDNA haplotypes in the *A. cursor* complex are not at drift–gene flow equilibrium, as there is no pattern of isolation by distance even though gene flow among distant localities is likely to be low. This suggests recent occupation of much of the Atlantic Forest region. Geise et al. (2001) rejected the peripheral isolate hypothesis for the origin of *A. aff. cursor* from within *A. cursor* because the two taxa show reciprocal monophyly of mtDNA (whereas the peripheral isolate hypothesis predicts that *A. aff. cursor* mtDNA should be nested within the haplotypes found in *A. cursor*). The recent divergence of *A. cursor* and *A. aff. cursor* is consistent with Quaternary speciation but whether speciation resulted from isolation in refuges or across ecological gradients or among mountain isolates remains to be determined.

Ecological Gradients Hypothesis

Recently, evidence for the Ecological Gradients Hypothesis (EGH) has come from observations that morphological differentiation among populations occupying adjacent but differing habitats occurs despite high rates of gene flow (Schneider et al., 1999; Smith et al., 1997) and in the direction predicted by selection. The demonstration of population divergence in response to selection across ecological gradients suggests that this may be an important process in speciation. However, analyses of species relationships across altitudinal gradients in separate river valleys in the eastern Andes of Peru (Patton and Smith, 1992) did not support the EGH. Rather than having sister species replace each other altitudinally in each valley, as predicted by the EGH, Patton and Smith found that sister species tended to occupy similar altitudinal ranges in each valley, suggesting divergence in allopatry (however as noted by Patton and Smith [1992}, divergence may have occurred once across the ecological gradient followed by dispersal of species among valleys and subsequent allopatric divergence).

The EGH has yet to be examined in the Atlantic Forest, except for one ongoing study (Pinheiro et al., 2001). The distribution of vegetation types, resulting largely from a rainfall gradient that declines from east to west in combination with orographic effects of mountain ranges, provides a clear ecological gradient across which disruptive selection may generate new diversity. Furthermore, the large altitudinal and corresponding habitat differences between lowland and upland forests provide another type of ecological gradient across which divergence could occur. At least one recent study (and possibly two others) indicates an intriguing relationship between divergence and habitat gradients in the Atlantic Forest spiny rat *Trinomys* where the distributions of its species coincide with the distribution of vegetation types and humidity gradients in the Atlantic Forest (Lara and Patton 2000). The Atlantic tree rat *Phyllomys* also shows a similar trend in that some species are restricted to semi-deciduous forest while others occur mainly in the coastal rainforests (Leite, 2002). In an ongoing study of morphometric differentiation among populations of *M. incanus*, Pinheiro et al. (2001) found highly significant differences between populations occupying the two forests types (evergreen and semi-deciduous forests). Within the wet, evergreen forest there were no differences among localities. However, within the semi-deciduous forest, there were significant differences among localities.

The association of species with different habitats is not sufficient to test the EGH. For example, the two major clades of *Trinomys* may have diverged as a result of vicariance induced by the uplift of the coastal mountain ranges with subsequent differentiation within each habitat type. The study of *M. incanus* is consistent with the prediction of the EGH that populations within species should show greater morphological divergence among rather than within habitats, but these observations need to be compared to levels of genetic distance among populations to determine the form of the relationship between genetic and phenotypic distance (Figure 1).

CONCLUSIONS

Studies of diversification (of most taxa, not only mammals) in the Atlantic Forest are in their infancy. In general, our knowledge of the evolutionary processes that generate phenotypic and species diversity in natural systems is remarkably poor given the number of studies that purport to address this important topic. Nearly 150 years after Darwin we still do not know how new species arise. As noted by Bush (1994), processes of speciation in tropical forests are likely to be complex and no single model is likely to be generally applicable. What is needed is a robust hypothesis-testing framework to distinguish among alternative speciation processes (see Patton et al., 2000; Patton and da Silva, 1998; Orr and Smith, 1998; Schluter and Nagel, 1995; Patton et al., 1994; Patton et al., 1990; Patton and Smith, 1992; Cadle and Patton, 1988; Patton, 1986; Endler, 1982b for examples). There are abundant opportunities to increase our understanding of the evolutionary processes involved in speciation in every region of the world and in almost every taxon. Laboratory studies of *Drosophila* have been informative regarding the genetics of speciation (see Rice and Hostert, 1993 and Wu, 2001 for reviews), but studies of laboratory populations can only go so far and tell us very little about how species interact with their evolving landscape during differentiation. We thank Jim Patton for his pioneering work in this area and for pointing the way to a better understanding of both the patterns and processes of speciation in natural populations.

ACKNOWLEDGMENTS

We thank the Shipwreck Committee for organizing the symposium and inviting us to participate. We also thank two anonymous reviewers and Phil Myers for insightful comments that improved this manuscript. We thank the Fundação de Amparo a Pesquisa do Rio de Janeiro (LG), the Brazilian Conselho Nacional de Desenvolvimento Científico e Tecnológico   (LG, ML), and National Science Foundation (DEB 997072 to CJS) for supporting our work on rainforest animal diversification.

LITERATURE CITED

Arbogast, B., and J. Slowinski
   1998   Pleistocene speciation and the mitochondrial DNA clock. Science 282: 1955.

Avila-Pires, T. C. S.
   1995   Lizards of Brazilian Amazonia (Reptilia: Squamata). Zoologische Verhandelingen 299:1-706.

Avise, J. C., and D. Walker
   1998   Pleistocene phylogeographic effects on avian populations and the speciation process. Proceedings Royal Society London, B 265:457-463.

Ayres, J. M., and T. H. Clutton-Brock
1992   River boundaries and species range size in Amazonian primates. American Nauralist 140:531-537.

Balmford, A., G. M. Mace, and J. R. Ginsberg
1999   The challenges to conservation in a changing world: putting processes on the map. Pp. 1-28 in Conservation in a changing world (G. M. Mace, A. Balmford, and J. R. Ginsberg, eds.). Cambridge Univ. Press, Cambridge.

Bonvicino, C. R., and L. Geise
1995   Taxonomic status of *Delomys collinus* Thomas, 1917 (Rodentia, Cricetidae) and description of a new karyotype. Zeitschrift fur Säugetierkunde 60:124-127.

Bonvicino, C. R., A. B. Langguth, P. Herskovitz, and A.C. Paula
1997   An elevational gradient study of small mammals at Caparaó National Park, Southeastern Brazil. Mammalia 61:547-560.

Brasil, Instituto Brasileiro de Geografia e Estatística
1992   Manual Técnico da Vegetação Brasileira. Série Manuais Técnicos Número 1. Departamento de Recursos Naturais e Estudos Ambientais, IBGE, Rio de Janeiro, 92pp.

Brown, K.
1987   Conclusions, synthesis, and alternative hypotheses. Pp. 175-196 in Quaternary History in Tropical America (T. C. Whitmore and G. T. Prance, eds.). Oxford University Press, Oxford.

Bush, M. B.
1994   Amazonian speciation: a necessarily complex model. Journal of Biogeography 21:5-17.

Cadle, J. E., and J. L. Patton
1988   Distribution patterns of some amphibians, reptiles, and mammals of the eastern Andean slope of southern Peru. Pp. 225-243 in Proceedings of a workshop on Neotropical distribution patterns (P. E. Vanzolini and W. R. Heyer, eds). Academia Brasileira de Ciências, Rio de Janeiro.

Câmara, I. D. G.
1991   Plano de ação para a Mata Atlântica. SOS Mata Atlântica, São Paulo.

Capparella, A. P.
    1992   Neotropical avian diversity and riverine barriers. Acta Congressus Internationalis Ornithologici 20:307-316.

Colinvaux, P. A., and P. E. de Oliveira
    2001   Amazon plant diversity and climate through the Cenozoic. Palaeogeography Palaeoclimatology Palaeoecology 166:51-63.

Colinvaux, P. A., P. E. de Oliveira, and M. B. Bush
    2000   Amazonian and neotropical plant communities on glacial time-scales: the failure of the aridity and refuge hypothesis. Quaternary Science Reviews 19:141-169.

Colinvaux, P. A., P. E. De Oliveira, J. E. Moreno, M. C. Miller, and M. B. Bush.
    1996   A long pollen record from lowland Amazonia: Forest and cooling in glacial times. Science 275:85-88.

Conservation International do Brasil, Fundação SOS Mata Atlântica, Fundação Biodiversitas, Instituto de Pesquisas Ecológicas, Secretaria do Meio Ambiente do Estado de São Paulo, and Instituto Estadual de Florestas-MG.
    2000   Avaliação e ações prioritárias para a conservacão da biodiversidade da Mata Atlântica e Campos Sulinos. Brasília: MMA/SBF.

Cracraft, J.
    1994   Species diversity, biogeography, and the evolution of biotas. American Zoologist 34:33-47.

Cracraft, J., and R. O. Prum
    1988   Patterns and processes of diversification: speciation and historical congruence in some neotropical birds. Evolution 42:603-620.

Crandall, K. A., O. R. P. Bininda-Emonds, G. M. Mace, and R. K. Wayne
    2000   Considering evolutionary processes in conservation biology. Trends in Ecology and Evolution 15:290-294.

da Silva, M. N. F., and J. L. Patton
    1993   Amazonian phylogeography: mtDNA sequence variation in arboreal echimyid rodents (Caviomorpha). Molecular Phylogeny and Evolution 2:243-255.

Darwin, C. R.
  1858   The autobiography of Charles Darwin. Collins, London.

Diamond, A. W., and A. C. Hamilton
  1980   The distribution of forest passerine birds and Quarternary climate change in tropical Africa. Journal of Zoology 191:379-402.

Diels, L.
  1908   Pfanzengeographie. Borntraeger, Berlin.

Duellman, W. E.
  1978   The biology of an equatorial herptofauna in Amazonian Ecuador. Miscellaneous Publications of the Museum of Natural History of the University of Kansas 65:1-352.

Endler, J. A.
  1977   Geographic variation, speciation, and clines. Princeton University Press, Princeton, NJ.

  1982a   Pleistocene forest refuges: fact or fancy? Pp. 641-657 in Biological diversification in the tropics (G. T. Prance, ed.). Colombia University Press, New York.

  1982b   Problems in distinguishing historical from ecological factors in biogeography. American Zoologist 22:441-452.

Erwin, T. L.
  1991a   An evolutionary basis for conservation strategies. Science 253:750-752.

  1991b   How many species are there? Revisited. Conservation Biology 5:330-333.

Fjeldså, J.
  1994   Geographical patterns for relict and young birds in Africa and South America and implications for conservation priorities. Biodiversity and Conservation 3:107-226.

Geise, L.
  1995   Os roedores Sigmondontinae do Estado do Rio de Janeiro (Rodentia: Muridae). Sistemática, citogenética, distribuição e variação geográfica. D.Sc., Univ. Fed. Rio de Janeiro.

Geise, L., M. F. Smith, and J. L. Patton
   2001   Diversification in the genus *Akodon* (Rodentia: Sigmodontinae) in southeastern South America: mitochondrial DNA sequence analysis. Journal of Mammalogy 82:92-101.

Haffer, J.
   1969   Speciation in amazonian forest birds. Science 165: 131-137.

   1974   Avian speciation in tropical South America. Publication of the Nuttall Ornithological Club 14.

   1993   Time's cycle and time's arrow in the history of Amazonia. Biogeographica 69:15-45.

Hershkovitz, P.
   1977   Living New World Monkeys (Platyrhini). Vol. 1. University of Chicago Press, Chicago.

Heyer, W. R., and L. R. Maxson
   1983   Relationships, zoogeography, and speciation mechanisms of frogs of the genus Cycloramphus (Amphibia, Leptodactylidae). Arquivos de Zoologia, São Paulo 30:341-373.

Jackson, J. F.
   1978   Differentiation in the genera *Enyalius* and *Strobilurus* (Iguanidae): implications for Pleistocene climatic changes in eastern Brazil. Arquivos de Zoologia, São Paulo 30:1-79.

Klicka, J., and R. M. Zink
   1997   The importance of recent ice ages in speciation: A failed paradigm. Science 277:1666-1669.

   1998   Pleistocene speciation and the mtDNA clock. Science 282:1955.

Lara, M. C., and J. L. Patton
   2000   Evolutionary diversification of spiny rats (genus *Trinomys*, Rodentia: Echimyidae) in the Atlantic Forest of Brazil. Zoological Journal of the Linnean Society 130:661-686.

Lara, M. C., J. L. Patton, and M. N. F. da Silva
    1996   The simultaneous diversification of South American Echimyid rodents (Hystricognathi) based on complete cytochrome b sequences. Molecular Phylogeny and Evolution 5:403-413.

Leite, Y. L. R.
    2002   Evolution and systematics of the Atlantic tree rats, genus *Phyllomys* (Rodentia, Echimyidae). University of California Press.

Mayr, E.
    1963   Animal species and evolution. Belknap Press, Cambridge, Massachusetts.

Mayr, E., and R. J. O'Hara
    1986   The biogeographical evidence supporting the Pleistocene forest refuge hypothesis. Evolution 40:55-67.

McNeely, J. A., K. R. Miller, W. V. Reid, R. A. Mittermeier, and T. B. Werner
    1990   Conserving the World's Biological Diversity. IUCN, Gland, Switzerland.

Miranda, E. E., and C. Mattos
    1992   Brazilian rain forest colonization and biodiversity. Agriculture, Ecosystems & Environment 40:275-296.

Moreau, R. E.
    1966   The bird faunas of Africa and its islands. Academic Press, New York.

Moritz, C., L. Joseph, M. Cunningham, and C. J. Schneider
    1997   Molecular perspectives on historical fragmentation of Australian tropical and subtropical rainforests: implications for conservation. Pp. 442-454 in Tropical Rainforest Remnants: Ecology, Management, and Conservation of Fragmented Communities (W. F. Laurance and R. O. Jr. Bierregaard, eds.). University of Chicago Press, Chicago.

Moritz, C., J. L. Patton, C. J. Schneider, and T. B. Smith
    2001   Tropical rainforest diversity: a molecular view. Annual Review of Ecology and Systematics 31:533-563.

Mustrangi, M. A., and J. L. Patton
  1997   Phylogeography and systematics of the slender mouse opossum *Marmosops* (Marsupialia: Didelphidae). University of California Press.

Myers, N., R. A. Mittermeier, C. G. Mittermeier, G. A. R. Da Fonseca, and J. Kent
  2000   Biodiversity hotspots for conservation priorities. Nature 403:853-858.

Nelson, B. W., C. A. C. Ferreira, M. F. da Silva, and M. L. Kawasaki
  1990   Endemism centers, refugia and botanical collection density in Brazilian Amazonia. Nature 345:714-716.

Orr, M. R., and T. B. Smith
  1998   Ecology and speciation. Trends in Ecology and Evolution 13:502-506.

Patton, J. L .
  1986   Patrones de distribucion y especiacion de fauna de mamiferos de los bosques nublados andinos del Peru. Anales del Museo de Historia Natural Valparaíso 17:87-94.

Patton, J. L., and L. P. Costa
  2003   Molecular phylogeography, species boundaries, and the biogeography of rainforest marsupial genera (Didelphimorphia: Didelphidae). Pp. 63-81 in Carnivores with Pouches. (M. Jones, C. Dickman, and M. Archer, eds.). CSIRO Press.

Patton, J. L., and M. N. F. da Silva
  1998   Rivers, refuges, and ridges: the geography of speciation of Amazonian mammals. Pp. 202-213 in Endless forms: models and mechanisms of speciation (D. Howard and S. Berlocher, eds.). Oxford University Press.

Patton, J. L., M. N. F. da Silva, M. C. Lara, and M. A. Mustrangi
  1997   Diversity, differentiation, and the historical biogeography of non-volant small mammals of the neotropical forests. Pp. 455-465 in Tropical Rainforest Remnants: Ecology, Management, and Conservation of Fragmented Communities (W. F. Laurance and R. O. Jr. Bierregaard, eds.). University of Chicago Press, Chicago.

Patton, J. L., M. N. F. da Silva, and J. R. Malcolm
  1994   Gene genealogy and differentiation among arboreal spiny rats (Rodentia, Echimyidae) of the Amazon Basin: a test of the riverine barrier hypothesis. Evolution 48:1314-1323.

Patton, J. L., M. N. F. da Silva, and J. R. Malcolm
    2000   Mammals of the Rio Juruá and the evolutionary and ecological diversification of Amazonia. Bulletin of the American Museum of Natural History 244:1-306.

Patton, J. L., P. Myers, and M. Smith
    1990   Vicariant versus gradient models of diversification; the small mammal fauna of eastern Andean slopes of Peru. Pp. 355-371 in Vertebrates in the tropics (G. Peters and R. Hutterer, eds.). Museum Alexander Koening.

Patton, J. L., and M. F. Smith
    1992   MtDNA phylogeny of Andean mice: a test of diversification across ecological gradients. Evolution 46:174-183.

Peres, C. A., J. L. Patton, and M. N. F. da Silva
    1996   Riverine barriers and gene flow in Amazonian saddle-back tamarins. Folia Primatologica 67:113-124.

Pinheiro, P. S., L. Geise, and C. E. V. Grelle
    2001   Differenciação morfométrica de *Marmosops incanus* (Didelphimorphia, Didelphidae) no sudeste do Brasil: um teste da hipótese dos gradientes ecológicos. I Congresso Brasileiro de Mastozoologia. resumo DP-14: 73.

Prance, G. T.
    1982   Biological Diversification in the Tropics. Columbia Univ. Press, New York.

Rice, W. R., and E. E. Hostert
    1993   Laboratory experiments on speciation: what have we learned in 40 years. Evolution 47:1637-1653.

Ron, S. R.
    2000   Biogeographic area relationships of lowland Neotropical rainforest based on raw distributions of vertebrate groups. Biological Journal of the Linnean Society 71:379-402.

Roy, M. S.
    1997   Recent diversification in African greenbuls (Pycnonotidae: *Andropadus*) supports a montane speciation model. Proceedingsof the Royal Society, London, B 264:1337-1344.

Rylands, A. B., R. A. Mittermeier, and E. Rodriguez-Luna
    1997   Conservation of neotropical primates: Threatened species and an analysis of primate diversity by country and region.   Folia Primatologica 68:161-180.

Schluter, D., and L. M. Nagel
    1995   Parallel speciation by natural selection. The American Naturalist 146: 292-301.

Schneider, C. J., M. Cunningham, and C. Moritz
    1998   Comparative phylogeography and the history of vertebrates endemic to the Wet Tropics rainforest of Australia. Molecular Ecology 7:487-498.

Schneider, C. J., and C. Moritz
    1999   Rainforest refugia and evolution in Australia's Wet Tropics. Proceedings of the Royal Society London B 266:191-196.

Schneider, C. J., T. B. Smith, B. Larison, and C. Moritz
    1999   A test of alternative models of diversification in tropical rainforests: Ecological gradients vs. rainforest refugia. Proceedings of the National Academy of Sciences USA 96:13869-13873.

Smith, T. B., M. W. Bruford, and R. K. Wayne
    1993   The preservation of process: the missing element of conservation programs. Biodiversity Letters 1:164-167.

Smith, T. B., R. K. Wayne, D. J. Girman, and M. W. Bruford
    1997   A role for ecotones in generating rainforest biodiversity. Science 276: 1855-1857.

Stebbins, G. L.
    1974   Flowering plants, evolution above the species level.   Harvard University Press.

Tuomisto, H., K. Ruokolainen, R. Kalliola, A. Linna, W. Danjoy, and Z. Rodriguez
    1995   Dissecting Amazonian biodiversity. Science 269:63-66.

Vanzolini, P. E.
   1988   Distributional patterns of South American lizards. Pp. 317-342 in Proceedings of a workshop on Neotropical distribution patterns (P. E. Vanzolini and W. R. Heyer, eds). Academia Brasileira de Ciências, Rio de Janeiro.

Vanzolini, P. E., and E. E. Williams
   1970   South American anoles: the geographic differentiation and evolution of the *Anolis chrysolepis* species group (Sauria: Iguanidae). Arquivos de Zoologia (São Paulo) 19:1-298.

   1981   The vanishing refuge: a mechanism for ecogeographic speciation. Papéis Avulsos de Zoologia, São Paulo 34:251-255.

Wallace, A. R.
   1852   On the monkeys of the Amazon. Proceedings of the Zoological Society London 20:107-110.

Whitmore, T. C., and G. T. Prance
   1987   Biogeography and Quaternary history in tropical America. Oxford University Press, New York.

Willis, J. C.
   1922   Age and Area. Cambridge University Press, Cambridge.

Wu, C-I.
   2001   The genic view of the process of speciation. Journal of Evolutionay Biology 14:851-865.

Zink, R. M.
   1997   Phylogeographic studies of North American birds. Pp. 301-324 in Avian Molecular Evolution and Systematics (D. P. Mindell, ed.). Academic Press, San Diego.

Small Mammal Communities in Upland and Floodplain Forests Along an Amazonian White Water River

Jay R. Malcolm, James L. Patton, and Maria Nazareth F. da Silva

Rivers are thought to exert an important influence on patterns of biological diversity in the Amazon. Unlike the Congo Basin, which straddles the equator, much of the Amazon Basin is in the southern hemisphere, which leads to synchronous rainy seasons throughout much of the Basin and strong annual and multi-annual flooding (Junk, 1984; Richey et al., 1989; Queiroz, 1995). Coupled with channel migration (Räsänen et al., 1992; Kalliola et al., 1992), this flooding generates a rich and dynamic mosaic of disturbances, nutrient regimes, and successional stages (Junk, 1984; Salo et al., 1986; Colinvaux, 1987; Furch and Klinge, 1989; Lamotte, 1990; Terborgh and Petren, 1991; Kalliola et al., 1991 and 1992; Worbes et al., 1992; Puhakka et al., 1992; see also Hughes, 1990). Extensive seasonally-flooded forests are common, including the *várzea* and *igapó* forests of the white- and black-water rivers (*sensu* Prance, 1979). These flooded forests cover large areas of the Amazon Basin. For example, between the channels of the Amazon (Solimões) and Japurá River in the western Brazilian Amazon, they are over 75 km in width. In eastern Peru far from the mouth of the Amazon, active floodplains comprised some 12% of 0.5 million km^2 surveyed using Landsat images (Salo et al., 1986). Presumably, the different ecological conditions of the floodplain and upland forests influence local species diversity and, because of possibilities for habitat differentiation, may lead to beta diversification (Erwin and Adis, 1982; Gentry, 1988). In the few comparisons undertaken between floodplain and adjacent non-flooded forests, striking differences in terrestrial species richness and community composition were observed. A high degree of habitat specificity among both plants and animals (Erwin and Adis, 1982; Remsen and Parker, 1983; Gentry, 1988; Dumont et al., 1990; Martius, 1994) and numerous adaptations to flooding (Gentry, 1983; Junk, 1984; Kubitzki and Ziburski, 1994) suggest that seasonal inundation has long been an important factor in the evolutionary history of the Amazon Basin.

Amazonian rivers also are thought to play fundamental roles in generating patterns of regional (gamma) diversity. Perhaps best-known is the "riverine barrier hypothesis" (Wallace, 1852; Gascon et al., 2000). Many bird and primate distributions are bounded by rivers and measures of across-river community and genetic divergence for several groups have been observed to vary with channel width, channel velocity, and the width of the meander belt (Hershkovitz, 1977,

Erwin and Adis, 1982; Beven et al., 1984; Capparella, 1987, 1988, and 1992; Haffer, 1993; Ayres and Clutton-Brock, 1992; Peres et al., 1996). Several other biogeographic roles of rivers have yet to be tested. Recent evidence suggests that much of western Amazonia, including upland areas far from contemporary river channels, has been repeatedly influenced by riverine and lacustrine processes (Campbell and Frailey, 1984; Colinvaux et al., 1985; Salo et al., 1986; Räsänen et al., 1987 and 1992; Salo and Kalliola, 1991; but see Dumont et al., 1990), raising the possibility that repeated formation and abandonment of floodplains due to tectonic events may lead to secondary contact between formerly separated river systems and, hence, to dynamics similar to the refuge pattern (Salo et al., 1986; Salo and Kalliola, 1991; see also Erwin and Adis, 1982). Dissection of the floodplain by meanders may lead to greater genetic differentiation within floodplain forests than in adjoining upland forests (see Terborgh et al., 1996). Alternately, just as gallery forests currently serve as corridors that allow rainforest taxa to inhabit xeric regions (Cerqueira, 1982; Redford and Fonseca, 1986), gene flow over time may have been higher in floodplain forests than in upland forests, especially during historical drier periods when upland forests were perhaps less widely distributed and contiguous (Haffer, 1969 and 1993).

Despite the potential influence of Amazonian rivers on patterns of alpha, beta, and gamma diversity, few field studies have systematically compared communities among riverine habitats, or have compared patterns of species richness and turnover along the lengths of Amazonian rivers. Studies of floodplain and upland mammal communities in the Amazon have compared only a single pair of sites (Ayres, 1986; Bodmer, 1990; Fleck and Harder, 1995; Woodman et al., 1995), with the exception of Peres (1997). Moreover, comprehensive distributional information that might be used to test for river effects is lacking (Gentry, 1988; Robinson and Terborgh, 1990; Tuomisto et al., 1995; Pimm and Gittleman, 1992). Even for primates, which are the best-known Neotropical mammal group (Patton et al., 1997), species distributions are poorly understood in the upper reaches of the Amazon Basin (Hershkovitz, 1977; Emmons and Feer, 1990).

In this paper, we report on a geographically replicated survey of non-volant small mammal communities in floodplain and upland forest along a large, white-water tributary in the western part of the Amazon Basin, the Juruá River. Our previous work focused on the riverine barrier hypothesis (Patton et al., 1994, 1996 and 2000; Gascon et al., 2000) and on systematics (da Silva and Patton, 1993; da Silva, 1995 and 1998; Patton et al., 2000). Herein, we expand on preliminary analyses in Patton et al. (2000) and focus on differences in community structure between upland and floodplain forests. We address four objectives. First, we compare small mammal community composition, abundance, and richness between upland and floodplain forests. Second, we examine correlations between small mammal community structure and various measurements of forest physiognomy. Third, we compare regional patterns of species richness and composition between upland and floodplain forests. We were particularly interested in testing whether gamma diversity was higher in floodplain than in upland forests, indicating a possible role of *in situ* barriers or frequent vicariant events in the floodplain. We also examined the possibility that certain stretches of the river might exhibit strong gradients of turnover in species composition. In its middle reaches, the Juruá River crosses the former location of the Jutaí (Iquitos) Arch, one of a series of ancient ridges that have been posited to play important biogeographic roles in the Amazon (Räsänen et al., 1987 and 1992). Our genetic analyses suggest that the Arch may have served as a barrier separating small mammal populations in the former upstream and downstream subsidence basins (Patton et al., 1994, 1996 and 2000). Fourth, our use of both terrestrial and canopy trapping allowed us to examine patterns of species composition, richness, and turnover in the two strata.

MATERIALS AND METHODS

Study Sites and Sampling Design

We sampled four regions spanning ca 960 km of the Rio Juruá over the course of one year, including upland (*terra firme*) and floodplain (*várzea*) forests on both sides of the river (Figure 1; see Patton et al., (2000) for detailed maps of sampling locations). Sixteen sites were sampled in total (four regions times two forest types times two river banks). Each site was sampled once. The floodplain varied in width from approximately 6 km in the headwaters region to nearly 25 km near the river's mouth (RADAM 1:250,000 maps). Two consecutive, 2-3 week long trapping sessions were conducted in each region. In each session, we simultaneously sampled both the terrestrial and canopy faunas at an upland and floodplain site (Table 1).

Upland sites were in the relatively hilly, dissected terrain characteristic of upland forest. They were chosen to be easily accessible from the river and, as a result, were close to places where the river channel reached the edge of the floodplain. Usually, the upland forest that we sampled abutted floodplain forest, but at one site (the headwaters site on the right [facing downstream] bank), an upland low-lying bamboo forest (*tabôca*) that was ca 2 km wide separated the two. Within upland forest we avoided swamps and secondary forests and all traps were set at least 150 m from the floodplain.

In the floodplain, we also sampled mature primary forest. Based on descriptions of floodplain forests along the Rio Manu (Foster, 1990; Terborgh and Petren, 1991; see also Worbes et al., 1992), the forests that we sampled were probably more than 300 years old. We avoided setting traps close to young successional forests (such as *Cecropia* stands), areas that flooded deeply (> 3 m), vine (liana) forests (Bodmer, 1990), or second growth forests. Traps were, in most cases, at least 100 m from early successional habitats along the channel margin, and, except at sites close to the river mouth (see below), were at least 150 m from upland forest. We avoided the inner portions of meander loops in case these features create a "peninsular effect" on small mammal communities.

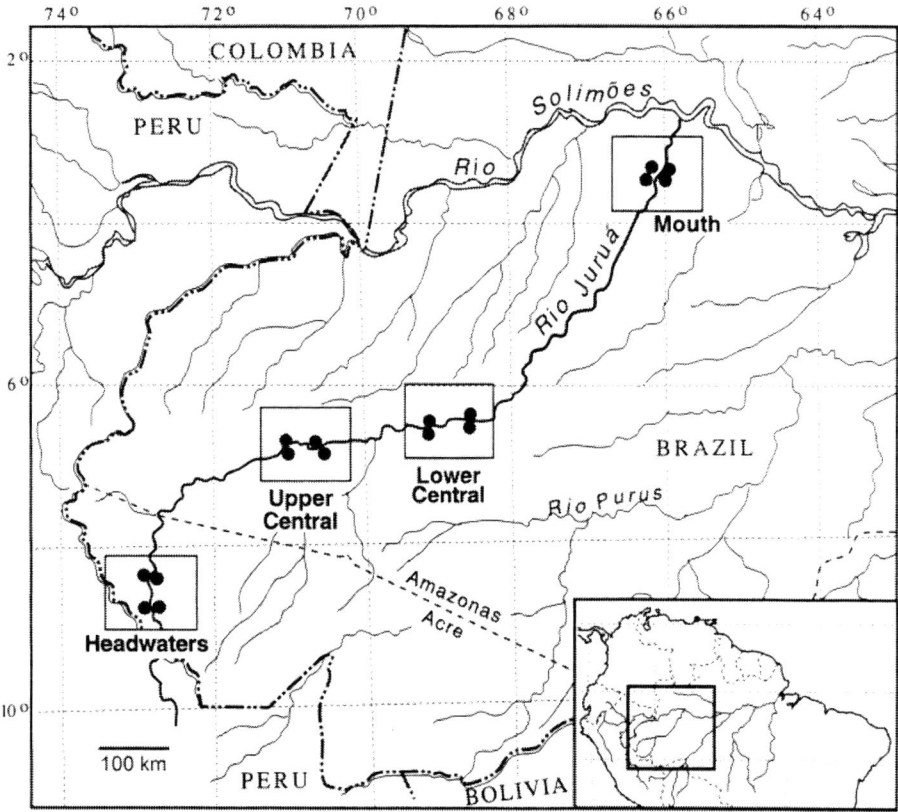

Figure 1. Study sites in four regions along the Juruá River in western Brazil sampled for non-volant small mammals with terrestrial and canopy live-trapping. Modified from Patton et al. (2000).

Table 1. Number of traplines and the dates that traps were set in four regions along the Juruá River in western Brazil. A trapline comprised 15 trap stations (one Sherman and one Tomahawk trap) spaced at 20-m intervals. Traps on terrestrial traplines were activated for 7 nights (105 station nights); those on arboreal traplines for 12 nights (180 station nights).

Region[1]	Trap session	Dates traps first set	Upland forest Right[2] bank Terrestrial	Canopy	Left Bank Terrestrial	Canopy	Floodplain forest Right bank Terrestrial	Canopy	Left Bank Terrestrial	Canopy
Headwaters	1	16 Feb. 92	5	2	0	0	0	0	5	2
	2	7 Mar. 92	0	0	5	3	4	3	1	0
Upper	3	23 Aug. 91	5	3	0	0	0	0	5	3
Central	4	18 Sep. 91	0	0	5	3	5	3	0	0
Lower	5	16 Oct. 91	0	0	5	3	3	2	2	2
Central	6	9 Nov. 91	5	3	0	0	2	2	3	2
Mouth	7	4 May 92	5	3	0	0	5	3	0	0
	8	27 May 92	0	0	5	3	0	0	5	3

[1] see Figure 1 for locations. [2] Facing downstream.

Because trapping in the central regions of the river was conducted during the low-water season, we were able to sample areas that flooded annually. As determined from water marks on trees, all trap locations in the lower central region had flooded during the previous high-water season (mean flood depth = 1.4 m, range 0.3 to 3.0 m, n = 120 trap stations). Similarly, all trap stations on the right bank in the upper central region had flooded the previous season (mean flood depth = 1.1 m, range 0.5 to 1.8 m; n = 45 trap stations). However, part of the left bank site in the upper central region was on slightly higher floodplain forest that had not flooded the previous season [17 of 45 trap stations]). Mean depth at the remaining stations at that site was only 0.4 m (range 0.1-1.1 m).

Because of a smaller catchment area, year-to-year variation in flood depth was more pronounced in the headwaters region than further downstream and was heavily influenced by local rainfall. As at Cocha Cashu (Foster, 1990; Terborgh and Petren, 1991), water levels could change by several meters overnight and large areas of the floodplain did not flood in all years. According to local residents, our floodplain sites in the headwaters region flooded approximately every five years.

Because of logistical constraints, we could visit the river's mouth only during peak flood. As a result, traplines were placed either in areas of the floodplain that flooded only occasionally and were dry at the time (*restinga*), or in areas of upland forest immediately adjacent to the inundated floodplain. On the right bank, two of five terrestrial traplines (and two of three arboreal traplines) were on an approximately 3-ha *restinga* island that apparently flooded only in exceptional years. During the trapping session, rising floodwaters eventually reduced the island to ca 2 ha. The nearest upland forest was approximately 100 m distant from the island. The rest of the traps on the right bank and all of the traps on the left bank were placed along the edge of the upland forest within 10 m of the flooded forest.

Small Mammal Trapping

At upland sites, we cut a central trail perpendicular to the floodplain and established five or six 280-m long perpendicular trails spaced at 150-m intervals. At floodplain sites, five or six 280-m long trails also were established; in most cases they were also parallel to each other and spaced at 150-m intervals. When we trapped in the floodplain on both sides of the river during a session (see Table 1), we established two or three parallel trails per side.

Traplines were established along the trails and consisted of fifteen trap stations spaced at 20-m intervals. We report here on captures from these standardized lines only; captures from non-systematic trapping are reported in Patton et al. (2000). Terrestrial trap stations consisted of a 14 x 14 x 40 cm wire mesh (Tomahawk) trap and a 8 x 8 x 23 cm folding aluminum (Sherman) trap placed 2 - 4 m apart. Two trap sizes were used to lessen biases to particular size classes of mammals. Canopy stations consisted of a Sherman trap fastened on top of a Tomahawk, which in turn was fastened to a wood frame (Malcolm, 1991a). Pulleys, attached to wood or plastic (PVC) frames that were transported between sites, were used to raise and lower traps. Canopy trap height measured with a rangefinder to the nearest 0.1 m averaged 11.9 m in upland forest ($N = 345$, SE = 0.13, range = 6.6-19.2 m) and 10.3 m in floodplain forest ($N = 375$, SE = 0.13, range = 2.5-20.7 m). Because canopy trapping was laborious, canopy stations were set on only 2-4 of the five or six parallel trails established at a site (Table 1).

For bait, we used ripe plantains and a mixture of ground peanuts, vegetable oil, and honey. Terrestrial traps were rebaited daily but, because this was prohibitively time consuming for canopy traps, the latter were rebaited only when an animal was captured. To compensate, canopy traps also were baited with a long-term bait source not easily removed by insects (a small cloth sack containing raisins and the ground peanut mixture).

Terrestrial traps were left open for seven and canopy traps for 12 consecutive nights (traps were sometimes left open longer, but we exclude those captures here). We left canopy traps open longer than terrestrial traps to help compensate for the lower effort and capture success in the canopy (Figure 2).

Individuals were identified to species based on external, cranial, karyotypic, and, in a few cases, mitochondrial DNA features (Patton et al., 2000). Specimens are deposited at the Instituto Nacional de Pesquisa da Amazônia and the Museum of Vertebrate Zoology, University of California. Taxonomy follows Patton et al. (2000).

Habitat Variables

Variation in small mammal abundance and diversity in the tropics is strongly influenced by the three-dimensional structure of the forest (e.g., Malcolm, 1995; Malcolm and Ray, 2000). Therefore, we measured vertical stratification of foliage, understory density, and the amount of downed woody debris along a subset of the traplines (53 terrestrial traplines and 47 arboreal traplines). At 5-m intervals along the trapline, we used a 2.5-m pole to make a vertical sighting into the canopy ($N = 57$ sightings per trapline). To avoid vegetation damage along the trail itself, the sightings were offset 5 m perpendicular to the trapline (always on the same side of the trapline). Using a rangefinder, along the sighting we estimated the proportion of vegetative cover in six height intervals: 0-2, 2-5, 5-10, 10-20, 20-30, and 30-40 m. The first three intervals were scored as: 0 (<0.25 coverage), 1 (0.25-<0.50 coverage), 2 (0.50-<0.75 coverage), or 3 (>0.75 coverage). The last three were scored as 0 (<0.10 coverage), 1 (0.10-<0.50 coverage), 2 (0.50-<0.75 coverage), or 3 (>0.75 coverage). Non-zero scores were transformed to "foliage thickness" by multiplying the mean proportion corresponding to a score by the number of meters in the interval (for example, a score of 1 [0.25-<0.50] in the second height interval [2-5 m] would be

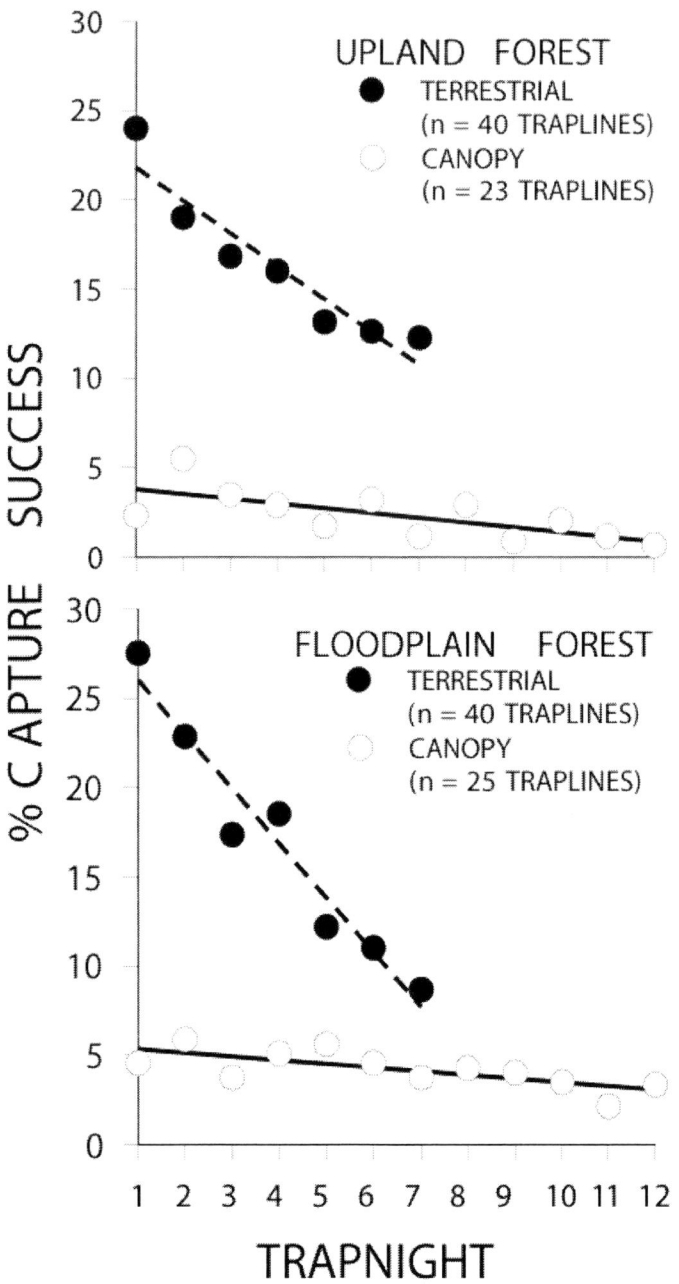

Figure 2. Nightly capture success (number of individuals per 100 station nights) averaged across traplines. A trapline comprised 15 trap stations set at 20-m intervals. Stations were either on the ground (terrestrial traplines) or at an average height of approximately 11 m (canopy traplines).

recorded as 0.375 x 3 m = 1.125 m). Following Malcolm (1995), these data were used to derive statistically uncorrelated measures of the amount, variability, and spatial grain of foliage in the lower story (0-10 m) and the overstory (10-40 m) for each transect. Understory density was quantified by use of a 2.5-m pole with 10-cm wide banding. At each trap station, the pole was positioned 20 m into the forest, perpendicular to the trail, and the number of bands that were at least partly obscured by vegetation was counted. Again, these data were used to derive statistically uncorrelated measurements of the amount, variability, and spatial grain of understory foliage. Finally, the amount of downed woody debris was quantified at the same locations that had vertical sightings into the canopy. Along two 5-m long perpendicular transects centered at the location (one parallel to and one perpendicular to the trail), the number of 10-cm intervals that were completely covered by downed wood was counted.

Data Analyses

Multivariate Patterns of Species Composition. To examine variation in community structure, we calculated terrestrial and canopy percent capture success (number of individuals per 100 station nights) for each species for each of the 16 sites. Detrended Correspondence Analysis (DCA; Jongman et al., 1995) was conducted on the resulting 32-row matrix (16 rows each for the terrestrial and canopy data). We focused on relatively common taxa by excluding species that did not occur in at least five of the 32 samples. Because of the strong contrast between the terrestrial and canopy faunas, we also conducted DCAs separately on data from the terrestrial and canopy traplines. These separate DCAs included all species, but included only those traplines with habitat measurements.

Relationships between small mammal community structure and habitat structure were investigated by use of Canonical Correspondence Analysis (CCA; Jongman et al., 1995). In the CCAs, Monte Carlo forward selection permutation tests were performed to determine the statistical significance of the habitat variables (reduced-model, unrestricted permutations repeated 9,999 times). All multivariate tests were conducted using CANOCO (v. 4.0). Other statistical tests were performed using SAS (v. 8.01). Parametric tests were used only where the assumptions of normality and equality of variances were reasonable (as judged by examination of residuals).

Comparisons of Species Abundances by Trap Stratum, Forest Type, Region, and River Bank. To compare differences in species abundances between terrestrial and canopy traps, we used paired t-tests on percent capture successes in the two strata across the 16 sites. When a species was absent at a site (i.e., caught neither on the ground nor in the canopy), the site was excluded from the test. Species were tested only if they were present at four or more sites.

We used three-way ANOVAs to test for forest type, region, and river bank effects across the 16 sites. Rather than conducting the tests separately on terrestrial and canopy success, the two were combined (by summing terrestrial and canopy percent trap success). When a species was absent from a region, the region was excluded from the test. The test was conducted on a species only if it was present in at least two regions. In a first set of tests, all possible two-way interactions were included in the model; in a second set, we dropped interaction terms that were not close to significant ($p > 0.15$).

Because of the relatively small number of sites that we sampled, we used a relatively liberal statistical procedure and did not correct probabilities in the above tests for the relatively large number of species tested.

Alpha (Site-specific) Diversity. For each trapline, we calculated total small mammal abundance and characterized diversity by computing three of Hill's "diversity numbers" (Ludwig and Reynolds, 1988):

$$N_0 = S,$$

where S is the total number of species,

$$N_1 = e^{H'},$$

where H' is Shannon's index (i.e., $-\Sigma p_i \ln p_i$), and

$$N_2 = 1/\lambda$$

where λ is Simpson's index (i.e., $1 - \Sigma p_i^2$). The last two differ in their tendency to include or ignore rarer species in a sample: N_2 is a measure of the number of "very abundant" species, while N_1, which is intermediate between N_2 and N_0, is a measure of the number of "abundant" species in the sample. As dominance in a community increases, both of these measures will tend toward a value of one (Ludwig and Reynolds, 1988). For each of the 16 sites, we calculated trapline means for each of these measures and conducted three-way ANOVAs (with all possible two-way interactions) on the log-transformed means. Examination of residuals indicated that the assumptions of normality and homogeneity appeared to be met.

Because species richness will increase asymptotically with the number of individuals captured, we further investigated species richness by statistically controlling for the number of individuals captured. We did this in two ways. First, we plotted stratum-specific mean trapline species richness against mean trapline abundance and used analysis of covariance on the log-log transformed data to test for forest type, region, and river bank effects. In these plots, variation in elevations (intercepts) of the richness-abundance regressions indicated variation in species richness independent of the number of individuals captured. By using a log-log transformation, we implicitly fitted a power function to the untransformed data, which, over the range of abundances that we recorded, captured the curvilinear relationship between species richness and abundance well. Second, we used extrapolation to estimate the asymptotic number of species for each of the 16 sites (Colwell and Coddington, 1994) based on the relationship between cumulative number of species and trap night. The Michaelis-Menten equation provided a smaller residual sums-of-squares on average than the negative exponential equation, and, hence, it was used to estimate species richness. Four canopy sites and one terrestrial site were discarded because the non-linear regressions either failed to converge or because residual sums of squares were inordinately large. Differences among asymptotes were tested via ANOVA.

Gamma (Among-site) Diversity. Small mammal richness in the two forest types at increasingly large spatial scales was compared by plotting species richness against abundance for traplines, sites, regions, pairs of regions, triplets of regions, and the total sample.

To compare terrestrial and arboreal small mammal species turnover along the course of the river within upland and floodplain forest, we used Jaccard's community similarity index:

$$JI = \frac{a}{a+b}$$

where a is the number of species found at both localities and b is the number of species found at only one locality (Ludwig and Reynolds, 1988). Similarity was calculated between all possible pairs of same-forest-type sites ($n = 28$, where a site was a unique combination of river bank, forest type, and region). Relationships between community similarity and inter-site distances were tested using Mantel tests with 9,999 permutations (Manly, 1991).

RESULTS

A total of 8,355 station nights on the ground and 8,546 station nights in the canopy yielded 1,661 individuals of 40 species from the families Didelphidae, Muridae, and Echimyidae (see Appendix 1 for capture success summarized according to forest type, region, and river bank). Ten species were new to science, illustrating the poorly known state of Amazonia's mammals (to date, nine have been described [da Silva and Patton, 1993; da Silva, 1995 and 1998; Patton et al., 2000]). Species (number of individuals) of Didelphidae captured were *Micoureus regina* (51), *Micoureus demerarae* (48), *Metachirus nudicaudatus* (33), *Didelphis marsupialis* (24), *Caluromys lanatus* (16), *Marmosops noctivagus* (14), *Philander mcilhennyi* (13),

Marmosops neblina (4), *Philander opossum* (3), *Marmosops impavidus* (3), *Monodelphis emiliae* (3), and *Marmosa murina* (1). Species of Muridae captured were *Oryzomys perenensis* (300), *Oryzomys yunganus* (90), *Oecomys roberti* (38), *Oecomys bicolor* (32), *Neacomys minutus* (23), *Neacomys spinosus* (21), *Scolomys juruaense* (17), *Oryzomys macconnelli* (12), *Rhipidomys leucodactylus* (6), *Oecomys trinitatis* (4), *Oecomys* sp. (2), *Oryzomys nitidus* (2), *Nectomys apicalis* (1), *Neacomys musseri* (1), *Oligoryzomys microtis* (1), and *Rhipidomys gardneri* (1). Species of Echimyidae captured were *Proechimys simonsi* (308), *P. steerei* (307), *Mesomys hispidus* (84), *Proechimys brevicauda* (51), *P. echinothrix* (46), *P. cuvieri* (26), *P. kulinae* (20), *P. pattoni* (18), *Isothrix bistriata* (17), *Proechimys gardneri* (14), *Makalata macrura* (3), and *Mesomys occultus* (3). Additional trapping that was either non-standardized or in other habitats yielded three more species (*Holochilus sciureus, Marmosops parvidens, Oecomys superans*; see Patton et al., 2000).

Multivariate Patterns of Variation

Of the 40 small mammal species captured, 17 were present in five or more of the 32 samples. The first two Detrended Correspondence axes accounted for 29.1% and 13.4% of the total variance (gradient lengths were 8.3 and 3.0 S.D., respectively). Because the third and fourth axes explained markedly less variance (5.8% and 2.4% of the total variance, respectively), we plotted only the first two axes (Figure 3). The first axis separated terrestrial from canopy samples. Slightly more than half of the 17 species were terrestrial (10 species scores were in the left half of the plot, seven were in the right half). The second axis separated sites according to forest type. Most terrestrial species were more abundant in upland forest than in floodplain forest, as evidenced by the fact that more species scores were in the lower left quadrant of the plot than in the upper left. The opposite appeared true of arboreal species, in that more species scores were in the upper right quadrant than in the lower right. The second axis also separated floodplain sites according to their position along the river, although patterns differed among the terrestrial and canopy samples. For terrestrial samples, floodplain headwaters sites showed similarities with upland sites, while mouth sites showed similarities with central floodplain sites. Conversely, for canopy samples, headwaters sites tended to be similar to central floodplain sites and mouth sites tended to be similar to upland sites. Separation according to river bank was not evident on either axis.

A DCA on terrestrial traplines alone showed strong separation between upland and floodplain sites along the first axis (16.5% of the total variance; gradient length of 4.2 S.D.; Figure 4a, b). Again, greater species richness was indicated for upland than for floodplain forest, as was variation among floodplain sites from the headwaters, mouth, and central regions. Despite the fact that sites in the headwaters floodplain were far from the nearest upland forest (>1 km), they approached upland sites in community composition. Community composition

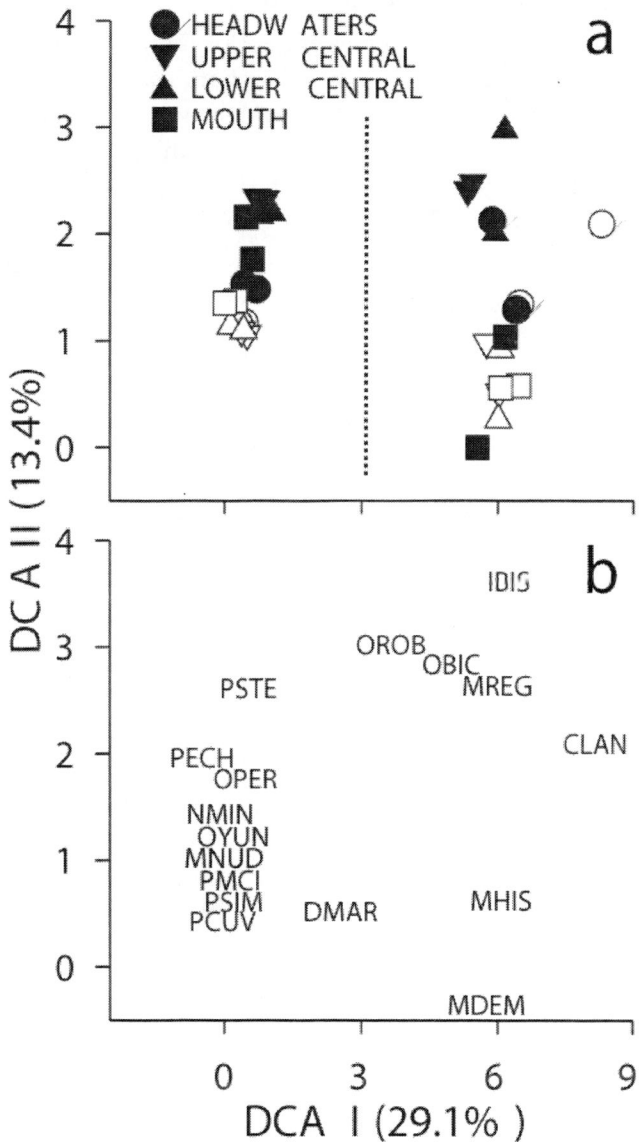

Figure 3. The first two axes from a detrended correspondence analysis on 17 relatively widespread small mammal species caught along the Juruá River (closed symbols = floodplain forest; open symbols = upland forest). In panel (a), site scores to the left of the dotted line are from terrestrial trapping; those to the right are from canopy trapping. Species scores in panel (b) are labeled with the first letter of the genus and the first three letters of the specific epithet.

varied greatly among floodplain traplines near the river's mouth. The two traplines in low-lying *restinga* island yielded communities typical of the central floodplain, while samples from the edge of upland forest showed similarities to those of upland forest. Interestingly, support for separation of upland communities above and below the Jutaí Arch was eivdent in a plot of the second and third DCA axes (9.6% and 7.6% of the total variance, respectively; gradient lengths of 3.4 and 2.7 S.D.; Figure 4c, d). Strong separation was shown between traplines in the river's upper half (headwaters and upper central regions) and those in the lower half (lower central and mouth regions). The means for the headwaters and upper central sites were more or less superimposed, as were those of the mouth and lower central sites, suggesting differentiation associated with the Arch rather than geographic isolation by distance. Floodplain sites showed little evidence of separation according to geographic position.

In the canopy community, separation according to forest type also was shown on the first DCA axis (18.4% of the total variance; gradient length 5.3 S.D.; Figure 5). Again, headwater sites tended to be intermediate between upland and central floodplain sites, but the mouth traplines (including those on the *restinga* island) tended to have communities similar to those of upland forest. Unlike the terrestrial fauna, there was little indication of differences in species richness along the first axis, with about equal numbers of species to the right and left of the origin. Neither the second nor third axes (not shown) showed patterns that were readily interpretable (10.3% and 4.1% of the variance, respectively, with gradient lengths of 3.1 and 2.9 S.D.).

Ordinations constrained by forest structure variables showed similar patterns for both the terrestrial and canopy communities (Figure 6). Differences between central floodplain and upland sites were found along the first axis and were significantly correlated with variation in forest structure. These axes accounted for 8.8 % (terrestrial) and 13.5% (canopy) of the total species variance. In both cases, the first four structural variables chosen in the forward selection were the variance and mean of overstory thickness, mean understory density, and abundance of downed wood. All were either significant or nearly so in the Monte Carlo tests (respective p values for terrestrial samples were < 0.01, 0.02, < 0.01, and 0.06; respective p values for canopy samples were 0.01, 0.06, 0.10, 0.07). In neither ordination was there evidence of differences among traplines from the floodplain headwaters, floodplain mouth, and upland forest.

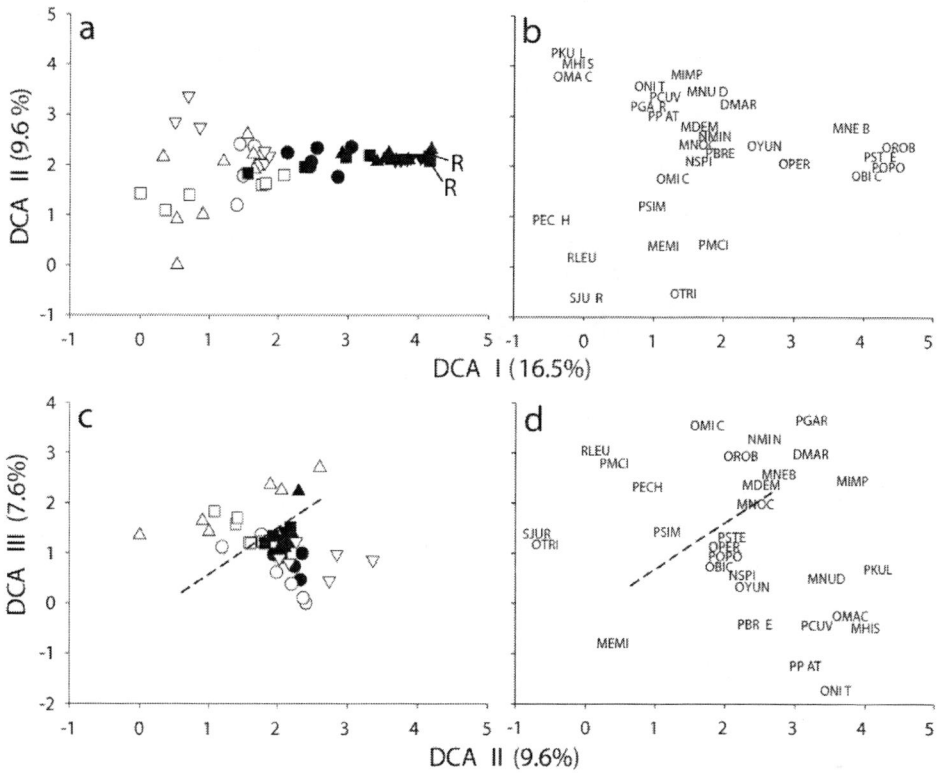

Figure 4. Detrended correspondence analysis of small mammal abundances from terrestrial traplines that were measured for habitat variables. Symbols are as in Figure 3. Axes one and two are shown in panels (a) and (b); axes two and three are shown in panels (c) and (d). Species scores shown in the right-hand-most plots are labeled with the first letter of the genus and the first three letters of the specific epithet. The dashed line in (c) and (d) highlights the dichotomy between upland traplines above and below the Jutaí Arch. In (a), the two right-hand-most mouth floodplain traplines (labeled "R") were from a floodplain island (*restinga*).

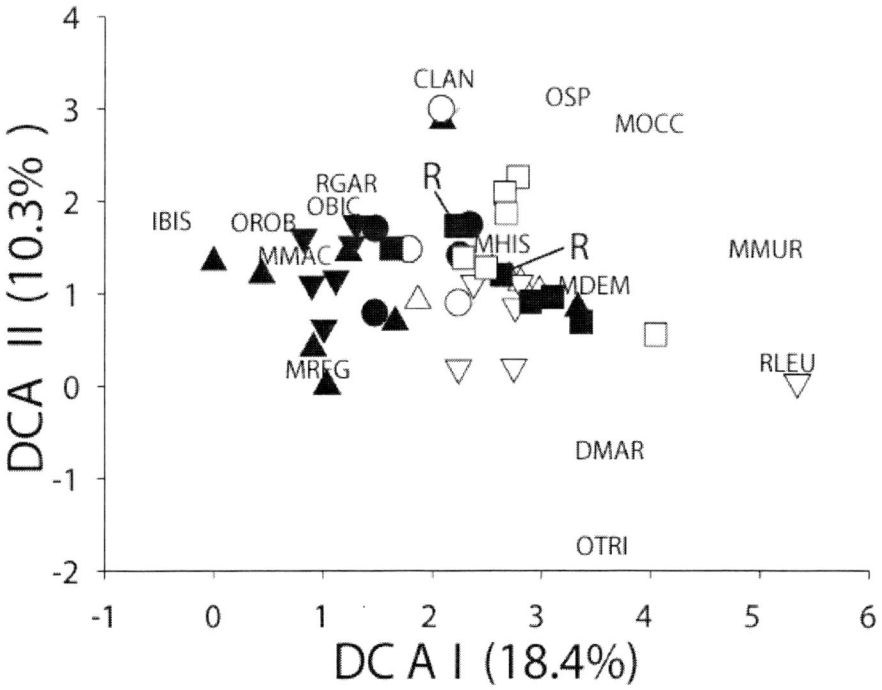

Figure 5. Detrended correspondence analysis of small mammal abundances from canopy traplines that were measured for habitat variables. Sites and species scores are shown on the same plot. Symbols are as in Figure 3. The mouth floodplain traplines that were second and third from the left (labeled "R") were from a floodplain island (*restinga*).

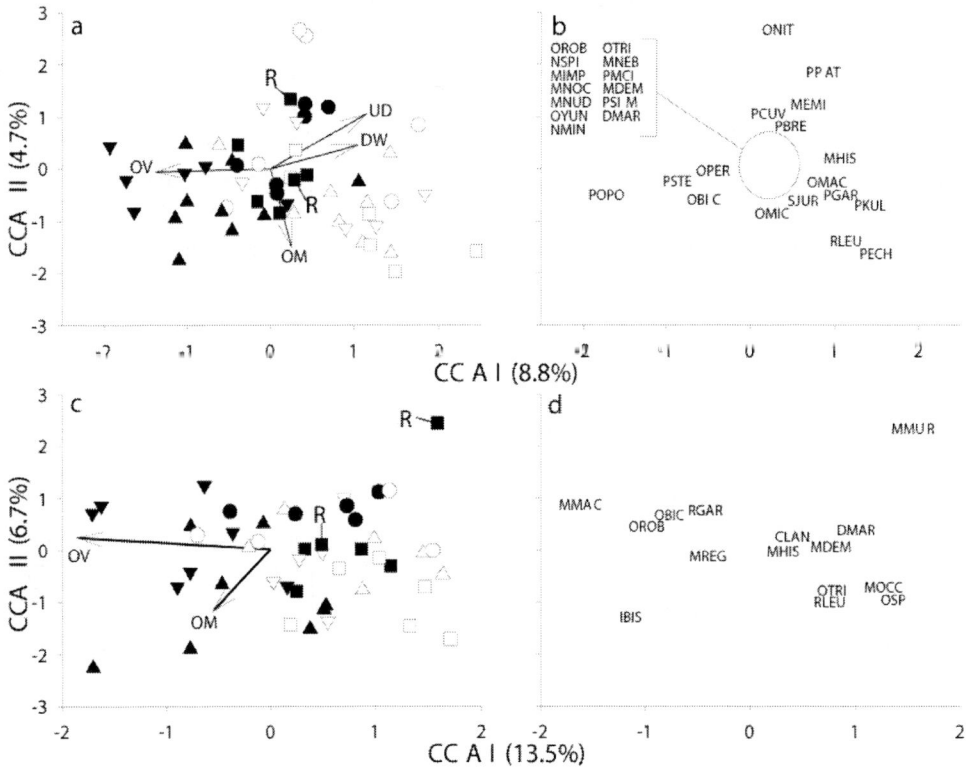

Figure 6.   Canonical correspondence analysis in which gradients in species composition among terrestrial (a,b) and canopy (c,d) traplines are constrained by variables measuring forest structure. Symbols are as in Figure 3; vectors are shown for habitat variables that were important ($p < 0.06$) in Monte Carlo tests (OM = mean overstory thickness; OV = mean variance in overstory thickness; UD = mean understory density; and DW = abundance of downed wood). Species scores shown in the right-hand-most plots are labeled with the first letter of the genus and the first three letters of the specific epithet. In (a), the mouth floodplain traplines fourth and fifth from the left were on a floodplain island (*restinga*); in (c), they are the first and fourth from the right (labeled "R").

Species Abundances According to Trap Stratum, Forest Type, Region, and River Bank

The importance of trap stratum and forest type was confirmed in species-specific tests. Most species were much more abundant either in terrestrial or canopy traps: of 21 species caught in four or more of the 16 habitat-by-region-by-bank sites, 16 differed significantly in abundance between the two strata (Figure 7a). Only two species were captured with approximately equal frequency on the ground and in the canopy (*Oecomys roberti* and *O. trinitatus*). Only three of the 28 species that were caught in two or more regions showed evidence of two-way interactions in the three-way ANOVAs. *P. steerei* was very abundant in all floodplain sites compared to upland sites, except in the headwaters region (forest type-by-region interaction p = 0.02). *Rhipidomys leucodactylus* was more abundant on the left (facing downstream) than right bank, but only in upland forest (forest type-by-bank interaction p = 0.03; see Appendix 1). *Mesomys hispidus* was more abundant on the left bank than the right bank in upland forest, but the converse was true for floodplain forest (forest type-by-bank interaction p = 0.04). In addition, this species was approximately equally abundant in floodplain and upland forest in the upper central and mouth regions; however the headwaters and lower central regions showed contrasting patterns of habitat-specific abundance (region-by-habitat interaction p = 0.02). Of the remaining 25 species, seven showed significant differences in abundance according to forest type. *P. simonsi* and *Scolomys juruaense* were significantly more common in upland than floodplain forest, whereas the converse was true for *Oryzomys perenensis, Oecomys bicolor, O. roberti, Micoureus regina,* and *Philander opossum* (Figure 7b). Only one species differed significantly in abundance among regions (*P. echinothrix*) and none differed significantly in abundance between river banks.

Alpha diversity

Terrestrial small mammal abundance did not differ between upland and floodplain sites (p = 0.85); however, relative to upland sites, floodplain sites had fewer species (p = 0.03) and showed significantly greater dominance (p < 0.01 for N1 and N2) (Table 2). The ANOVAs also indicated significant differences in terrestrial species richness and dominance among regions (p < 0.05 for both). Region-specific means ranked in accordance with position along the river, with upstream sites having higher diversity and lower dominance than downstream sites. When region was entered into the ANOVA as a quantitative rather than a nominal variable, N_0, N_1, and N_2 showed significant regional effects (p < 0.01; region was coded as 1, 2, 3, and 4 from the headwaters to the mouth), supporting the existence of a systematic geographic trend in these variables along the length of the Juruá.

Canopy-dwelling small mammals, on the other hand, were significantly more abundant in the floodplain than in upland forest (p = 0.03) and, although not significant, differences in richness and diversity in the canopy were opposite to those observed on the ground. That is, compared to upland forest, floodplain forest tended to have more species and lower dominance (Table 2). Regional effects for canopy abundance, richness, and diversity were not significant. River bank effects and two-way interactions were not significant in any of the tests.

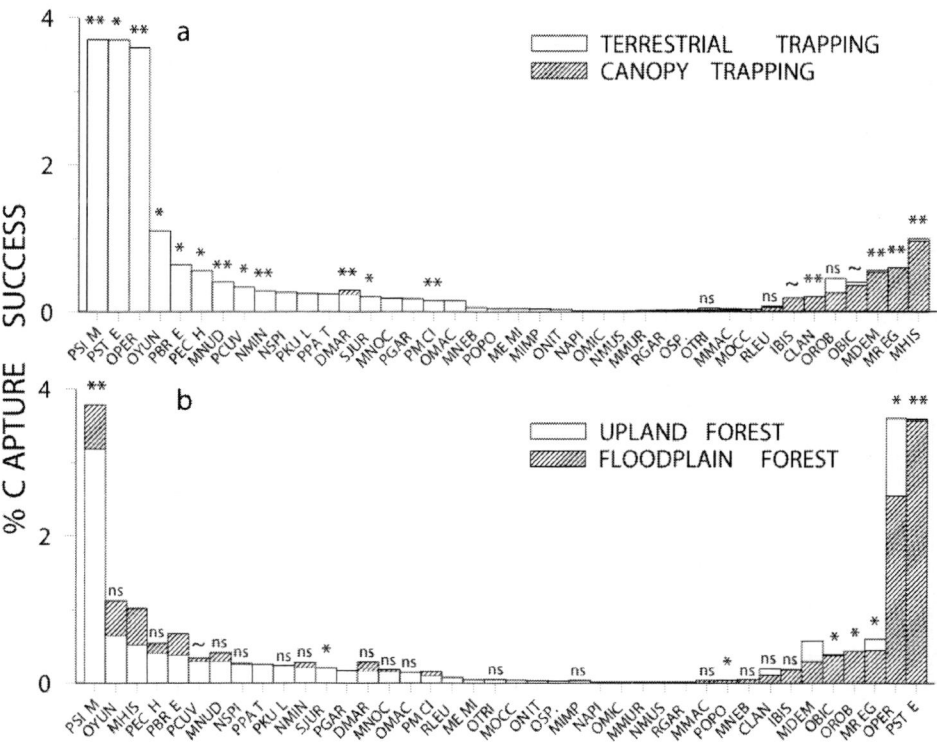

Figure 7. Mean capture success (per 100 station nights) compared between canopy and terrestrial live-trapping (a) and between upland and floodplain forest (b). Symbols represent results from significance tests (<u>ns</u> = not significant; ~ = 0.05 < p ≤ 0.10; * = p < 0.05; ** = p < 0.01; see text for details of tests). Taxa are labeled along the abscissa with the first letter of the genus and the first three letters of the species epithet.

Table 2a. Mean trapline small mammal abundance, richness, and diversity and site-specific asymptotic species richness for 16 sites along the Juruá river. Asymptotic richness is based on 4-6 terrestrial traplines per site (see Table 1).

	Upland forest								Floodplain forest							
	HE		UC		LC		MO		HE		UC		LC		MO	
	LB	RB	LB	RB	LB	RB	LB	RB	LB	RB	LB	RB	LB	RB	LB	RB
Abundance[1]	13.6	21.4	17.2	35.5	8.2	12.4	19.6	9.6	15.2	19.0	36.2	23.8	9.2	10.8	13.6	14.0
N0[2]	5.6	6.8	6.2	8.2	3.6	4.4	4.0	3.0	4.8	6.5	4.2	3.4	2.6	3.2	3.8	2.8
N1[3]	0.25	0.19	0.23	0.16	0.39	0.34	0.32	0.44	0.31	0.21	0.43	0.45	0.53	0.39	0.37	0.57
N2[4]	3.63	4.66	3.50	5.21	2.69	2.97	3.12	2.24	2.77	4.13	1.87	2.06	1.64	2.47	2.54	1.66
Smax[5]	12.6	14.2	14.8	14.7	-	15.9	6.4	8.7	11.2	17.6	7.2	6.7	8.9	6.7	8.1	5.4

[1] Abundance = mean percent capture success (number of individuals per 100 station nights) per trapline.

[2] N0 = mean number of species per trapline.

[3] N1 = mean relative number of abundant species per trapline.

[4] N2 = mean relative number of very abundant species per trapline.

[5] Smax = site-specific asymptotic species richness estimated from Michaelis-Menten equations fit to the cumulative number of species plotted against nightly trap effort.

Table 2b. Mean trapline small mammal abundance, richness, and diversity and site-specific asymptotic species richness for 16 sites along the Juruá river. Asymptotic richness is based on 2-4 canopy traplines per site (see Table 1).

	Upland forest								Floodplain forest							
	HE		UC		LC		MO		HE		UC		LC		MO	
	LB	RB	LB	RB	LB	RB	LB	RB	LB	RB	LB	RB	LB	RB	LB	RB
Abundance[1]	1.0	3.5	3.7	9.1	2.3	3.7	7.3	3.3	9.2	5.3	16.2	11.0	5.8	3.0	9.0	4.0
N0	0.7	3.0	2.3	4.0	1.7	2.0	4.3	1.7	5.0	3.0	4.3	4.3	2.0	1.5	3.0	2.7
N1	1.00	0.34	0.39	0.31	0.53	0.53	0.27	0.70	0.27	0.37	0.31	0.27	0.51	0.77	0.42	0.45
N2	1.00	2.83	2.50	3.17	1.80	1.84	3.22	1.49	3.40	2.85	3.10	3.39	1.86	1.40	2.41	2.32
Smax		5.7	9.1		3.5		2.2	9.7	5.4		8.8	7.8	6.2	4.9	4.6	6.5

[1] Abundance = mean percent capture success (number of individuals per 100 station nights) per trapline.

[2] N0 = mean number of species per trapline.

[3] N1 = mean relative number of abundant species per trapline.

[4] N2 = mean relative number of very abundant species per trapline.

[5] Smax = site-specific asymptotic species richness estimated from Michaelis-Menten equations fit to the cumulative number of species plotted against nightly trap effort.

Comparisons of species richness while controlling for abundance confirmed the lower diversity of terrestrial small mammals in floodplain compared to upland forests, but suggested that the slightly greater species richness of canopy small mammals in floodplain compared to upland forest was an artifact of higher abundances (Figure 8a, b). For terrestrial richness, forest type and regional effects were significant both without (type I $p < 0.03$) and with the covariate (type III $p < 0.047$). In contrast, canopy richness varied significantly across both of these factors when the covariate was excluded (type I $p < 0.03$), but neither was significant when the covariate was included in the model (type III $p > 0.18$). River bank effects were never significant (p always > 0.39).

These contrasting patterns in the terrestrial and canopy faunas were confirmed in analyses of asumptotic small mammal richness. Compared to upland forest, floodplain forest had lower terrestrial richness (means for floodplain and upland traplines, respectively, were 9.0 and 12.5; three-way main effects ANOVA, forest type effect $p = 0.06$). However, there was little evidence of any difference in canopy richness between the two forest types (respective means were 6.3 and 6.1, three-way main effects ANOVA, forest type effect $p = 0.85$). Terrestrial richness tended to be higher at upstream than downstream sites (three-way main effects ANOVA, region effect $p = 0.08$; two-way ANOVA with region as a (quantitative) covariate, region effect $p = 0.01$), but for canopy richness there was little evidence of any regional variation (three-way main effects ANOVA, region effect $p = 0.23$; two-way ANOVA with region as a (quantitative) covariate, region effect $p = 0.67$). Again, river bank was never significant (p always > 0.42).

Gamma Diversity

Patterns of diversity observed at super-regional levels were similar to those observed at the site level. On the ground, even though we captured approximately equal numbers of individuals in the two forest types, we captured fewer species in the floodplain than in upland forests (Figure 8c). We captured more individuals in the floodplain canopy than in the canopy of upland forests, but the observed increase in canopy species richness with number of individuals captured was similar in the two forest types (Figure 8d).

For both forest types and both trap strata (ground or canopy), the decrease in inter-site community similarity with increasing distance between sites was significant ($p < 0.001$). However, for both strata, the decrease was close to linear for upland forest, but was curvilinear for floodplain forest (Figure 9). In the floodplain, the nearest and furthest-apart sites tended to show similarities that were slightly above those from upland forest, whereas sites at intermediate distances tended to show similarities that were below those from upland forest. Community similarities that excluded the headwaters and mouth regions (triangles in Figure 9) tended to be similar in floodplain and upland sites.

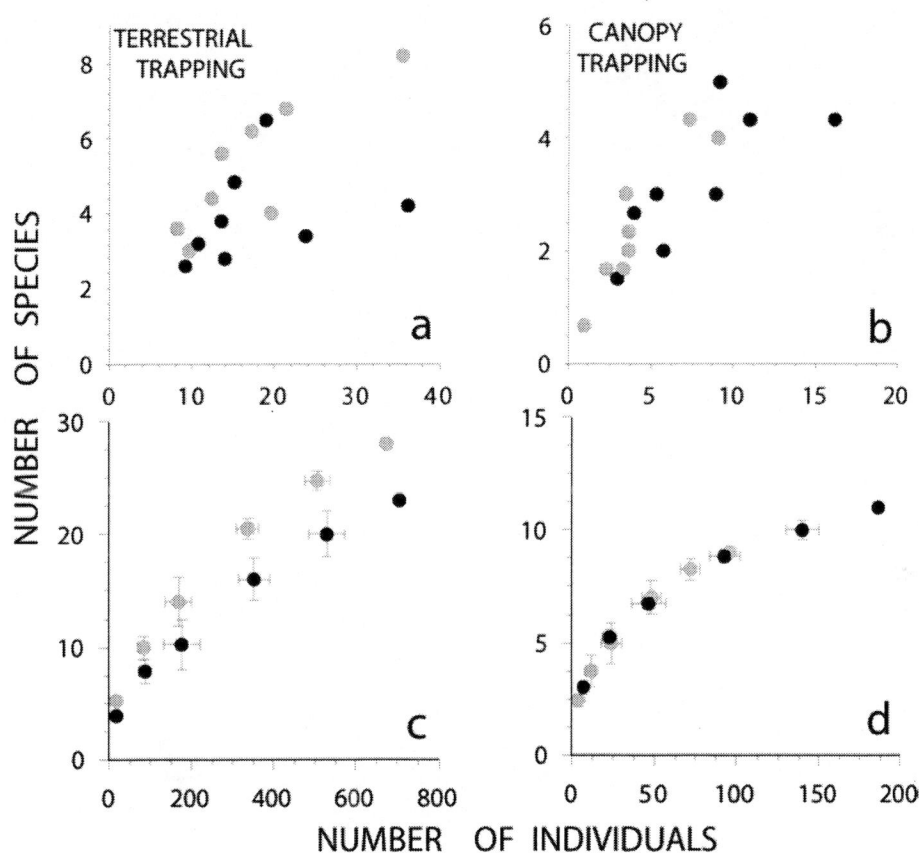

Figure 8. Mean trapline species richness plotted against mean trapline abundance for terrestrial (a) and canopy (b) trapping (open symbols = upland forest; closed symbols = floodplain forest). In the lower panels, six means (± S.E.) are shown for terrestrial (c) and canopy (d) trapping: traplines, sites (n = 8), regions (n = 4), all possible pairs of regions (n = 6), all possible combinations of three regions (n = 4), and the total sample (n = 1).

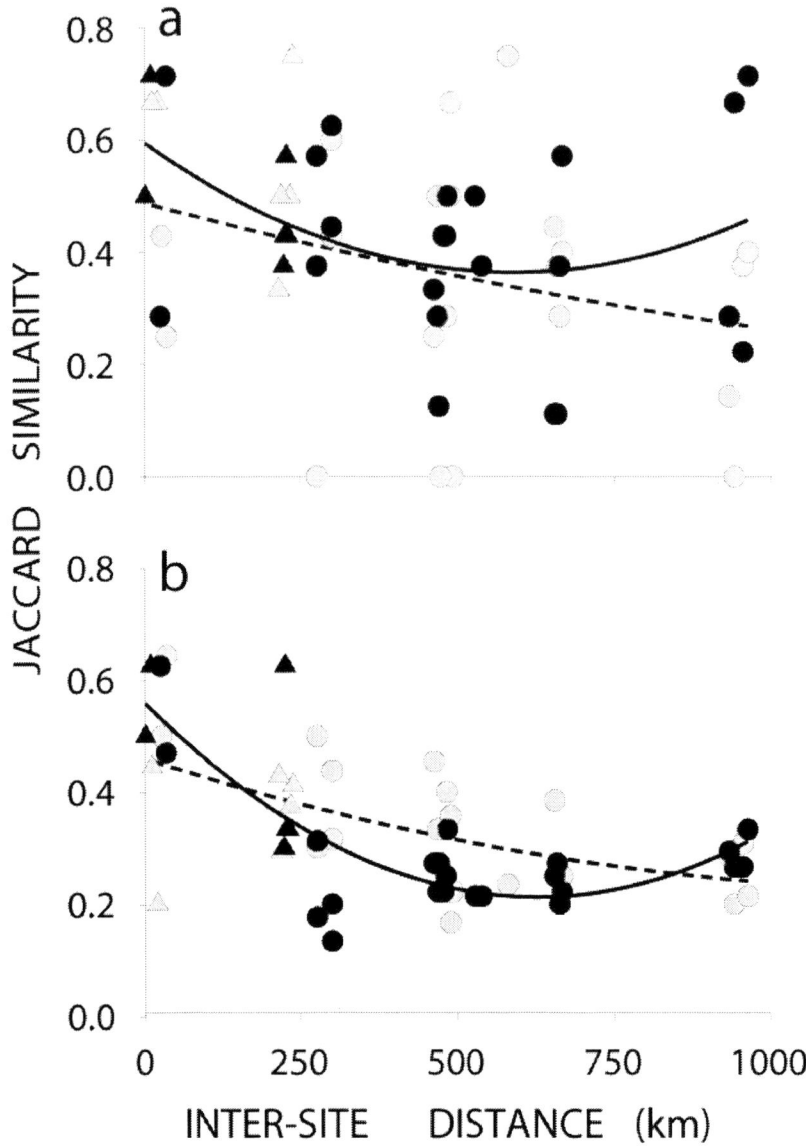

Figure 9. Values of the Jaccard similarity index calculated for site pairs of the same forest type plotted against inter-site distance. Results for canopy trapping (a) and terrestrial trapping (b) are shown (open symbols = upland forest; closed symbols = floodplain forest). Lines are quadratic least-squares regressions (dashed = upland forest; solid = floodplain forest). Triangles indicate site pairs that did not include either headwaters or mouth sites.

DISCUSSION

There are only four studies that have intensively sampled canopy small mammals in tropical rainforests (Charles-Dominique et al., 1981; Malcolm, 1991b; Malcolm and Ray, 2000; the present study) and comparisons among them provide for some interesting parallels and contrasts. All agree that most species tend to be relatively much more abundant either on the ground or in the canopy – specialization is the rule. Presumably, this reflects design constraints: locomotory adaptations for effective arboreal living are ill-suited to terrestrial living and vice-versa. The studies also agree that terrestrial specialists outnumber canopy specialists; for example, if one assigns species to one stratum or the other based on relative abundances, terrestrial species outnumbered canopy species in both the central Amazon (Malcolm, 1991b) and the Central African Republic (Malcolm and Ray, 2000). Similarly, along the Juruá River, asymptotic alpha species richness in the canopy averaged only about 60% of terrestrial richness, although this comparison is not exact because canopy specialists eventually will be caught on the ground (and hence will inflate terrestrial richness) while the converse is not generally true.

Our samples along the Juruá are the first to permit simultaneous examination of species turnover across sites (i.e., gamma diversity) for canopy and terrestrial faunas and they suggest that the relative paucity of canopy species within a site is true at larger spatial scales as well. To illustrate, across all four sampling regions (a linear distance of c. 960 km), we obtained 11 species in the floodplain canopy sample of 186 individuals. Although sample size was smaller in the canopy of upland forest, similar results are expected given the nearly identical species accumulation curves in the two forest types. If the terrestrial fauna was equally diverse, one would expect fewer species within a single sampling region. However, terrestrial samples of ca 190 individuals resulted in an average of at least 10 species within a region (Figure 8). This lower diversity in the canopy may reflect fundamental differences in the biology of the two groups of organisms. Arboreal species are faced with complex locomotory requirements, which may result in relatively "K-selected" life-history parameters. Also, variation in resource availability can be expected to be greater for terrestrial organisms than arboreal ones (Peres, 1999) because most rainforest primary productivity takes place in the canopy. These differences may, in turn, influence population persistence and, ultimately, rates of speciation. Alternatively, possibilities for niche differentiation may be lower in the canopy. Unfortunately, autecological studies have been undertaken for few tropical small mammal species and dimensions of resource partitioning are poorly understood.

While studies are in agreement concerning species richness in the canopy and terrestrial faunas, they yield rather dissimilar results with respect to relative abundances in the two strata. Our markedly lower trap success in canopy- vs. ground-based trapping came as a surprise in light of results from the central

Amazon, where trapping in continuous forest for nearly four years consistently showed greater success in the canopy (Malcolm 1991b, Malcolm 1995). The two most abundant small mammal species in the central Amazon were both canopy specialists (*Caluromys philander* and *Micoureus demerarea*) and *C. philander* was more than three times as abundant as the most abundant terrestrial genus (*Proechimys*). It is not yet clear which pattern – relatively higher abundance in the canopy compared to on the ground or vice-versa – is typical of tropical rainforests, nor is it clear to what extent abundances in two strata may vary from year to year. Comparison of Charles-Dominique et al. (1981) with Guillotin (1982) suggests that terrestrial and canopy densities and biomasses were similar at their respective sites in French Guiana. However, Voss et al.'s (2001) study, also in French Guiana, resulted in much lower success in the canopy than on the ground, although his canopy sampling effort was small. In the Central African Republic, trap success was considerably lower in the canopy than on the ground (6.8 vs. 15.7 individuals per 100 trapnights; Malcolm and Ray, 2000).

Comparisons between floodplain and upland forests showed strongly contrasting patterns of small mammal abundance and diversity for the two vertical strata. Compared to upland forest, terrestrial diversity was lower in the floodplain, but approximately equal numbers of captures were obtained in the two habitats. In contrast, diversity in the canopy was similar in the two habitats, but canopy small mammal abundance was relatively higher in floodplain than upland forest. Lower species diversity in floodplain compared to upland forests has been reported for many taxa, including plants (Dumont et al., 1990; Klinge et al., 1990; Patton et al., 2000), termites (Martius, 1994), and ants (Wilson, 1987; Majer and Delabie, 1994) and has been attributed to the harsh conditions created by periodic flooding. This is particularly important for organisms that use the ground and lower strata of the forest (August, 1984; Junk, 1984; Gentry, 1988; Peres, 1997; see also Fleck and Harder, 1995). The contrasting patterns of diversity that we observed for the ground and canopy small mammal fauna support the importance of flooding. In addition, terrestrial species richness in the floodplain was highest in the headwaters floodplain, where widespread floods were least frequent, also suggesting that flooding is important. Peres (1997) found that one of the most floodplain-averse primate taxa in the lower reaches of the river (*Saguinus* spp.) was present in the headwaters region. Ferreira (1997) reported that tree species richness decreased with mean water levels and flood duration. It is also of interest in this regard to note that the only terrestrial species that is essentially restricted to the floodplain (*Proechimys steerei*) is the largest of the *Proechimys* species (see Patton et al., 2000). Fleck and Harder (1995) also reported larger body size in small mammals from a floodplain compared to an upland site. Large body size in small mammals is thought to impart physiological and competitive advantages under harsh environmental conditions (Emmons, 1984; Malcolm, 1995) and, hence, may also indicate the importance of flooding in driving patterns of terrestrial diversity. The arboreal species experience the effects of flooding only indirectly; hence any impacts on species richness should be reduced in the canopy, a result that we obtained for small mammals and also was observed for primates (Peres, 1997) and ants (Majer and Delabie, 1994).

The net effect of (1) roughly equal terrestrial abundances in the two forest types but (2) greater canopy abundances in floodplain than upland forests was to generate greater overall small mammal abundances in floodplain than in upland forests, a result also obtained for primates (Peres, 1997). Given the large body size of *P. steerei*, it is very likely that terrestrial small mammal biomass in the floodplain also exceeded that in upland forest. The most obvious explanation for high biomass in the floodplain is high productivity due to seasonal deposition of nutrient-rich alluvium (see Peres, 1997). An example of the same phenomenon may be the seasonally extremely productive grassy areas found along the river channel (Piedade et al., 1994), which yielded extremely high abundances of *Oligoryzomys microtis* (Patton et al., 2000). The higher abundance of the canopy small mammal fauna in the floodplain compared to upland forests can be similarly explained. Although diets of Amazonian small mammal species are poorly known, key food resources include arthropods and fruit. High folivore biomass in floodplain forests has been attributed to high foliage productivity and low concentrations of secondary compounds (Peres, 1999), which would presumably also give rise to high arthropod productivity. More frequent canopy openings in the floodplain compared to upland forest and the resulting increased structural variability of the canopy stratum may also result in rapid seasonal vegetation growth and more accessible foliage nutrients (Peres, 1997). Similar effects were hypothesized to lead to increased insect and small mammal biomass in edge-modified and secondary habitats in the central Amazon (Malcolm, 1995, 1997a and 1997b). In support of this idea, Terborgh (1985) found that while nectarivorous and frugivorous bird species were strongly dependent on the availability of appropriate food resources (and hence on floristic composition), richness of insectivores increased monotonically along a floodplain successional gradient. Small mammals tend to be more insectivorous than larger mammalian taxa (reviewed in Malcolm, 1995) and, hence, the lower tree species diversity found in floodplain vs. upland forest may be largely irrelevant for them. Some evidence suggests that floodplain forests produce more fruit than upland forests, although the proportion of gut-dispersed species appears to be lower (see Peres, 1997). Higher productivity of fruits could also result in more abundant small mammal populations, although pulp-eaters might be at a relative disadvantage compared to granivorous species.

Our results also suggest that flooding plays an important role in generating beta diversity along the Juruá. The gradient in species composition between the two forest types was approximately one third as strong as that between the terrestrial and canopy faunas (gradient lengths in the Detrended Correspondence Analysis were 3.0 and 8.3 S.D., respetively). Specialization on one habitat or the other was found among both terrestrial and arboreal species and included some of the most abundant species in the sample, most notably *Proechimys steerei* and *P. simonsi*. As suggested in the constrained ordinations, part of this differentiation may be due to differences in forest structural features, such as the greater overstory variability, less dense understory, and lower quantities of downed wood in floodplain compared to upland forests. In addition, at least for the terrestrial fauna, the flooding regime presumably leads to strong selection for particular life history characteristics, notably the ability to rapidly colonize newly available habitats and to persist in the face of extensive habitat loss. The potential for population turnover in the floodplain appears to be astonishing. A very rough estimate of the number of *Proechimys steerei* in the Juruá floodplain at low water suggests perhaps 1.8 million animals (assuming that one half of the floodplain is available for habitat, that the floodplain is 900 km long with a width of 6 km at the headwaters and 25 km at the mouth, and that one of our traplines exhaustively sampled an area of 280 x 100 m). Our sampling at the mouth suggested that the great majority of these animals are displaced and that large numbers perish. Although some persisted on *restinga* islands, these islands appeared to be rare (pers. obs.). In addition, although we expected to find a superabundance of terrestrial floodplain taxa along the edge of the flooded forest (due to an influx of animals from the adjacent, recently-submerged floodplain), densities appeared to be comparable to those in the floodplain's central reaches. Our sampling sites at the mouth appeared to have been invaded by terrestrial floodplain taxa, as evidenced by the fact that both the canopy fauna and forest structure were more typical of upland forest. Nevertheless, they presumably offered a small sanctuary compared to the originally extensive habitat.

The possibility of different life history strategies in the two forest types is supported by a comparison of reproductive parameters between two closely related pairs of abundant floodplain- and upland-specializing species. *Oryzomys perenensis* was significantly more abundant in the floodplain, and, of 185 females, 42% were gravid with a mean litter size of 3.5 (S.E. = 0.12, range = 1-6). In contrast, for *O. yunganus*, which was somewhat more abundant in upland forest, only 31% of 59 females were gravid and mean litter size was 2.6 (S.E. = 0.20, range = 1-4). Similar results were obtained for the two most common small mammal species along the Juruá, *Proechimys steerei* and *P. simonsi*, both of which were highly restricted to one forest type or the other. Of 216 female *P. steerei*, a floodplain specialist, 39% were gravid and mean litter size was 3.0 (S.E. = 0.12, range = 1-7). For *P. simonsi*, an upland specialist, only 29% of 230 females were gravid and mean litter size was 2.0 (S.E. = 0.08, range = 1-3). The dynamics of these floodplain populations, including the roles of source and sink habitats, competitive interactions with upland taxa, resource partitioning, and adaptations to flooding, all deserve detailed study.

Patterns of species turnover along the Juruá were unexpected in several ways. Our previous work found little evidence of a riverine barrier effect for either communities (Gascon et al., 2000) or species (Patton et al., 1994, 1996 and 2000), a result that is corroborated by the additional tests undertaken here. The main gradients in species composition were unrelated to river bank: we found no significant differences in small mammal abundance, richness, or diversity between river banks; and species-specific tests also showed little evidence of river bank effects.

Instead of occurring along the river as expected under the riverine barrier hypothesis, distributional boundaries generally tended to be perpendicular to the river, at least within the level of resolution provided by our sampling (Gascon et al., 2000). Our genetic studies also revealed gradients perpendicular to the course of the river rather than along it. Interestingly, the steep gradients of mtDNA haplotype diversity did not appear to be evenly distributed along the river, but were concentrated in the central reaches of the river in the vicinity of the Jutaí Arch. Our ordinations of upland communities also indicated a contrast between communities above and below the Arch, supporting a historical role of the Andean thrust-and-fold belts in generating species diversity in the Amazon Basin.

Unfortunately, because the mouth floodplain was inundated during our visit to that region, our attempt to compare species turnover between floodplain and upland forests was unsuccessful. Rather than sampling the floodplain itself, we usually sampled the edge of upland forest, which resulted in captures of both floodplain and upland species. The net effect was a spurious curvilinear relationship in the floodplain between community similarity and inter-site distance. Specifically, because floodplain samples from both the headwaters and mouth contained upland species, they exhibited high similarity with each other. For the same reason, they both showed low similarity with the "true" floodplain communities of the river's central reaches. Our comparison of gamma diversity within the central regions alone showed little evidence of a contrast between upland and floodplain forests, although our sample size was small. More conclusive tests await more sites and a greater range of inter-site distances.

Small mammal communities along the Juruá showed evidence of the east-to-west increase in species diversity reported for several Amazonian taxa (e.g., Remsen and Parker, 1983; Emmons, 1984; Duellman, 1988; Gentry, 1988; Peres, 1999). Several explanations of this phenomenon have been put forward, including the following: (1) differences in soil fertility and ecosystem productivity (Emmons, 1984; Gentry and Emmons, 1987; Gentry, 1988), (2) habitat diversity (Gentry, 1988; Robinson and Terborgh, 1990; Cohn-Haft et al., 1997), and (3) historical factors (Patton et al., 1997). Concerning the first, despite higher soil fertility in the floodplain and increased abundances of small mammals, we did not observe increased species diversity, even in the canopy. This comparison is not definitive, however, because it is complicated by numerous other differences between these forest types, including differences in forest structure and relatively lower floristic diversity in the floodplain. The potential importance of the second factor, habitat diversity, was clearly demonstrated in the headwaters floodplain, where a single habitat type contained a mixture of upland and floodplain specialists. Several other authors have noted the contribution of Amazonian habitat diversity to within-site faunal diversity, especially at headwaters sites (Gentry, 1988; Robinson and Terborgh, 1990; Redford and Fonseca, 1986). On the basis of flooding alone, increased diversity is expected within the floodplain as one proceeds up the river because of a less regular flooding regime and the gradual appearance of upland taxa. No such riverine effects are expected among our upland sites, however, which were far enough from the floodplain to exclude floodplain species. These sites nonetheless exhibited the east-west gradient. Cohn-Haft et al. (1997) suggest that even within upland forests, topographical diversity in the peripheral areas of the Basin creates a finer grained mixture of habitats and hence leads to higher local diversity. Our sample may provide support for the importance of historical factors independent of present ecological conditions in that most of the observed increase in species richness from east to west occurred across the former location of the Jutaí Arch. An alternative hypothesis is that the gradient in richness across the Arch is related to the contrasting soil formations of the pre-Pleistocene Acre and Central Amazon fluvial deposition systems (Räsänen et al., 1987, 1990 and 1992), as reflected in different phytogeographic units present on either side of the Arch (Projeto RadamBrasil, 1997; Daly and Prance, 1989).

In conclusion, our work along the Juruá clearly supports a role for the flooding regime in influencing patterns of small mammal alpha, beta, and gamma diversity. Further investigations of floodplain populations are needed, as are comparisons among a larger set of Amazonian rivers, which show a remarkable diversity of sizes, landform types, and floodwater chemistry. In addition, our investigations of small mammal communities are suggestive of a role of late Cenozoic foreland dynamics in influencing ecological patterns in western Amazonia.

ACKNOWLEDGEMENTS

This research was funded by the Wildlife Conservation Society, the National Geographic Society, the Museum of Vertebrate Zoology, the Natural Sciences and Engineering Research Council of Canada (postdoctoral fellowship to JRM), and the Conselho Nacional de Desenvolvimento Científico e Technológico (CNPq; doctoral fellowship to MNFS). We are indebted to CNPq and the Instituto Brasileiro do Meio Ambiente e dos Recursos Naturais Renovavais (IBAMA) for permission to conduct the research, and to M. Ayres and J. Robinson for their support in initiating the project. We especially thank Drs. C. Gascon and C. A. Peres, who co-led the project along with the authors and assisted in the field. We also are indebted to D. Novaes, C. Patton, A. Percequillo, O. Pereira da Cunha, L. Pereira da Cunha, A. Pinheiro, and J. Ray for their able field assistance. D. Kelt and an anonymous reviewer made helpful comments on an earlier draft of the manuscript.

LITERATURE CITED

August, P. V.
  1984   Population ecology of small mammals in the Llanos of Venezuela. Special Publications of the Museum, Texas Tech University 22:1-234.

Ayres, J. M.
  1986   Uakaris and Amazonian flooded forest. Ph.D. dissertation, Cambridge University, Cambridge, United Kingdom.

Ayres, J. M., and T. H. Clutton-Brock.
  1992   River boundaries and species range size in Amazonian primates. American Naturalist 140:531-537.

Beven, S., E. F. Connor, and K. Beven.
  1984   Avian biogeography in the Amazon basin and the biological model of diversification. Journal of Biogeography 11:383-399.

Bodmer, R. E.
  1990   Responses of ungulates to seasonal inundations in the Amazon floodplain. Journal of Tropical Ecology 6:191-201.

Campbell, K. E., and D. Frailey.
  1984   Holocene flooding and species diversity in southwestern Amazonia. Quaternary Research 21:369-375.

Capparella, A. P.
  1987   Effects of riverine barriers on genetic differentiation of Amazonian forest undergrowth birds. Ph.D. dissertation. Louisiana State University, Baton Rouge.

  1988   Genetic variation in Neotropical birds: implications for the speciation process. Acta Congressus Internationalis Ornithologici 19:1658-1664.

  1992   Neotropical avian diversity and riverine barriers. Acta Congressus Internationalis Ornithologici 20:307-316.

Cerqueira, R.
    1982   South American landscapes and their mammals. Pp. 53-75 in
            Mammalian biology in South America (M. A. Mares and H. H.
            Genoways, eds.). University of Pittsburgh, Pittsburgh.

Cohn-Haft, M., A. Whittaker, and P. C. Stouffer.
    1997   A new look at the "species-poor" central Amazon: the avifauna north of
            Manaus, Brazil. Ornithological Monographs 48:205-235.

Colwell R. K., and J. A. Coddington.
    1994   Estimating terrestrial biodiversity through extrapolation. Philosophical
            Transactions of the Royal Society B 345:101-118.

Charles-Dominique, P. M., M. Atramentowicz, M. Charles-Dominique, H. Gerard,
C. M. Hladik, and M. F. Prevost.
    1981   Les mammifères frugivores arboricoles nocturnes d'une forêt
            Guyanaise: inter-relations plantes-animaux. Revue d'Ecologie (Terre et
            Vie) 35:341-435.

Colinvaux, P.
    1987   Amazon diversity in light of the paleoecological record. Quaternary
            Science Review 6:93-114.

Colinvaux, P. A., M. C. Miller, K. Liu, M. Steinitz-Kannan, and I. Frost.
    1985   Discovery of permanent Amazon lakes and hydraulic disturbance in
            the upper Amazon Basin. Nature 313:42-45.

Daly, D.C., and G. T. Prance.
    1989   Brazilian Amazon. Pp. 401-426 in Floristic inventory of tropical
            countries: the status of plant systematics, collections, and vegetation,
            plus recommendations for the future (D. G. Campbell and H. D.
            Hammond, eds.). New York Botanical Garden, New York.

da Silva, M. N. F.
    1995   Systematics and phylogeography of Amazonian spiny rats of the genus
            Proechimys (Rodentia: Echimyidae). Ph.D. Dissertation. University of
            California, Berkeley.

    1998   Four new species of spiny rats of the genus *Proechimys* (Rodentia:
            Echimyidae) from the western Amazon of Brazil. Proceedings of the
            Biological Society of Washington 111:436-471.

da Silva, M. N. F., and J. L. Patton.
    1993   Amazonian phylogeography: mtDNA sequence variation in arboreal Amazonian echimyid rodents (Caviomorpha). Molecular Phylogenetics and Evolution 2:243-255.

Duellman, W. E.
    1988   Patterns of species diversity in anuran amphibians in the American tropics. Annals of the Missouri Botaniccal Garden 75:79-104.

Dumont, J. F., S. Lamotte, and F. Kahn.
    1990   Wetland and upland forest ecosystems in Peruvian Amazonia: plant species diversity in the light of some geological and botanical evidence. Forest Ecology and Management 33/34:125-139.

Emmons, L. H.
    1984   Geographic variation in densities and diversities of non-flying mammals in Amazonia. Biotropica 16:210-222.

Emmons, L. H., and F. Feer.
    1990   Neotropical rainforest mammals: a field guide. University of Chicago Press, Chicago, 281 pp.

Erwin, T. L., and J. Adis.
    1982   Amazonian inundation forests: their role as short-term refuges and generators of species richness and taxon pulses. Pp. 358-371 in Biological Diversification in the tropics (G. T. Prance, ed.). Columbia University Press, New York.

Ferreira, L. V.
    1997   Effects of the duration of flooding on species richness and floristic composition in three hectares in the Jaú National Park in floodplain forests in central Amazonia. Biodiversity and Conservation 6:1353-1363.

Fleck, D. W., and J. D. Harder.
    1995   Ecology of marsupials in two Amazonian rain forests in northeastern Peru. Journal of Mammalogy 76:809-818.

Foster, R. B.
　1990　The floristic composition of the Rio Manu floodplain forest. Pp. 99-111 in Four Neotropical rainforests (A. H. Gentry, ed.). Yale University Press, New Haven.

Furch, K., and H. Klinge.
　1989　Chemical relationships between vegetation, soil and water in contrasting inundation areas of Amazonia. Pp. 189-204 in Mineral nutrients in tropical forest and savanna ecosystems (J. Proctor, ed.). Blackwell Scientific Pub., Oxford, United Kingdom.

Gascon, C., J. R. Malcolm, J. L. Patton, M. N. F. Da Silva, J. P. Bogart, S. C. Lougheed, C. A. Peres, S. Neckel, and P. T. Boag.
　2000　Riverine barriers and the geographic distribution of Amazonian species. Proceedings of the National Academy of Sciences 97:13672-13677.

Gentry, A. H.
　1988　Changes in plant community diversity and floristic composition on environmental and geographic gradients. Annals of the Missouri Botanical Garden 75:1-34.

　1983　Dispersal ecology and diversity in neotropical forest communities. Sonderbd. Naturwiss. Verh. Hamburg 7:303-314.

Gentry, A. H., and L. H. Emmons.
　1987　Geographic variation in fertility, phenology, and composition of Neotropical forests. Biotropica 19:216-227.

Guillotin, M.
　1982　Place de Proechimys cuvieri (Rodentia, Echimyidae) dans les peuplements micromammaliens terrestres de la forêt Guyanaise. Mammalia 46:299-318.

Haffer, J.
　1969　Speciation in Amazonian forest birds. Science, 165:131-137.

　1993　Time's cycle and time's arrow in the history of Amazonia. Biogeographica 69:15-45.

Hershkovitz, P.
   1977   Living New World monkeys (Platyrrhini) with an introduction to Primates. Vol. 1. University of Chicago Press, Chicago.

Hughes, F. M. R.
   1990   The influence of flooding regimes on forest distribution and composition in the Tana River Floodplain, Kenya. Journal of Applied Ecology 27:475-491.

Jongman, R.H.G., C. J. F. ter Braak, and O. F. R. van Tongeren.
   1995   Data Analysis in Community and Landscape Ecology. Cambridge University Press, Cambridge, United Kingdom.

Junk, W. J.
   1984   Ecology of the várzea floodplain of Amazonian white-water rivers. Pp. 215-243 in The Amazon (H. Sioli, ed.). W. Junk, Dordrecht, Germany.

Kalliola, R., J. Salo, M. Puhakka, and M. Rajasilta.
   1991   New site formation and colonizing vegetation in primary succession on the western Amazon floodplains. Journal of Ecology 79:877-901.

Kalliola, R., J. Salo, M. Puhakka, M. Rajasilta, T. Häme, R. J. Neller, M. E. Räsänen, and W. A. Danjoy Arias.
   1992   Upper Amazon channel migration: implications for vegetation perturbance and succession using bitemporal Landsat MSS images. Naturissenschaften 79:75-79.

Klinge, H., W. J. Junk, and C. J. Revilla.
   1990   Status and distribution of forested wetlands in tropical South America. Forest Ecology and Management 33/34:81-101.

Kubitzki, K., and A. Ziburski.
   1994   Seed dispersal in flood plain forests of Amazonia. Biotropica 26:30-43.

Lamotte, S.
   1990   Fluvial dynamics and succession in the Lower Ucayali River basin, Peruvian Amazonia. Forest Ecology and Management 33/34:141-156.

Ludwig, J. A., and J. F. Reynolds.
   1988   Statistical ecology: a primer on methods and computing. John Wiley and Sons, New York.

Majer, J. D., and J. H. C. Delabie.

1994   Comparison of the ant communities of annually inundated and terra firme forests at Trombetas in the Brazilian Amazon. Insectes Sociaux, 41:343-359.

Malcolm, J. R.

1991a   Comparative abundances of Neotropical small mammals by trap height. Journal of Mammalogy 72:188-192.

1991b   The small mammals of Amazonian forest fragments: pattern and process. Ph.D. dissertation, University of Florida, Gainesville.

1995   Forest structure and the abundance and diversity of Neotropical small mammals. Pp. 179-197 in Forest canopies (M. D. Lowman and N. M. Nadkarni, eds.). Academic Press, San Diego.

1997a   Insect biomass in Amazonian forest fragments. Pp. 510-533 in Canopy arthropods (N. E. Stork, J. Adis, and R. K. Didham, eds.). Chapman & Hall, London, United Kingdom.

1997b   Biomass and diversity of small mammals in Amazonian forest fragments. Pp 207-221 in Tropical forest remnants: ecology, management and conservation of fragmented communities (W. F. Laurance and R. O. Bierregaard, Jr., eds.). University of Chicago Press, Chicago.

Malcolm, J. R., and J. C. Ray.

2000   Influence of timber extraction routes on central African small mammal communities, forest structure, and tree diversity. Conservation Biology 14:1623-1638.

Manly, B. F. J.

1991   Randomization and Monte Carlo methods in biology. Chapman and Hall, New York.

Martius, C.

1994   Diversity and ecology of termites in Amazonian forests. Pedobiologia 38:407-428.

Patton, J. L., M. N. F. da Silva, M. C. Lara, and M. A. Mustrangi.
1997   Diversity, differentiation, and the historical biogeography of nonvolant small mammals of the neotropical forests. Pp. 455-465 in Tropical forest remnants: ecology, management and conservation of fragmented communities (W. F. Laurance, and R. O. Bierregaard, Jr., eds.). University of Chicago Press, Chicago.

Patton, J. L., M. N. F. da Silva, and J. R. Malcolm.
1994   Gene genealogy and differentiation among arboreal spiny rats (Rodentia: Echimyidae) of the Amazon basin: a test of the riverine barrier hypothesis. Evolution 48:1314-1323.

1996   Hierarchical genetic structure and gene flow in three sympatric species of Amazonian rodents. Molecular Ecology 5:229-238.

2000   Mammals of the Rio Juruá and the evolutionary and ecological diversification in Amazonia. Bulletin of the American Museum of Natural History 244:1-306.

Peres, C. A.
1997   Primate community structure at twenty western Amazonian flooded and unflooded forests. Journal of Tropical Ecology 13:381-405.

1999   The structure of nonvolant mammal communities in different Amazonian forest types. Pp. 564-581 in Mammals of the Neotropics. Vol. 3 (J. F. Eisenberg, and K. H. Redford, eds.). University of Chicago Press, Chicago.

Peres, C. A., J. L. Patton, and M. N. F. da Silva.
1996   Riverine barriers and gene flow in Amazonian saddle-back tamarins. Folia Primatologica 67:113-124.

Piedade, M. T. F., S. P. Long, and W. J. Junk.
1994   Leaf and canopy photosynthetic CO_2 uptake of a stand of *Echinochloa polystachya* on the Central Amazon floodplain. Oecologia 97:193-201.

Pimm, S. L., and J. L. Gittleman.
1992   Biological diversity: where is it? Science 255:940.

Prance, G. T.
  1979   Notes on the vegetation of Amazonia III.   The terminology of Amazonian forest types subject to inundation. Brittonia 31:26-38.

Projeto RadamBrasil.
  1977   Levantamento de recoursos naturais: folha SB.19 Juruá. Departamento Nacional da Produção Mineral, Rio de Janeiro, Brazil.

Puhakka, M., R. Kalliola, M. Rajasilta, and J. Salo.
  1992   River types, site evolution and successional vegetation patterns in Peruvian Amazonia. Journal of Biogeography 19:651-665.

Queiroz, H. L.
  1995   Preguiças e guaribas: os mamíferos folívoros arborícolas do Mamirauá. Conselho Nacional de Desenvolvimento Científico e Tecnológico, Brasília, Brazil.

Räsänen, M., R. Neller, J. Salo, and H. Jungner.
  1992   Recent and ancient fluvial deposition systems in the Amazonian Foreland basin, Peru. Geological Magazine 129:293-306.

Räsänen, M. R., J. S. Salo, H. Jungner, and L. R. Pittman.
  1990   Evolution of the western Amazon lowland relief: impact of Andean foreland dynamics. Terra Research 2: 320-332.

Räsänen, M., J. Salo, and R. J. Kalliola.
  1987   Fluvial perturbance in the western Amazon basin: regulation by long term sub-Andean tectonics. Science 238:1398-1401.

Redford, K. H., and A. B. Fonseca.
  1986   The role of gallery forests in the zoogeography of the cerrado's non-volant mammalian fauna. Biotropica 18:126-135.

Remsen, J. V., Jr., and T. A. Parker III.
  1983   Contribution of river-created habitats to bird species richness in Amazonia. Biotropica 15:223-231.

Richey, J. E., C. Nobre, and C. Deser.
  1989   Amazon River discharge and climate variability: 1903 to 1985. Science 246:101-103.

Robinson, S. K., and J. Terborgh.
  1990   Bird communities of the Cosha Cashu biological station in Amazonian Peru. Pp. 199-216 in Four Neotropical Rainforests (A. H. Gentry, ed.). Yale University Press, New Haven.

Salo, J., R. Kalliola, I. Häkkinen, Y. Mäkinen, P. Niemelä, M. Puhakka, and P. D. Coley.
  1986   River dynamics and the diversity of Amazon lowland forest. Nature 322:254-258.

Salo, J. S., and R. J. Kalliola.
  1991   River dynamics and natural forest regeneration in the Peruvian Amazon. Pp. 245-256 in Rain forest regeneration and management (A. Gómez-Pompa, T. C. Whitmore, and M. Hadley, eds.). UNESCO, Paris, France.

Terborgh, J.
  1985   Habitat selection in Amazonian birds. Pp. 311-338 in Habitat selection in birds (M. L. Cody, ed.). Academic Press, New York.

Terborgh, J., R. B. Foster, and P. Nuñez V.
  1996   Tropical tree communities: a test of the nonequilibrium hypothesis. Ecology 77:561-567.

Terborgh, J., and K. Petren.
  1991   Development of habitat structure through succession in an Amazonian floodplain forest. Pp. 28-46 in Habitat structure: the physical arrangement of objects in space (S. S. Bell, E. D. McCoy, and H. R. Mushinsky, eds.). Chapman and Hall, London, United Kingdom.

Tuomisto, H., K. Roukolainen, R. Kalliola, A. Linna, W. Danjoy, and Z. Rodriguez.
  1995   Dissecting Amazonian biodiversity. Science 269:63-66.

Voss, R. S., D. P. Lunde, and N. B. Simmons.
  2001   The mammals of Paracou, French Guiana: a Neotropical lowland rainforest fauna. Part 2. Nonvolant species. Bulletin of the American Museum of Natural History 253:1-236.

Wallace, A. R.
  1852   On the monkeys of the Amazon. Proceedings of the Zoological Society of London 20:107-110.

Wilson, E. O.

    1987   The arboreal ant fauna of Peruvian Amazon forests: a first assessment. Biotropica 19:245-251.

Woodman, N., N. A. Slade, R. M. Timm, and C. A. Schmidt.

    1995   Mammalian community structure in lowland, tropical Peru, as determined by removal trapping. Zoological Journal of the Linnean Society 113:1-20.

Worbes, M., H. Klinge, J. D. Revilla, and C. Martius.

    1992   On the dynamics, floristic subdivision and geographical distribution of várzea forests in Central Amazonia. Journal of Vegetation Science 3:553-564.

Appendix 1. Percent capture success of small mammal species along the Rio Juruá, summarized according to forest type, region, and river bank. Values represent the sum of terrestrial percent capture success (per 100 station nights) and canopy percent capture success (per 100 station nights). Trap effort is detailed in Table 1.

	Upland forest								Floodplain forest							
	HE		UC		LC		MO		HE		UC		LC		MO	
	LB	RB	LB	RB	LB	RB	LB	RB	LB	RB	LB	RB	LB	RB	LB	RB
Caluromys lanatus	0.56	0.28	0	0	0	0	0.56	0	0.29	0.74	0	0.19	0.28	0	0	0.19
Didelphis marsupialis	0.38	0.19	0.19	1.03	0	0.76	0	0	0	0.48	0	0	0.19	0.19	1.32	0
Marmosa murina	0	0	0	0	0	0	0	0	0	0	0	0	0	0	0.19	0
Marmosops impavidus	0	0	0.19	0	0.19	0	0	0	0	0	0	0	0	0	0.19	0
M. neblina	0	0	0	0	0	0	0	0	0	0.48	0.19	0	0	0.19	0	0
M. noctivagus	0	0	0.19	2.29	0	0	0	0	0	0.48	0	0	0	0	0	0
Metachirus nudicaudatus	0.38	0.38	0.38	2.08	0.38	0	0.95	0.19	0.48	0.71	0	0	0	0	0.19	0.38

Appendix 1. Continued.

	Upland forest								Floodplain forest							
	HE		UC		LC		MO		HE		UC		LC		MO	
	LB	RB	LB	RB	LB	RB	LB	RB	LB	RB	LB	RB	LB	RB	LB	RB
Micoureus demerarae	0	0.28	0.56	1.42	0.37	0.74	0.93	0.19	0.29	0.79	0	0	0.14	0.28	2.78	0.37
M. reginus	0	0.56	0.19	1.41	0.37	0	0	0	1.15	0.37	2.39	1.30	0.83	1.11	0	0
Monodelphis emiliae	0.19	0.38	0	0	0	0	0	0	0	0	0	0	0	0	0	0
Philander mcilhennyi	0.38	0.38	0	0.63	0	0.19	0	0	0.48	0.24	0	0	0	0	0	0
P. opossum	0	0	0	0	0	0	0	0	0	0.24	0.19	0.19	0	0	0	0
Neacomys musseri	0	0	0	0	0	0	0	0	0.16	0	0	0	0	0	0	0
N. minutus	0	0	0	0	0.19	0.76	1.33	0	0	0	0.38	0.38	0	0	0.38	0
N spinosus	0.38	0	0	3.75	0	0	0	0	0.16	0	0	0	0	0	0	0
Nectomys apicalis	0	0	0.19	0	0	0	0	0	0	0	0	0	0	0	0	0

Appendix 1. Continued.

	Upland forest								Floodplain forest							
	HE		UC		LC		MO		HE		UC		LC		MO	
	LB	RB	LB	RB	LB	RB	LB	RB	LB	RB	LB	RB	LB	RB	LB	RB
Oecomys bicolor	0.19	0	0	0	0	0	0	0	1.72	0.37	1.59	2.42	0.19	0	0	0.19
O. roberti	0	0	0	0	0	0	0	0	0	0	2.96	2.44	0.42	1.28	0	0
O. species	0	0	0	0	0	0	0.19	0.19	0	0	0	0	0	0	0	0
O. trinitatus	0	0	0	0.2	0.19	0	0.19	0	0.16	0	0	0	0	0	0	0
Oligoryzomys nicrotis	0	0	0	0	0	0.19	0	0	0	0	0	0	0	0	0	0
Oryzomys maccounnelli	0	0	1.71	0	0	0	0	0.57	0	0	0	0	0	0	0	0
O. nitidus	0	0.38	0	0	0	0	0	0	0	0	0	0	0	0	0	0
O. perenensis	1.9	3.24	1.14	3.33	0.38	0.95	5.33	0.19	7.62	5.24	5.52	8.00	3.62	3.81	5.9	0.76

Appendix 1. Continued.

	Upland forest								Floodplain forest							
	HE		UC		LC		MO		HE		UC		LC		MO	
	LB	RB	LB	RB	LB	RB	LB	RB	LB	RB	LB	RB	LB	RB	LB	RB
O. yunganus	0.38	0.95	0.57	6.46	0	0.38	1.52	0	1.11	2.14	3.24	0	0.38	0.76	0	0
Scolomys juruaense	0.76	0	0.38	0.83	1.33	0	0	0	0	0	0	0	0	0	0	0
Rhipidomys gardneri	0	0	0	0	0	0	0	0	0.29	0	0	0	0	0	0	0
R. leucodactylus	0	0	0.37	0.2	0	0	0.37	0.19	0	0	0	0	0	0	0	0
Isothrix bistriata	0	0	0	0	0	0	0	0	0.29	0	0.40	0.37	1.53	0	0	0.19
Makalata macrura	0	0	0	0	0	0	0	0	0	0	0.40	0	0.19	0	0	0

Appendix 1. Continued.

	Upland forest								Floodplain forest							
	HE		UC		LC		MO		HE		UC		LC		MO	
	LB	RB	LB	RB	LB	RB	LB	RB	LB	RB	LB	RB	LB	RB	LB	RB
Mesomys hispidus	0	0.83	1.31	1.62	0.56	1.3	1.3	1.48	1.15	1.16	1.79	1.11	0	0.14	1.48	1.30
M. occultus	0	0	0	0	0	0	0.56	0	0	0	0	0	0	0	0	0
Proechimys brevicauda	2.1	3.43	0	0	0	0	0	0	1.27	3.33	0	0	0	0	0	0
P. cuvieri	0	1.33	0.95	2.08	0.19	0	0	0	0.16	0.48	0	0	0	0	0	0
P. echinothrix	0	0	0	0	0.76	0	2.48	3.43	0	0	0	0	0	0	0.19	1.90
P. gardneri	0	0	0	0	0	2.67	0	0	0	0	0	0	0	0	0	0
P. kulinae	0	0	3.62	0	0.19	0	0	0	0	0	0	0	0	0	0	0
P. pattoni	0.19	3.24	0	0	0	0	0	0	0	0	0	0	0	0	0	0

Appendix 1. Continued.

	Upland forest								Floodplain forest							
	HE		UC		LC		MO		HE		UC		LC		MO	
	LB	RB	LB	RB	LB	RB	LB	RB	LB	RB	LB	RB	LB	RB	LB	RB
P. simonsi	5.71	6.48	6.29	10.6	4.00	5.71	7.05	4.57	2.54	2.14	0	0	0	0	2.29	2.10
P. steerei	0	0	0.19	0	0	0.19	0	0	0.32	1.67	24.4	12.38	4.19	4.19	3.05	8.19

Margaret Oliveira Corrêa
Departamento de Zoologia
Instituto de Biologia
Universidade Federal do Rio de
    Janeiro
Rio de Janeiro, RJ, Brazil
Email: margaret@biologia.ufrj.br

Guillermo D'Elía
Laboratorio de Evolución
Facultad de Ciencias
Iguá 4225
Montevideo 11400, Uruguay
Email: guillermo@fcien.edu.uy

Louise H. Emmons
Division of Mammals
Smithsonian Institution
P.O. Box 37012
Washington, DC 20013-7012 USA
Email: Emmons.Louise@nmnh.si.edu

Thales Renato O. de Freitas
Departamento de Genética
Univ. Federal do Rio Grande do Sul
Caixa Postal 15053, 91501-970
Porto Alegre, RS, Brazil
Email: thales.freitas@ufrgs.br

John Carlos Garza
Southwest Fisheries Science Center
Santa Cruz Laboratory
Santa Cruz, CA 95060 USA
Email: carloz.garza@noaa.gov

Lena Geise
Univ. do Estado do Rio de Janeiro
Setor Zoologia
Dept. de Biologia Animal e Vegeta
Rio de Janeiro, RJ, Brazil
Email: geise@ufrj.br

Pablo Rodrigues Gonçalves
Departamento de Vertebrados
Museu Nacional
Univ. Federal do Rio de Janeiro
Rio de Janerio, RJ, Brazil
Email: prg@acd.ufrj.br

Harry W. Greene
Dept. Ecology & Evolutionary
    Biology
Cornell University
Ithaca, NY 14853-2701 USA
Email: hwg5@cornell.edu

Hopi E. Hoekstra
Ecology, Behavior & Evolution
Division of Biological Sciences
University of California, San Diego
La Jolla, CA 92093-0116 USA
Email: hoekstra@ucsd.edu

Patrick A. Kelly
California State University, Stanislaus
Endangered Species Recovery Prog.
1900 N. Gateway Blvd., Suite 101
Fresno, CA 93727 USA
Email: patrickk@esrp.csustan.edu

Eileen A. Lacey
Museum of Vertebrate Zoology
3101 VLSB
University of California
Berkeley, CA 94720-3160 USA
Email: ealacey@berkeley.edu

Márcia C. Lara
The Broad Institute of MIT and
    Harvard
The Genome Center
320 Charles Street
Cambridge, MA 02141 USA
Email: mlara@broad.mit.edu

Enrique P. Lessa
Laboratorio de Evolución
Facultad de Ciencias
Iguá 4225
Montevideo 11400, Uruguay
Email: lessa@fcien.edu.uy

Barbara L. Lundrigan
Michigan State University Museum
Department of Zoology
Michigan State University
East Lansing, MI 48824 USA
Email: lundriga@msu.edu

Jay R. Malcolm
33 Willcocks Street
Faculty of Forestry
University of Toronto
Toronto, Ontario, Canada M5S 3B3
Email: jay.malcolm@utoronto.ca

Philip Myers
Mammal Division
Museum of Zoology
University of Michigan
Ann Arbor, MI 48109 USA
Email: pmyers@umich.edu

Michael W. Nachman
Dept. Ecology & Evolutionary
    Biology
University of Arizona
Tucson, AZ 85745 USA
Email: nachman@u.arizona.edu

João Alves de Oliveira
Departamento de Vertebrados
Museu Nacional
Univ. Federal do Rio de Janeiro
Rio de Janeiro, RJ, Brazil
Email: jaoliv@mn.ufrj.br

Ulyses F. J. Pardiñas
Centro Nacional Patagonico
CC 128
9120 Puerto Madryn
Chubut, Argentina
Email: ulyses@cenpat.edu.ar

James L. Patton
Museum of Vertebrate Zoology
3101 VLSB
University of California
Berkeley, CA 94720 USA
Email: patton@uclink.berkeley.edu

Leila Maria Pessôa
Departamento Zoologia
Instituto de Biologia
Universidade Federal do Rio de
    Janeiro
Rio de Janeiro, RJ, Brazil
Email: pessoa@acd.ufrj.br

Scott E. Phillips
California State University, Stanislaus
Endangered Species Recovery Prog.
1900 N. Gateway Blvd., Suite 101
Fresno, CA 93727 USA
Email: sphillips@esrp.csustan.edu

Javier A. Rodríguez-Robles
Department of Biological Sciences
University of Nevada, Las Vegas
Las Vegas, NV 89154–4004 USA
Email:
javier.rodriguez@ccmail.nevada.edu

Christopher J. Schneider
Boston University
Biology Department
Boston, MA 02215 USA
Email: cschneid@bio.bu.edu

Maria Nazareth F. da Silva
Departamento de Ecologia
Instituto Naional de Pesquisa da
    Amazônia
Manaus, AM, Brazil 69011
Email: mnfs@inpa.gov.br

Robert E. Vande Kopple
Univ. of Michigan Biological Station
9008 Biological Road
Pellston, MI 49769 USA
Email: bvk@umich.edu

Gabriela Wlasiuk
Dept. Ecology & Evolutionary
    Biology
University of Arizona
Tucson, AZ 85721 USA
Email: wlasiuk@email.arizona.edu

Daniel F. Williams
California State University Stanislaus
Endangered Species Recovery Prog.
1900 Gateway Blvd., Suite 101
Fresno, CA 93727 USA
Email: dwilliam@esrp.csustan.edu